Advances in Intelligent Systems and Computing 192

For further volumes:
http://www.springer.com/series/11156

Ivan Zelinka, Otto E. Rössler, Václav Snášel,
Ajith Abraham, and Emilio S. Corchado (Eds.)

Nostradamus: Modern Methods of Prediction, Modeling and Analysis of Nonlinear Systems

Springer

Editors
Ivan Zelinka
Department of Computer Science
Faculty of Electrical Engineering and
Computer Science VŠB-TUO
Ostrava-Poruba
Czech Republic

Prof. Dr. Otto E. Rössler
Institute of Physical and
Theoretical Chemistry
University of Tuebingen
Tuebingen
Germany

Václav Snášel
Department of Computer Science
Faculty of Electrical Engineering
and Computer Science VŠB-TUO
Ostrava-Poruba
Czech Republic

Dr. Ajith Abraham
Machine Intelligence Research
Labs (MIR Labs)
Scientific Network for Innovation
and Research Excellence
Auburn, Washington
USA

Dr. Emilio S. Corchado
Departamento de Informática
y Automática
Faculty of Science
University of Salamanca
Salamanca
Spain

ISSN 2194-5357
e-ISSN 2194-5365
ISBN 978-3-642-33226-5
e-ISBN 978-3-642-33227-2
DOI 10.1007/978-3-642-33227-2
Springer Heidelberg New York Dordrecht London

Library of Congress Control Number: 2012946262

Printed on acid-free paper

Springer is part of Springer Science+Business Media (www.springer.com)

Preface

This proceeding book of Nostradamus conference (http://nostradamus-conference.org) contains accepted papers presented at this event in 2012. Nostradamus conference was held in the one of the biggest and historic city of Ostrava (the Czech Republic, http://www.ostrava.cz/en), in September 2012.

Conference topics are focused on classical as well as modern methods for prediction of dynamical systems with applications in science, engineering and economy. Topics are (but not limited to): *prediction by classical and novel methods, predictive control, deterministic chaos and its control, complex systems, modelling and prediction of its dynamics* and much more.

Prediction of the behaviour of the dynamical systems and modelling of its structure is vitally important problem in engineering, economy and science today. Examples of such systems can be seen in the world around us and of course in almost every science discipline, including such "exotic" domains like the earth's atmosphere, turbulent fluids, economies (exchange rate and stock markets), population growth, physics (control of plasma), chemistry and complex networks .

The main aim of the conference is to create periodical possibility for students, academics and researchers to exchange their ideas and novel methods. This conference will establish a forum for the presentation and discussion of recent trends in the area of applications of various predictive methods for researchers, students and academics.

The conference Nostradamus was officially established in 1998 and has been repeated successfully each year for 5 years. Due to administrative and technical conditions it has been cancelled since now. By this year 2012 we are opening this conference permanently with support of wide spectra of publishers (Springer, Computers & Mathematics with Applications, Journal of Unconventional Computing, International Journal of Energy Optimization and Engineering and Springer series Emergence Complexity and Computation), societies and organizations, one of the biggest university in the Czech Republic and with financial support of EU projects. Conference is supposed to be repeated every year to share knowledge and create new useful relations between researchers.

The selection of papers was extremely rigorous and we did it in order to maintain the high quality of the conference. We would like to thank the members of the Program Committees and reviewers for their hard work - each contribution has got 6 reviews. We believe that Nostradamus conference creates a high standard conference in the domain of prediction and modelling of complex systems.

Nostradamus 2012 enjoyed outstanding keynote speeches by distinguished guest speakers: **Otto E. Rössler** and **Ali Sanayei** (Germany), **Hendrik Richter** (Germany), **Mohammed Chadli** (France), **Oldřich Zmeškal** (Czech Republic) and **Juan Carlos Burguillo-Rial** (Spain). Also a few tutorials were done by Ivan Zelinka (Czech Republic), Roman Šenkeřík (Czech Republic) and Petr Dostal (Czech Republic).

Historically the first Nostradamus conference 1998

For this special edition, as a follow-up of the conference, we anticipate further publication of selected papers in a special issue of the prestigious journal Computers & Mathematics with Applications, special book in Studies in Computational Intelligence series and more.

Particular thanks go as well to the Workshop main Sponsors, IT4Innovations, VŠB-Technical University of Ostrava, MIR labs (USA), Centre for Chaos and Complex Networks (Hong Kong), Journal of Unconventional Computing (UK). Special thanks belong to Ministry of Education of the Czech Republic. This conference was supported by the Development of human resources in research and development of latest soft computing methods and their application in practice project, reg. no. CZ.1.07/2.3.00/20.0072 funded by Operational Programme Education for Competitiveness, co-financed by ESF and state budget of the Czech Republic.

We would like to thank all the contributing authors, as well as the members of the Program Committees and the Local Organizing Committee for their hard and highly valuable work. Their work has helped to contribute to the success of the Nostradamus conference.

The Editors

Ivan Zelinka
Otto E. Rössler
Václav Snášel
Ajith Abraham
Emilio S. Corchado

This conference was supported by the Development of human resources in research and development of latest soft computing methods and their application in practice project, reg. no. CZ.1.07/2.3.00/20.0072 funded by Operational Programme Education for Competitiveness, co-financed by ESF and state budget of the Czech Republic.

european social fund in the czech republic

EUROPEAN UNION

MINISTRY OF EDUCATION, YOUTH AND SPORTS

OP Education for Competitiveness

INVESTMENTS IN EDUCATION DEVELOPMENT

IT4Innovations

Centrum excelence

Committees

International Conference Committee

Edward Ott (USA)
Ivan Zelinka (Czech Republic)
Guanrong Chen (Hong Kong)
Otto E. Rössler (Germany)
Sergej Čelikovsky (Czech Republic)
Mohammed Chadli (France)
Ajith Abraham (MIR Labs, USA)
Vaclav Snasel (Czech Republic)
Emilio Corchado (Spain)
Andy Adamatzky (UK)
Jiří Pospíchal (Slovakia)
Jouni Lampinen (Finland)
Juan Carlos Burguillo-Rial (Spain)
Pandian Vasant (Malysia)
Petr Dostál (Czech Republic)
Davendra Donald (Fiji, Czech Republic)
Bernabé Dorronsoro (Luxembourg)
Oplatková Zuzana (Czech Republic)
Linqiang Pan (China)
Šenkeřík Roman (Czech Republic)
Fečkan Michal (Slovakia)
Jašek Roman (Czech Republic)
Joanna Kolodziej (Poland)
Radek Matoušek (Czech Republic)
Hendrik Richter (Germany)
Zdeněk Beran (Czech Republic)
Ana Peleteiro (Spain)
Vadim Strijov (Russia)
Oldřich Zmeškal (Czech Republic)
Masoud Mohammadian (Australia)

Local Conference Committee

Jan Martinovič
Lenka Skanderová
Jan Platoš
Eliška Odchodková
Martin Milata
Pavel Krömer
Michal Krumnikl
Miloš Kudělka
Pavel Moravec
Jiří Dvorský
Tilkova Ludmila
Kvapulinska Petra

Contents

Keynote Speech

Tutorials

Regular Participations

Is Hot Fusion Made Feasible by the Discovery of Cryodynamics?

Otto E. Rössler[1], Ali Sanayei[1], and Ivan Zelinka[2]

[1] Institut für Physikalische und Theoretische Chemie, University of Tübingen, Auf der Morgenstelle 8, D-72076 Tübingen, Germany
otto.rossler@iptc.uni-tuebingen.de, sanayeiali8@gmail.com
[2] VŠB-Technical University of Ostrava, Faculty of Electrical Engineering and Computer Science, Department of Computer Science, 17. listopadu 15, 708 33 Ostrava-Poruba, Czech Republic
ivan.zelinka@vsb.cz

Abstract. In a recent numerical experiment, the previously conjectured new fundamental physical discipline of cryodynamics, sister discipline to thermodynamics, was tentatively confirmed. Cryodynamics governs the statistical hyperchaotic behavior of attractively interacting particles. A terrestrial example is a hot plasma. The predicted phenomenon of paradoxical cooling (hotter particles cool the cooler ones in a process of anti-equipartition) can be used in principle to locally cool, and thereby stabilize, a thermonuclear fusion reactor of the ITER type.

1 Introduction

Statistical mechanics was recently proposed to contain a new subfield besides statistical thermodynamics, called statistical cryodynamics, which possesses the same rank [1]. It has a long history. About seven decades after Clausius introduced the basic notions of statistical thermodynamics, Chandrasekhar proved the existence of "dynamical friction" in a gas of mutually attractive rather than repulsive particles [2]. This notwithstanding, the paradoxical cooling effect described by him went unrecognized in its fundamental significance for another 7 decades. The first groping paper, received with benevolence by the late Ilya Prigogine [3], was very much less sophisticated than Chandrasekhar's. Nonetheless the basic idea was the same and goes back to a 1929 paper of Zwicky [4] (which contained an error as pointed out by its author [5]), Chandrasekhar used partial differential equations of stochastic diffusion type in the footsteps of Smoluchowski and Einstein to demonstrate the "cooling" of a fast, potentially lower-in-kinetic energy, particle in a gas of attractive particles (a globular star cluster). The fact that the energy can be lower without qualitatively changing the effect was only mentioned in a footnote.

I. Zelinka et al. (Eds.): Nostradamus: Mod. Meth. of Prediction, Modeling, AISC 192, pp. 1–4.
springerlink.com

A deterministic description of the same phenomenon was first obtained as a numerical exercise in a recent doctoral dissertation devoted to describing a numerically robust symplectic integration algorithm of the fourth order [6]. The sample calculations performed on a constrained two-particle system in formally one dimension were time consuming. Six out of 11 pertinent runs with differing parameters turned out to show the new behavior, which fact came as a surprise to the maximally meticulous and skeptical author.

2 A Numerical Impasse

Numerical reasons make it virtually impossible to try studying the same braking effect when it is exerted by higher-in-kinetic energy lower-in-mass particles on more massive much slower particles. The predicted symmetry to thermodynamics [1] calls for the effect to persist in this case. Think of injected hot electrons to interactively cool less hot protons in a plasma.

Of course, there are many technical questions to consider in this physical situation. Hot fusion reactors are a science of their own. Here electromagnetism is bound to strongly modify the attractive Newtonian conditions that would be predominant at low mutual velocities of the involved particles. Specifically, both radiation and quantum effects are bound to interfere. Nevertheless, the new effect if valid in the pure Newtonian limiting case is bound to survive in some form also under more complicated conditions.

Therefore, the idea is to for once use a physical analog in order to help numerical mathematics out of an impasse, rather than vice versa. While obviously beset with its own problems, an empirical study of this real-life type may for once be the best chance to come closer to the solution of a theoretical problem that has been simmering for the better part of a century.

3 A Proposed Experiment

Since the prediction made by cryodynamics – that hotter lighter particles statistically cool more massive particles of lower kinetic energy – cannot be tested numerically, an "analog model" needs to be found. The example that comes to mind is *hot fusion*.

Hot fusion, in the form of a Tokamak reactor, is a field in which cryodynamics is of direct concern. There are two classes of hot particles involved in the plasma, both endowed with highly unequal masses differing by a factor of 2.000 or more: protons or helium nuclei on the one hand and electrons on the other. Their temperatures are – as holds true also inside the sun – markedly different for reasons that may turn out to have something to do with the existence of cryodynamics, which is not our current problem, however. Note that both particle classes form a classical thermodynamic equilibrium of their own locally since they repel their own kin.

The problem at stake is the fact that hot fusion does not work so far. In case of the simplest reactor type, the Tokamak, the plasma is being confined magnetically so it cannot touch the walls of the necessarily much cooler container which even will immediately collapse the machine's temperature and thereby interrupt any on-going fusion process.

What would be needed is a "benevolent" cooling mechanism that allows a region that is about to get out of control and touch the wall, to be "called to order" again in a tender fashion as it were.

Cryodynamics offers itself for the purpose – if it is true that even lower-mass particles of a higher temperature can "cool" higher-mass particles, despite the fact that the numerics has currently no chance to show the effect.

4 The Concrete Proposal

Electrons can easily be made very "hot" in their kinetic energy content in so-called synchrotrons. An electron of one electron volt (eV) of kinetic energy has a temperature, kinetic-energy-wise, of about 1.000 Kelvin. Therefore, even a one-million degree hot plasma – as in the sun's core – can easily be imbued locally by concentric beams of electrons of a considerably higher temperature. Fairly simple electron synchrotrons suffice for this purpose. Their beams can be electronically controlled to rapidly converge on any "instability spot" that is detected on the surface of the magnetically confined hot plasma. If this idea is not misleading, empirically testing the concept of an "anti-thermodynamics" (cryodynamics) existing under attractive conditions, in order to overcome a numerical impasse, appears not totally unreasonable. Especially so as at the same time a technological impasse – the elusive solution of humankind's most pressing technological problem – can be hoped to be overcome.

5 Discussion

A numerical proposal has been offered: to recruit a most prestigious high-energy technology for the task of making life easier for a handful of numerical aficionados who call themselves chaos theorists. As a side effect, a technological impasse of decades-long standing would find its unexpected solution. This is, of course, wishful thinking. Hard problems usually remain hard problems. On the other hand, cryodynamics, if it really represents a new fundamental physical science, can be expected to have a certain clout. So the proposal to combine two big topics – a new fundamental insight and an old big technical problem – appears less outrageous than meets the eye.

To conclude, a new "analog computer" was proposed to be recruited for a numerical mathematical experiment. The excuse which lies behind – the hope that the whole of physics might be blessed with a second evolutionary principle besides far-from-equilibrium thermodynamics to which we all owe our own existence – will perhaps motivate the readers to forgive the authors for having tickled their imagination – most likely in vain since hopes have a tendency not to come true.

Acknowledgments. We thank Klaus Sonnleitner, Nils Schopohl, Jürgen Parisi, Christophe Letellier, Joachim Peinke, Peter Kloeden, Frank Kuske, Hans Diebner, Henry Gebhardt, Thimo Böhl, Boris Hagel and Tobias Müller for discussions. For J.O.R. This work was also supported by the European Regional Development Fund in the IT4Innovations Centre of Excellence project (CZ.1.05/1.1.00/02.0070) and by the Development of human resources in research and development of latest soft computing methods and their application in practice project, reg. no. CZ.1.07/2.3.00/20.0072 funded by Operational Programme Education for Competitiveness, co-financed by ESF and state budget of the Czech Republic.

References

[1] Rössler, O.E.: Hun Tun versus Big Bang: How classical chaos implies both thermodynamics and cryodynamics. International Journal of Bifurcation and Chaos 22(2), 1230007-1-1230007-9 (2012), Abstract: http://www.worldscientific.com/doi/abs/10.1142/S0218127412300078

[2] Chandrasekhar, S.: Dynamical friction I, General considerations: the coefficient of dynamical friction. Astrophys. J. 97, 255–263 (1943), http://repository.ias.ac.in/21144/1/326.pdf

[3] Rössler, O.E., Frohlich, D., Kleiner, N.: A Time-symmetric Hubble-like law: Light rays grazing randomly moving galaxies show distance-proportional redshift. Z. Naturforsch. 58a, 807–809 (2003), http://www.znaturforsch.com/aa/v58a/s58a0807.pdf

[4] Zwicky, F.: On the red shift of spectral lines through interstellar space. Proceedings of the National Academy of Sciences 15, 773–779 (1929), http://www.pnas.org/content/15/10/773

[5] Zwicky, F.: On the Possibilities of a gravitational drag of light. Phys. Rev. 34, 1623–1624 (1929), http://authors.library.caltech.edu/5559/1/ZWIpr29.pdf

[6] Sonnleitner, K.: StV4: Ein symplektisches zeitreversibles Störmer–Verlet-Verfahren vierter Ordnung für Hamiltonsche Mehrteilchensysteme mit zwei Anwendungsbeispielen (Gas, T-Rohr-Anordnung)-. Ph.D. dissertation (2010), http://www.wissensnavigator.com/documents/StV4-universell.pdf

Fitness Landscapes and Evolutionary Dynamics

Hendrik Richter

HTWK Leipzig University of Applied Sciences, Faculty of Electrical Engineering and Information Technology, D–04251 Leipzig, Germany
richter@eit.htwk-leipzig.de

Abstract. This key note considers fitness landscapes and their use for understanding evolutionary dynamics in natural and artificial biological systems. Landscape paradigms are meanwhile ubiquitous in several branches of science. This introductory overview discusses concepts, issues and application with a main focus on evolutionary biology and evolutionary computation.

Introductory Overview

The origin of the complexity and beauty of living structures and processes is the product of evolutionary dynamics. The fundamental Darwinian ideas of evolution in connection with genetic coding offer an explanatory framework for these developments of biological systems. Main elements of this framework are an inheritable genetic coding that allows to pass on abilities and features of the living beings, their survival and reproduction success depending on these abilities and features and an uneven distribution of the success due to (possible changing) environmental conditions. In this context, one of the core questions is how the genetic make-up (roughly to the equated with genotype), the abilities and features (approximated as phenotype) and the survival and reproduction success (expressed as fitness) interrelated with each other. An attempt to capture these interrelations are fitness landscapes. In evolutionary biology they are used as a mathematical framework for understanding evolutionary dynamics. In engineering and computer science the Darwinian ideas of evolution found application in optimization and modelling tools using evolutionary computation techniques. Also here, for studying and analyzing these algorithms fitness landscapes are useful. So, the main focus is on finding and/or creating realistic fitness landscapes, on analyzing and visualization tools for fitness landscapes, on how fitness landscapes can help to understand working principles of evolutionary computation techniques, and on how these landscapes can be used to reflect complex evolutionary dynamics.

A (static) fitness landscape Λ_S can be expressed by [4, 13]

$$\Lambda_S = (\mathbb{X}, n, f), \tag{1}$$

I. Zelinka et al. (Eds.): Nostradamus: Mod. Meth. of Prediction, Modeling, AISC 192, pp. 5–8.
springerlink.com

where $\mathbb{X}$ is a configuration space, $n(x)$ is a neighborhood structure that assigns to every $x \in \mathbb{X}$ a set of (more or less distant) neighbors, and $f(x) : \mathbb{X} \rightarrow \mathbb{R}$ is a fitness function that gives to every $x \in \mathbb{X}$ a proprietary quantity to be interpreted as a 'quality' information. In other words, the configuration space in connection with the neighborhood structure expresses a (possibly multi–dimensional) 'location', while the fitness is an orthogonal projection from location, defining an 'elevation' or 'height' and at the same time giving a location its most important property. Fitness is usually considered a single parameter but it seems perfectly possible to have a height measure with several dimensions.

The origin and character of configuration space, neighborhood structure and fitness function differ, naturally, in evolutionary biology and evolutionary computation. In evolutionary biology, the configuration space is made up by the genotypes of the biological system under study [17, 3]. The genotype characterizes the genetic make–up of a generic individual. It comprises of the sum (or union) of all genetically possible individuals and hence is the total genetic information. The neighborhood of a genotypical location is usually defined by the property of which genotypes can mutate from one to another [3]. Assigning fitness to each element of the genotypical space requires additional considerations. Up until recently, this question was answerable only purely theoretical and also requires to define an intermediate level between genotype and fitness, the phenotypical space. The reason for that is that it is complicated or even infeasible to assign a fitness value to the 'microscopic' genotype. Fitness, at least in any sensible biological sense, is connected to longevity and fertility and ultimately to reproduction success of a specific individual. Such a phenotypical individual can be thought of as an instance of the generic individual specified by a genotype. Hence, such a fitness landscape Λ_S is, strictly speaking, the product of a genotype–to–phenotype–to–fitness mapping and such landscapes have been the subject of much theoretical work on evolutionary dynamics [2, 15, 14, 8, 9]. However, some recent studies have shown that a direct experimental approach to construct fitness landscapes and analyze possible evolutionary pathways is possible [7, 5]. These results have led to a renewed interest in the framework of fitness landscapes as for the first time the question of the predictability of real evolutionary processes became addressable.

There is a fine but important conceptional difference in the approach to fitness landscapes in evolutionary biology and evolutionary computation. The main focus in evolutionary biology is to look for what fitness landscapes origin if we employ methods to extract its structure and topology from real biological data and what conclusions about the working and the outcome of the evolutionary process can be drawn from these landscapes. In evolutionary computation the main focus is on search for topological features in a landscape that is given by the optimization problem under study. This can be related to the question of how an evolutionary algorithm (an artificial model mimicking a simplified version of natural evolutionary processes) interacts with the landscape and what behavior and performance can be expected in the search.

Fig. 1 Static fitness landscape in $\mathbb{R}^2$ as mountainous region with peaks, valleys, ridges and plateaus

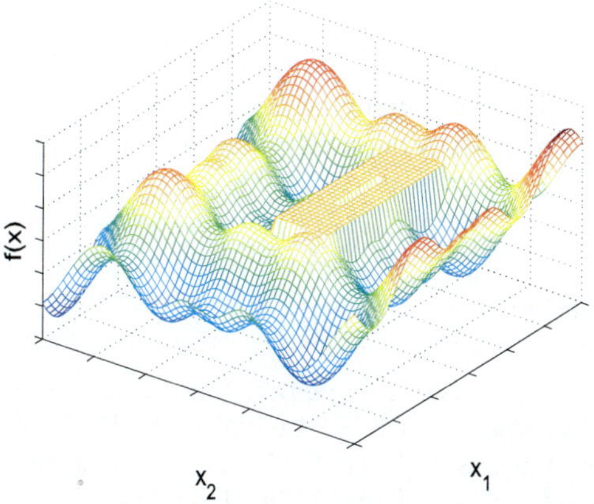

In evolutionary computation, therefore, the configuration space is made up by the search space obtained from encoding all possible solutions of the optimization problem. The neighborhood structure is a consequence of the search space and hence the objects to be optimized over, but also of the genetic operators the evolutionary search employs [4]. If the search space is metric (as for instance if the search space elements are real or integer numbers, and the genetic operators act on these numbers), then the neighborhood structure is inherent by the ordering of numbers. If the search space is not metric (or can have several different kinds of metrics), the neighborhood structure needs to be defined additionally. Examples are binary coding, where the neighborhood structure can be a Hamming distance of different length, or tree representation where the neighbors of a branch differ by a (smaller or larger) variation in a subtree.

As discussed so far, fitness landscapes in both evolutionary biology and evolutionary computation have the same base and are an attempt to answer similar or highly related questions. If, as a special case, we consider the configuration space as two–dimensional and the neighborhood structure continuously metric, we end up with the fitness landscape metaphor frequently depicted: that of a mountainous region with peaks, valleys, ridges and plateaus, see Fig. 1. It appears almost a little surprising that such a rather naive picture has meaning in branches of sciences as illustrious as physics, biology and informatics. Interestingly, landscape paradigms are closely related to conceptualization of behavior that is usually related to complexity [10, 1]. The main motivation to employ a landscape approach is that it offers a framework for a computational treatment. This treatment becomes geometrically interpretable in a meaningful way for the aforementioned simple two–dimensional case, but there is a multitude of ways to employ landscape measures or visualization methods that are applicable for any given configuration space dimension [8, 9, 16, 6].

The fitness landscape approach presented here has been applied to different kinds of problems in both evolutionary biology and evolutionary computation, and yields understanding of evolutionary dynamics. I think that the recent progresses and findings might be the beginnings for further developments that promise to address even more fundamental questions about the working and the outcome of evolutionary processes.

References

1. Frauenfelder, H., Leeson, D.T.: The energy landscape in non–biological and biological molecules. Nature Structural Biology 5, 757–759 (1998)
2. Kauffman, S.A., Levin, S.: Towards a general theory of adaptive walks on rugged landscapes. J. Theor. Biology 128, 11–45 (1987)
3. Kauffman, S.A.: The Origin of Order: Self–Organization and Selection in Evolution. Oxford University Press, New York (1993)
4. Kallel, L., Naudts, B., Reeves, C.R.: Properties of fitness functions and search landscapes. In: Kallel, L., Naudts, B., Rogers, A. (eds.) Theoretical Aspects of Evolutionary Computing, pp. 177–208. Springer, Heidelberg (2001)
5. Lobkovsky, A.E., Wolf, Y.I., Koonin, E.V.: Predictability of evolutionary trajectories in fitness landscapes. PLoS Comput. Biol. 7(12), e1002302 (2011), doi:10.1371/journal.pcbi.1002302
6. McCandlish, D.M.: Visualizing fitness landscapes. Evolution 65, 1544–1558 (2011)
7. Poelwijk, F.J., Kiviet, D.J., Weinreich, D.M., Tans, S.J.: Empirical fitness landscapes reveal accessible evolutionary paths. Nature 445, 383–386 (2007)
8. Richter, H.: Coupled map lattices as spatio–temporal fitness functions: Landscape measures and evolutionary optimization. Physica D237, 167–186 (2008)
9. Richter, H.: Evolutionary Optimization and Dynamic Fitness Landscapes: From Reaction-Diffusion Systems to Chaotic CML. In: Zelinka, I., Celikovsky, S., Richter, H., Chen, G. (eds.) Evolutionary Algorithms and Chaotic Systems. SCI, vol. 267, pp. 409–446. Springer, Heidelberg (2010)
10. Sherrington, D.: Landscape paradigms in physics and biology: Introduction and Overview. Phys. D 107, 117–121 (1997)
11. Smith, T., Husbands, P., Layzell, P., O'Shea, M.: Fitness landscapes and evolvability. Evolut. Comput. 10, 1–34 (2002)
12. Stadler, P.F.: Landscapes and their correlation functions. J. Math. Chem. 20, 1–45 (1996)
13. Stadler, P.F., Stephens, C.R.: Landscapes and effective fitness. Comm. Theor. Biol. 8, 389–431 (2003)
14. Vassilev, V.K., Fogarty, T.C., Miller, J.F.: Information characteristics and the structure of landscapes. Evolut. Comput. 8, 31–60 (2000)
15. Weinreich, D.M., Watson, R.A., Chao, L.: Sign epistasis and constraint on evolutionary trajectories. Evolution 59, 1165–1174 (2005)
16. Wiles, J., Tonkes, B.: Hyperspace geography: Visualizing fitness landscapes beyond 4D. Artificial Life 12, 211–216 (2006)
17. Wright, S.: The roles of mutation, inbreeding, crossbreeding and selection in evolution. In: Jones, D.F. (ed.) Proc. of the Sixth International Congress on Genetics, pp. 356–366 (1932)

Observer Design for Polytopic Systems: Application to Chaotic System Reconstruction

Mohammed Chadli

University of Picardie Jules Verne, 7, Rue du Moulin Neuf, 80000, Amiens, France
mohammed.chadli@u-picardie.fr

Abstract. Many studies concerning design of controllers and observers for a class of nonlinear systems described in polytopic representation are carried out. Such representation includes Takagi-Sugeno models, LPV models, switching models, PLDI... Particularly, T-S models are obtained by interpolation of M local LTI (linear time invariant) models throughout convex functions. The choice of the number of local models may be intuitively chosen by considering some operating regimes. Each LTI model can be obtained by using a direct linearization of an a priori nonlinear model around operating points, or alternatively by using an identification procedure. Based on the Lyapunov method and Linear Matrix Inequalities (LMI) formulation, sufficient conditions have been derived for controllers and observers design. Recently, systems subject to unknown inputs are considered for measurable and immeasurable decision variables. Unknown inputs can result either from model uncertainty, faults or due to the presence of unknown external excitation. These different results have been widely applied in the field of fault diagnosis (FDI), fault tolerance (FTC) and also for secure communications. Indeed, the increasing need of secure communications leads to the development of many techniques which make difficult the detecting of transmitted message. Based on unknown inputs observer design, many works have been carried out on secure communication and chaotic system reconstruction problem. In this framework, unknown inputs Takagi-Sugeno fuzzy observer has been exstensively used. The design of such observers is considered based on LMI and Lyapunov methods. The pole placement in an LMI region is also considered to improve the observer performances. Examples are given to illustrate a chaotic cryptosystem procedure where the plaintext (message) is encrypted using chaotic signals at the drive system side and the plaintext is retrieved via the designed unknown input observer.

Keywords: Fuzzy model, unknown inputs, state estimation, Lyapunov method, linear matrix inequalities (LMI), chaotic system reconstruction.

1 Introduction

In last two decades, many studies concerning stability analysis and design of controllers and observers for a class of systems described by multiple model approach

I. Zelinka et al. (Eds.): Nostradamus: Mod. Meth. of Prediction, Modeling, AISC 192, pp. 9–24.
springerlink.com

[2] are carried out. Such representation results from the interpolation of M local LTI (linear time invariant) models throughout convex functions. These functions can be viewed as a weighted sum of local LTI models and quantify the relative contribution of each local model to the global model. The choice of the number of local models may be intuitively chosen by considering some operating regimes. Each LTI model can be obtained by using a direct linearization of an a priori nonlinear model around operating points, or alternatively by using an identification procedure [2, 4]. From a practical point of view, LTI model describes the system's local behavior around the ith, $i : 1...M$ regime. This approach includes Takagi-Sugeno fuzzy models [5] and PLDI representation [1]. Based on the Lyapunov method and Linear Matrix Inequalities (LMI) formulation, sufficient conditions have been derived for stability analysis, controllers and observers design (see among others [3, 32, 8, 9, 6]). Recently, systems subject to unknown inputs are extensively considered in the literature. Unknown inputs can result either from model uncertainty, faults or due to the presence of unknown external excitation. This problem, usually referred as the unknown input observer design, has been considered actively for linear systems [10, 11, 12, 13, 14], for descriptor and nonlinear systems (see among other [15, 16, 17] and for T-S model approach (see for example [3, 7] and references therein).

The increasing need of secure communications leads to the development of many techniques which make difficult the detecting of transmitted message. Indeed, the problem we are faced with consists of transmitting some coded message with a signal broadcasted by a communication channel. At the receiver side, the hidden signal is recovered by a decoding system. A rich amount of literature, working with synchronization, exist (see for example [18], [19], [20], [21], [22], [36], [37], [38], [39], [40]). Evolutionary algorithms (EA) are also capable to synchronize simple chaotic systems, without knowledge of internal system structure, see for example [23], [24], [25], [26], [27], [28], [29] where applications of EAs on chaotic dynamic control and synchronization are given.

The goal of this study is to show how to use chaotic T-S model and how to design the proposed observers for chaotic system reconstruction. Then in section 2 and 3, a considered unknown inputs T-S model in continuous-time case and his corresponding observer are given. Synthesis conditions for the proposed observer are given in LMI terms. To improve the performances of the proposed observer, the pole assignment in a LMI region is also studied. Unknown input estimation is given in section 3. Then these design conditions are extended to unknown inputs discrete-time T-S model in section 4. To illustrate the given synthesis LMI conditions, numerical examples and applications dealing with the chaotic system reconstruction for both continuous-time and discrete-time T-S model are proposed.

Notation. Throughout this paper, $\mathbf{R}^n$ and $\mathbf{R}^{n\times m}$ denote, respectively, the n dimensional Euclidean space and the set of all $n \times m$ real matrices. The notation $X > Y$ where X and Y are symmetric matrices, means that $X - Y$ is positive definite. $\otimes$ is the Kronecker product, $\mathbf{I}$ is the identity matrix with compatible dimensions, the symbol $(*)$ denotes the transpose elements in the symmetric positions and $I_M = \{1, 2, \cdots, M\}$.

2 Unknown Input T-S Fuzzy Model Representation

Consider a continuous-time fuzzy models with unknown inputs defined by

$$\begin{cases} \dot{x}(t) = \sum_{i=1}^{M} \mu_i(\xi(t))(A_i x(t) + B_i u(t) + R_i v(t)) \\ y(t) = Cx(t) + Fv(t) \end{cases} \tag{1}$$

with

$$\mu_i(\xi(t)) \geq 0, \sum_{i=1}^{M} \mu_i(\xi(t)) = 1 \tag{2}$$

M being the number of sub-models, $x(t) \in \mathbf{R}^n$ the state vector, $u(t) \in \mathbf{R}^m$ the input vector, $v(t) \in \mathbf{R}^q$, the unknown input and $y \in \mathbf{R}^p$ the measured outputs. $A_i \in \mathbf{R}^{n\times n}$, $B_i \in \mathbf{R}^{n\times m}$, $D_i \in \mathbf{R}^n$ and $C \in \mathbf{R}^{p\times n}$ define the ith local model. Matrices $R_i \in \mathbf{R}^{n\times q}$ and $F \in \mathbf{R}^{p\times q}$ represent the influence of the unknown inputs. We assume that $q < p$ and, without loss of generality, that

Assumption 1: $rank(F) = q$ and $rank(R_i) = q$, i.e. F and R_i are full column ranks.

Assumption 2: $rank(C) = p$, i.e. C is full row rank.

The *activation functions* $\mu_i(.)$ depend on the so-called decision vector $\xi(t)$ assumed to depend on measurable variables.

We are concerned by the reconstruction of state variable $x(t)$ of unknown inputs T-S model (1) using only the available information, namely known input $u(t)$ and measured output $y(t)$.

3 Unknown Input T-S Fuzzy Observer Design

In order to estimate the state of the unknown input T-S fuzzy model (1), the considered unknown input T-S observer structure has the following form

$$\begin{cases} \dot{z}(t) = \sum_{i=1}^{M} \mu_i(\xi) \left(N_i z(t) + G_i u(t) + L_i y(t) \right) \\ \hat{x}(t) = z(t) - Ey(t) \end{cases} \tag{3}$$

The considered observer only uses known variables ($u(t)$ and $y(t)$) and the same activation functions $\mu_i(.)$ as used for the T-S model (1). The unknown inputs $v(t)$ are considered non available. The variables $N_i \in \mathbf{R}^{n\times n}$, $G_i \in \mathbf{R}^{n\times m}$, $L_i \in \mathbf{R}^{n\times p}$ and $E \in \mathbf{R}^{n\times p}$ are the observer gains to be determined in order to estimate the state of the unknown input T-S model (1). For that purpose, let us define the state estimation error:

$$\tilde{x}(t) = x(t) - \hat{x}(t) \tag{4}$$

The following subsections give LMI conditions satisfying $\tilde{x}(t) \to 0$ when $t \to \infty$. Poles placement in LMI region to improve the performance of the designed T-S observer finish this section.

3.1 LMI Synthesis Conditions

The following result gives sufficient LMI conditions guaranteeing the global asymptotic convergence of state estimation error (4).

Theorem 1. *The state estimation error between T-S observer (3) and unknown input T-S model (1) converges globally asymptotically towards zero, if there exists matrices $X > 0$, S and W_i such that the following conditions hold $\forall\, i \in I_M$:*

$$A_i^T X + XA_i + A_i^T C^T S^T + SCA_i - W_iC - C^T W_i^T < 0 \tag{5a}$$
$$(X + SC)R_i = W_iF \tag{5b}$$
$$SF = 0 \tag{5c}$$

Then T-S observer (3) is completely defined by:

$$E = X^{-1}S \tag{6a}$$
$$G_i = (\mathbf{I} + X^{-1}SC)B_i \tag{6b}$$
$$N_i = (\mathbf{I} + X^{-1}SC)A_i - X^{-1}W_iC \tag{6c}$$
$$L_i = X^{-1}W_i - N_iE \tag{6d}$$

Proof. From estimation error (4) with the expression of $\hat{x}(t)$ given by T-S observer (3) and T-S model (1), we obtain the following expression:

$$\tilde{x}(t) = (\mathbf{I} + EC)x(t) - z(t) + EFv(t) \tag{7}$$

The dynamic of state estimation error is then given by

$$\dot{\tilde{x}}(t) = \sum_{i=1}^{M} \mu_i(\xi)\Big(N_i\tilde{x}(t) + \big(TA_i - K_iC - N_i\big)x(t) + \big(TB_i - G_i\big)u(t) + \big(TR_i - K_iF\big)v(t)\Big) + EF\dot{v}(t) \tag{8}$$

with

$$T = \mathbf{I} + EC, K_i = N_iE + L_i \tag{9}$$

Using (9) and the following change of variables

$$W_i = XK_i, \;\; S = XE \tag{10}$$

with $X > 0$, the above expression can be rewritten as follows

$$\dot{\tilde{x}}(t) = \sum_{i=1}^{M} \mu_i(\xi)\Big(N_i\tilde{x}(t) + \big(TA_i - K_iC - N_i\big)x(t) + \big((\mathbf{I} + X^{-1}SC)B_i - G_i\big)u(t) + X^{-1}SF\dot{v}(t) + X^{-1}\big((X + SC)R_i - W_iF\big)v(t)\Big) \tag{11}$$

Then taking account (5b-c) and (6b-c), we get

$$\dot{\tilde{x}}(t) = \sum_{i=1}^{M} \mu_i(\xi) N_i \tilde{x}(t) \tag{12}$$

with $N_i = TA_i - K_iC$. Then, the state estimation error (12) converges asymptotically to zero if there exist $X > 0$ such that $\forall\, i \in I_M$:

$$XN_i + N_i^T X < 0 \tag{13}$$

With the same variable change (10), inequalities (13) are equivalent to

$$(X + SC)A_i - W_iC + ((X + SC)A_i - W_iC)^T < 0 \tag{14}$$

This completes the proof.

Remark 1. Classical numerical tools may be used to solve LMI problem (5a) on variables $X > 0$, S, W_i and scalars ε_i with the linear equality constraints (5b). The developed results can easily be solved using numerical tools such as the LMITOOL [42].

Remark 2. The proposed result deals with the case of common output matrix $C_i = C, \forall i \in I_M$. The case of different multiple output matrices leads to non convex constraints not easy to resolve with existing numerical tools.

3.2 Unknown Inputs Estimation

A lot of works have been considered for the unknown input estimation problem (see for example [34, 41]). The method proposed here is based on the hypothesis of the good estimation of the state variables [34]. Indeed, when the state estimation error is equal to zero, we get the following approximation:

$$\hat{y} = C\hat{x} + F\hat{v} \tag{15}$$

Since the assumption 1 holds, i.e. the matrix F is of full column rank, an estimation of unknown inputs can be carried out in a simpler way by

$$\hat{v} = F^+(y - \hat{y}) \tag{16}$$

where F^+ is any generalized inverse of F such that $FF^+F = F$.

3.3 Performance Improvement

In this part, we investigate how to improve the performances of the proposed observer (3) for T-S fuzzy model (1). In order to achieve a desired transient performance, a pole placement should be considered. For many problems, exact pole assignment may not be necessary; it suffices to locate the pole in a sub-region of the complex left half plane [35]. This section discusses a pole assignment in LMI regions $S(\alpha,\beta)$.

Since a prescribed LMI region will be added as supplementary constraint to these of theorem 1, it suffices to locate the poles of matrix $\sum_{i=1}^{M} \mu_i(\xi(t))N_i$ in prescribed LMI regions. Indeed, estimation error (12) is D-stable if there exists a matrix $X > 0$ such that [35]:

$$\alpha \otimes X + \beta \otimes (N_i X) + \beta^T \otimes (N_i X)^T < 0 \tag{17}$$

With the same changes of variables (10) applied to inequalities (17), we obtain the following result.

Corollary 1: If there exit matrices $X > 0$, S and W_i such that the following conditions hold $\forall\, i \in I_M$:

$$\alpha \otimes X + \beta \otimes (XA_i + SCA_i - W_i C) + \beta^T \otimes (XA_i + SCA_i - W_i C)^T < 0 \tag{18a}$$

$$(X + SC)R_i = W_i F \tag{18b}$$

$$SF = 0 \tag{18c}$$

Then, T-S observer (3) is globally asymptotically convergent with the performance defined by complex region $S(\alpha,\beta)$. The observer gains are defined by (6).

3.4 Numerical Example

To show the validness of the proposed results, a numerical example is proposed. The example deals also with poles placement. Now, consider the T-S model (1) with the following data:

$$A_1 = \begin{bmatrix} -3 & 1.5 & 1.5 \\ 1.5 & -4 & 0 \\ 3 & 1.5 & -9 \end{bmatrix},\ A_2 = \begin{bmatrix} -6 & 4 & 5 \\ 6 & -4 & 0 \\ 1 & 2 & -5 \end{bmatrix}$$

$$B_1 = \begin{bmatrix} 1 \\ -1 \\ -0.5 \end{bmatrix},\ B_2 = \begin{bmatrix} -1 \\ 2 \\ -0.5 \end{bmatrix},\ F = \begin{bmatrix} 1 \\ 1.2 \end{bmatrix}$$

$$R_1 = \begin{bmatrix} 1 \\ -1 \\ 1.2 \end{bmatrix},\ R_2 = \begin{bmatrix} 1 \\ 1.5 \\ -1 \end{bmatrix},\ C = \begin{bmatrix} 1 & 1 & 1 \\ 1 & 0 & 1 \end{bmatrix}$$

In order to guarantee some performances of the observer to be designed, a region $S_r(\alpha,\beta)$ defined as the intersection between a circle, of center $(0,0)$ and of radius β, and the left half plane limited by a vertical straight line of x-coordinate equal to $(-\alpha)$, $\alpha > 0$, is considered. The corresponding LMI formulation is given by the following LMI conditions:

$$\begin{bmatrix} -\beta X & XA_i + SCA_i - W_iC \\ (*) & -\beta X \end{bmatrix} < 0 \tag{19a}$$

$$XA_i + SCA_i - W_iC + (XA_i + SCA_i - W_iC)^T + 2\alpha X < 0 \tag{19b}$$

$$(X + SC)R_i = W_iF \tag{19c}$$

$$SF = 0 \tag{19d}$$

With $\alpha = 2.5$ and $\beta = 12$, the resolution of conditions (19) gives:

$$\mathrm{X} = 10^{-2}\begin{bmatrix} 1 & 0 & 0 \\ 0 & 1 & 0 \\ 0 & 0 & 1 \end{bmatrix}, W_1 = \begin{bmatrix} 0.0298 & -0.0153 \\ 0.0688 & -0.0612 \\ 0.0397 & -0.0194 \end{bmatrix}$$

$$W_2 = \begin{bmatrix} 0.0517 & -0.0375 \\ 0.0643 & -0.0516 \\ 0.0452 & -0.0548 \end{bmatrix}, S = 10^{-3}\begin{bmatrix} -2.2 & 1.8 \\ -8.4 & 7.0 \\ -7.1 & 5.9 \end{bmatrix}$$

From (6), the corresponding T-S observer is defined by:

$$E = \begin{bmatrix} -0.22 & 0.18 \\ -0.84 & 0.70 \\ -0.71 & 0.59 \end{bmatrix} \quad L_1 = \begin{bmatrix} 1.65 & -0.43 \\ 0.21 & -0.57 \\ -2.95 & 3.83 \end{bmatrix} \quad L_2 = \begin{bmatrix} 5.39 & -3.94 \\ -1.06 & 1.07 \\ 0.95 & -2.51 \end{bmatrix}$$

$$G_{11} = \begin{bmatrix} 1.20 \\ -0.22 \\ 0.15 \end{bmatrix}, G_{21} = \begin{bmatrix} -1.38 \\ 0.52 \\ -1.74 \end{bmatrix}$$

4 Extension to Discrete-Time T-S Observer Design

Consider a discrete-time T-S system subject to unknown inputs represented by

$$\begin{cases} x(t+1) = \sum_{i=1}^{M} \mu_i(\xi(t))(A_ix(t) + B_iu(t) + R_iv(t)) \\ y(t) = Cx(t) + Fv(t) \end{cases} \tag{20}$$

where $x(t) \in \mathbf{R}^n$ is the state vector, $u(t) \in \mathbf{R}^m$ the input vector, $v(t) \in \mathbf{R}^q$, $q < n$, contains the unknown inputs and $y(t) \in \mathbf{R}^p$ the measured outputs. Matrices $A_i \in \mathbf{R}^{n\times n}$ and $B_i \in \mathbf{R}^{n\times m}$. Matrices $R_i \in \mathbf{R}^{n\times q}$ and $F \in \mathbf{R}^{p\times q}$ are assumed to satisfy assumptions 1 and 2.

The considered structure of the T-S observer is

$$\begin{cases} z(t+1) = \sum_{i=1}^{M} \mu_i(\xi(t))\big(N_iz(t) + G_iu(t) + L_iy(t) \\ \hat{x}(t) = z(t) - Ey(t) \end{cases} \tag{21}$$

where $N_i \in \mathbf{R}^{n\times n}$, $G_i \in \mathbf{R}^{n\times m}$, $E \in \mathbf{R}^{n\times p}$, $L_i \in \mathbf{R}^{n\times p}$ are the observer gains to be determined. The following result gives sufficient conditions for the global asymptotic convergence of the T-S observer (21).

Theorem 1. *The state estimation error between T-S model (20) and unknown input T-S observer (21) converges globally asymptotically towards zero if there exists matrices $X > 0$, S and W_i such that the following conditions hold $\forall\, i \in I_M$:*

$$\begin{pmatrix} X & * \\ XA_i + SCA_i - W_iC & X \end{pmatrix} > 0 \tag{22a}$$

$$(X + SC)R_i = W_iF \tag{22b}$$

$$SF = 0 \tag{22c}$$

T-S observer (21) is then completely defined by:

$$E = X^{-1}S \tag{23a}$$

$$G_i = (\mathbf{I} + X^{-1}SC)B_i \tag{23b}$$

$$N_i = (\mathbf{I} + X^{-1}SC)A_i - X^{-1}W_iC \tag{23c}$$

$$L_i = X^{-1}W_i - N_iE \tag{23d}$$

Pole assignment - To improve performances of the T-S observer for better estimation of system state, dynamics of the T-S observer are constrained to be faster than that of the T-S model. As stated above, it is possible to assign the poles to a specific sub-region is the complex plane [35]. For example if the prescribed region $S_r(\sigma, r)$ is a disk centered at $(\sigma, 0)$ and radius r as shown in figure (1), the LMI formulation of the previous problem is expressed by the following corollary.

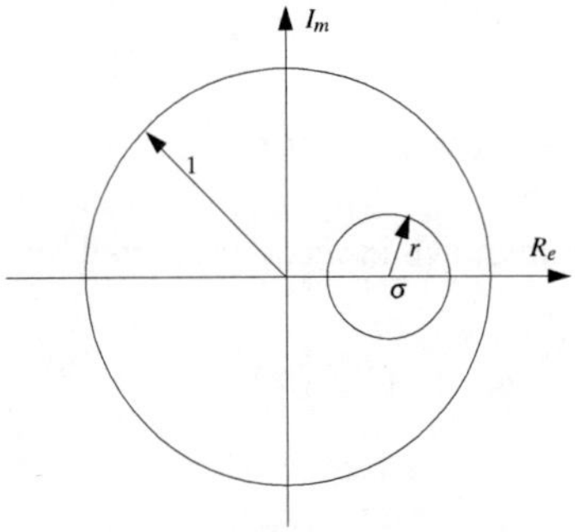

Fig. 1 Sub-region $S_r(\sigma, r)$ for poles location

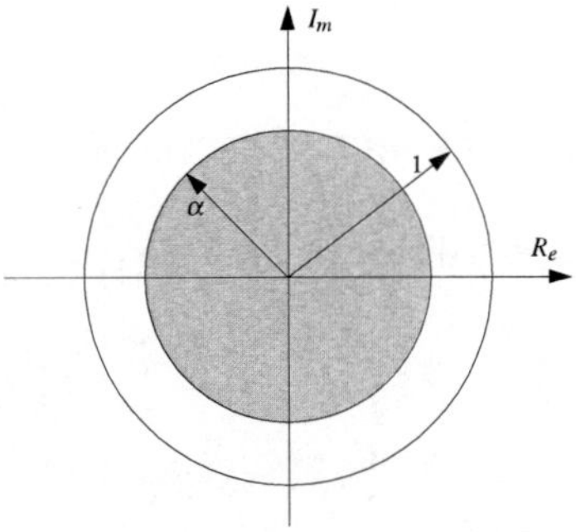

Fig. 2 Sub-region $S_0(0, \alpha)$ for poles location

Corollary 1: If there exist matrices X, S and W_i such that the following conditions hold $\forall\, i \in I_M$:

$$\begin{pmatrix} rX & * \\ X(A_i - \sigma\mathbf{I}) + SC(A_i - \sigma\mathbf{I}) - W_iC & rX \end{pmatrix} > 0 \tag{24a}$$

$$(X + SC)R_i = W_iF \tag{24b}$$

$$SF = 0 \tag{24c}$$

then the T-S observer (21) is globally asymptotically convergent with the performance defined by the complex region $S(\sigma, r)$. The observer parameters are as defined by (6).

Remark 3. Note that the LMI constraints (24) can be obtained from (22) by simply replacing the matrices A_i by $(A_i - \sigma\mathbf{I})/r$. Moreover if we are interested by the region $S_0(0, \alpha)$ as shown in figure (2) it suffices to chose $\sigma = 0$ and $r = \alpha$.

Summarizing the estimation procedure, the design of T-S observer and the estimation of unknown inputs can be implemented as follows:

i) Solve the linear constraints (22) with numerical tools such as the LMITOOL software,

ii) Deduce the observer parameters N_i, G_i, L_i and E of the T-S observer (21) using the equations (6).

iii) Under the assumption 1, estimate unknown input estimation using equation (16).

Remark 4. Recently, design conditions are developped in [43]. These conditions take into account of the case of unmeasurable decision variables. Moreover they give more relaxed design conditions by including addional variables and using polyquadratic lyapunov functions.

5 Application to Chaotic System Reconstruction

In this section, the proposed T-S observer is used to reconstruct states of chaotic systems and can be exploited in secure communication scheme. The message to be encoded is the unknown input of the chaotic T-S model.

5.1 Example 1: Continuous-Time Case

Results developed in section 3 can be applied to reconstruct states of chaotic system and also for a secure communication system. Indeed, the problem we are faced with consists of transmitting some coded message with a signal broadcasted by a communication channel. At the receiver side, the hidden signal is recovered by a decoding system. The increasing need of secure communications leads to the development of many techniques which make difficult the detecting of transmitted message (se for example [36], [37], [38], [39], [40]). In this section, our goal to show how the designed observer could be used in chaotic system reconstruction and in a secure communication scheme. For this purpose we use the nonlinear Lorenz model as chaotic systems represented by his equivalent chaotic T-S model. Consider the non linear Lorenz equation:

$$\begin{cases} \dot{x}_1(t) = -ax_1(t) + ax_2(t) \\ \dot{x}_2(t) = cx_1(t) - x_2(t) - x_1(t)x_3(t) \\ \dot{x}_3(t) = x_1(t)x_2(t) - bx_3(t) \end{cases} \tag{25}$$

Which can be rewritten as follows

$$\dot{x}(t) = A(x(t))x(t) \tag{26}$$

with:

$$x = \begin{bmatrix} x_1 \\ x_2 \\ x_3 \end{bmatrix}, \quad A(x(t)) = \begin{bmatrix} -a & a & 0 \\ c & -1 & -x_1(t) \\ 0 & x_1(t) & -b \end{bmatrix}$$

and a, b, and c are constants. Assume that $x_1(t) \in [-d,d]$ with $d > 0$. Then, we can write $x_1(t) = -d.\mu_1(x_1(t)) + d.\mu_2(x_1(t))$ with $\mu_1(x_1(t)) + \mu_2(x_1(t)) = 1$ which leads to the following T-S model:

$$\dot{x}(t) = (\mu_1(x_1(t))A_1 + \mu_2(x_1(t))A_2)\,x(t) \tag{27}$$

where

$$A_1 = \begin{bmatrix} -a & a & 0 \\ c & -1 & -d \\ 0 & d & -b \end{bmatrix}, \quad A_2 = \begin{bmatrix} -a & a & 0 \\ c & -1 & d \\ 0 & -d & -b \end{bmatrix}$$

and

$$\mu_1(x_1(t)) = \frac{1}{2}\left(1 + \frac{x_1(t)}{d}\right), \quad \mu_2(x_1(t)) = \frac{1}{2}\left(1 - \frac{x_1(t)}{d}\right)$$

Note that the obtained T-S model exactly represents the nonlinear Lorenz model under $x_1(t) \in [-d,d]$.

In the following, we consider the chaotic T-S model (27) with $a = 10$, $b = 8/3$, $c = 28$ and $d = 30$ in his general form:

$$\begin{cases} \dot{x} = \sum_{i=1}^{2} \mu_i(y_1)\left(A_i x + R_i v(t)\right) \\ y = Cx + Fv(t) \end{cases} \tag{28}$$

with:

$$A_1 = \begin{bmatrix} -10 & 10 & 0 \\ 28 & -1 & -30 \\ 0 & 30 & -{}^{8}\!/_{3} \end{bmatrix}, A_2 = \begin{bmatrix} -10 & 10 & 0 \\ 28 & -1 & 30 \\ 0 & -30 & -{}^{8}\!/_{3} \end{bmatrix}$$

$$B_1 = \begin{bmatrix} 0 \\ 0 \\ 0 \end{bmatrix} \; B_2 = \begin{bmatrix} 0 \\ 0 \\ 0 \end{bmatrix} \; C = \begin{bmatrix} 1 & 0 & 0 \\ 0 & 1 & 0 \end{bmatrix} \; F = \begin{bmatrix} 1 \\ 1 \end{bmatrix}$$

The message to be encoded constitutes the so-called unknown input of the T-S model which plays the role of the encoder. The output of this model is transmitted using a public channel. On the receiving side, an unknown input T-S observer serves as a decoder in order to re-build the message. The goal of the proposed example is only to show the feasibility of the proposed design in chaotic system reconstruction and in secure communication procedure. The hidden transmitted message is given in figure 3.

The corresponding T-S observer is

$$\begin{cases} \dot{z} = \sum_{i=1}^{2} \mu_i(y_1)\left(N_i z + L_i y\right) \\ \hat{x} = z - Ey \end{cases} \tag{29}$$

The resolution of conditions (5) with $B_1 = B_2 = (0,0,0)^\top$ lead to the following result:

$$X = \begin{bmatrix} 1.750 & 1.650 & -0.003 \\ 1.650 & 1.750 & -0.003 \\ -0.003 & -0.003 & 0.195 \end{bmatrix}, E = \begin{bmatrix} -3.05 & 3.05 \\ 3.99 & -3.99 \\ -0.004 & 0.004 \end{bmatrix}$$

$$N_1 = \begin{bmatrix} 33.66 & 47.89 & -91.72 \\ -36.18 & -45.78 & 89.96 \\ 61.12 & -32 & -2.79 \end{bmatrix}, N_2 = \begin{bmatrix} 35.06 & 49.54 & 91.72 \\ -37.82 & -47.14 & -89.96 \\ -62.08 & 31.20 & -2.53 \end{bmatrix}$$

$$L_1 = \begin{bmatrix} -16.44 & 17.44 \\ -14.94 & 15.94 \\ 253.9 & -252.9 \end{bmatrix} \quad L_2 = \begin{bmatrix} -19.38 & 17.32 \\ -13.67 & 17.67 \\ -252.38 & 253.37 \end{bmatrix}$$

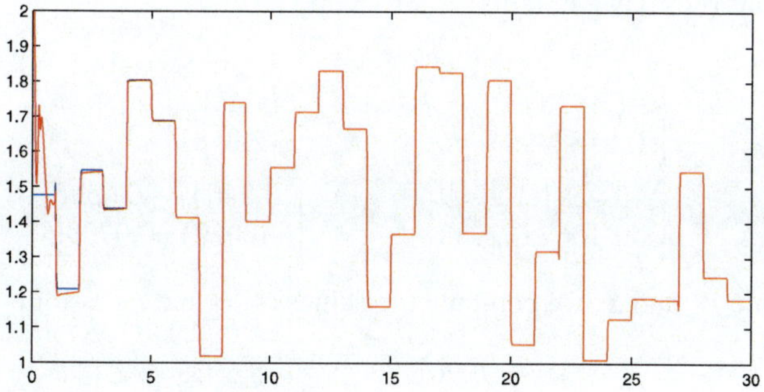

Fig. 3 Hidden message v and its estimate

Figure 3 displays the hidden transmitted message and its estimate.

5.2 *Example 2: Discrete-Time Case*

Consider a chaotic discrete-time T-S model that results from the interpolation of two local models:

$$\begin{cases} x(t+1) = \sum_{i=1}^{2} \mu_i(\xi(t))\big(A_i x(t) + R_i v(t)\big) \\ y(t) = Cx(t) + Fv(t) \end{cases} \tag{30}$$

The functions $\mu_i(.)$ depend on the T-S model output and expressed by

$$\mu_1(\xi(t)) = \frac{1-\xi(t)}{2}, \mu_2(\xi(t)) = 1 - \mu_1(\xi(t)) \tag{31}$$

with $\xi(t) = x_2(t)$. The numerical values of matrices are as follows:

$$A_1 = \begin{bmatrix} 0 & 1 & -0.1 \\ 1 & 0 & 0 \\ 0 & 1 & 0 \end{bmatrix}, \quad A_2 = \begin{bmatrix} 0 & -1 & -0.1 \\ 1 & 0 & 0 \\ 0 & 1 & 0 \end{bmatrix}, \quad D = \begin{bmatrix} 1.76 \\ 0 \\ 0 \end{bmatrix}, \quad C = \begin{bmatrix} 0 & 1 & 0 \end{bmatrix}, \quad F = 5$$

From the structure of T-S model (30), we can deduce the following values:

$$S = 0, \;\; E = 0, \;\; G_i = 0$$

For this example, since the encoding system (30) can be conceived at the same time as the decoding system (observer), the computation matrices R_i is then free. Thus, LMI (22a) can be solved without taking into account equalities (22b-c). The resolution of LMI (22a) gives

$$X = \begin{bmatrix} 1.6718 & -2.0563 \\ -2.0563 & 7.7169 \end{bmatrix} \quad W_1 = \begin{bmatrix} -3.9158 \\ 9.0362 \end{bmatrix} \quad W_2 = \begin{bmatrix} -2.3610 \\ 13.5810 \end{bmatrix}$$

Using equations (6d-e), we obtain

$$L_1 = \begin{bmatrix} -1.3418 \\ 0.8134 \end{bmatrix} \qquad L_2 = \begin{bmatrix} 1.1192 \\ 2.0581 \end{bmatrix}$$

$$N_1 = \begin{bmatrix} -0.4291 & 1.1709 \\ -0.1067 & 0.2933 \end{bmatrix} \quad N_2 = \begin{bmatrix} 0.2404 & -0.6596 \\ -0.0291 & 0.0709 \end{bmatrix}$$

from (30b) with W_i and X, we compute the values of R_1 and R_2 as follows

$$R_1 = \begin{bmatrix} -6.7090 \\ 4.0671 \end{bmatrix}, \quad R_2 = \begin{bmatrix} 5.5959 \\ 10.2906 \end{bmatrix}$$

The designed unknown T-S observer can be applied in chaotic system reconstruction and also in a secure communication procedure. In this context, the problem consists of transmitting a resulting ciphertext embedded to the output by a communication channel. At the receiver side, the hidden signal (plaintext) is retrieved via the synthesis approach, i.e. the designed unknown input T-S [36, 37, 39]. For simulation example, consider the message to be transmitted given by figure 4. Figures 5 and 6 show the state variable of the chaotic system and its estimation. Finally, figure 7 presents the estimated unknown input (message estimate) where the message is perfectly estimated.

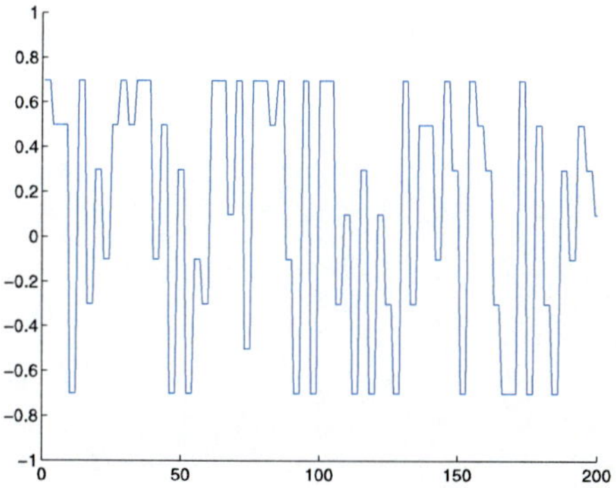

Fig. 4 Message $v(t)$

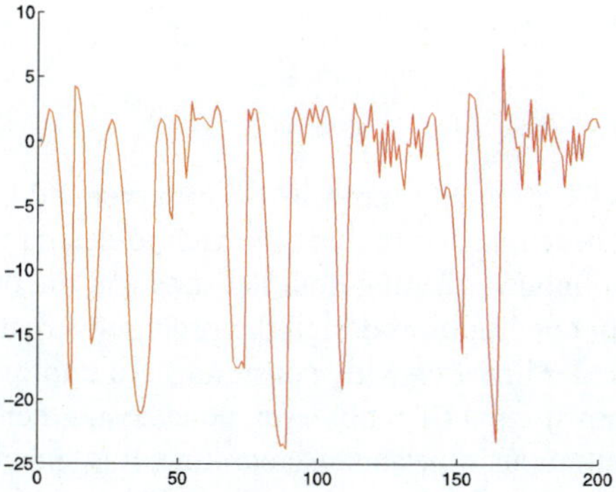

Fig. 5 $x_1(t)$ of chaotic T-S model and its estimate

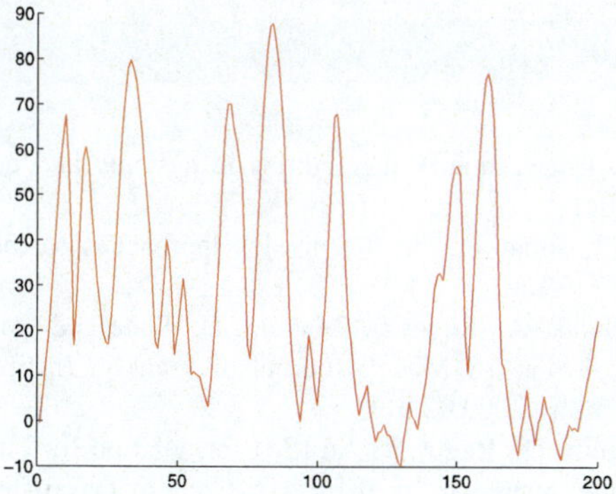

Fig. 6 $x_2(t)$ of chaotic T-S model and its estimate

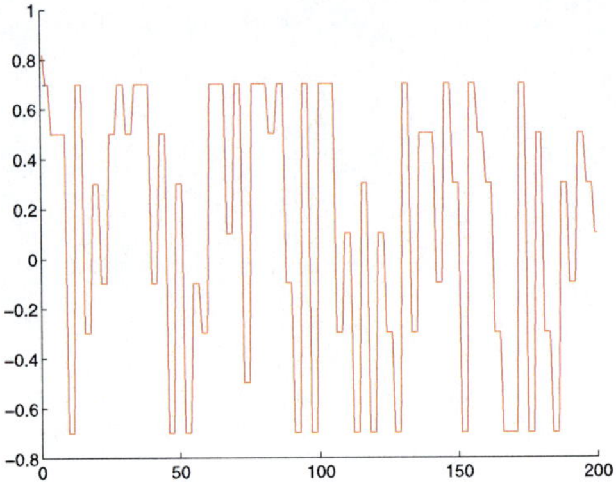

Fig. 7 Estimated message $\hat{v}(t)$

6 Conclusion

It is shown how to use T-S model approach, T-S observer and LMI formulation for chaotic system reconstruction. Indeed, an approach to design an observer for T-S models with unknown inputs affecting both the state and the output of the system is proposed. Sufficient conditions to design the proposed structure of observer are given in LMI terms under linear equality constraints. To improve the performances of the proposed unknown inputs T-S observer, poles assignment in LMI regions is also addressed for continuous-time and discrete-time. It is shown that this approach can be used in chaotic communications in the sense of signal masking and encryption. The designed unknown inputs T-S observer is shown to be satisfactory for message (unknown input) estimation and for chaotic system reconstruction.

References

[1] Boyd, S., et al.: Linear matrix inequalities in systems and control theory. SIAM, Philadelphia (1994)

[2] Murray-Smith, R., Johansen, T.: T-S model approaches to modelling and control. Taylor & Francis (1997)

[3] Chadli, M., Akhenak, A., Ragot, J., Maquin, D.: State and Unknown Input Estimation for Discrete Time T-S Model. Journal of Franklin Institute (2009) (in press), doi:10.1016/j.jfranklin.2009.02.011

[4] Chadli, M., Maquin, D., Ragot, J.: An LMI formulation for output feedback stabilisation in T-S model approach. In: IEEE 41st Conference on Decision Control, USA, December 10-13 (2002)

[5] Takagi, T., Sugeno, M.: Fuzzy identification of systems and its application to modelling and control. IEEE Trans. on Systems, Man, Cybernetics. 15(1), 116–132 (1985)

[6] Tanaka, K., Wang, H.O.: Fuzzy Control Systems Design and Analysis: A linear Matrix Inequality Approach. John Wiley & Sons, Inc. (2001)

[7] Chadli, M., Akhenak, A., Ragot, J., Maquin, D.: On the Design of Observer for Unknown Iinputs Fuzzy Models. International Journal of Automation and Control 2(1), 113–125 (2008)
[8] Kim, E., Lee, H.: New approaches to relaxed quadratic stability condition of fuzzy control systems. IEEE Trans. on Fuzzy Sets 8(5), 523–534 (2000)
[9] Xiaodiong, L., Qingling, Z.: New approach to H_∞ controller designs based on observers for T-S fuzzy systems via LMI. Automatica 39, 1571–1582 (2003)
[10] Guan, Y., Saif, M.: A novel approach to the design of unknown input observers. IEEE Transactions on Automatic Control 36(5), 632–635 (1991)
[11] Floquet, T., Barbot, J.P.: A sliding mode approach of unknown input observers for linear systems. In: IEEE Conference on Desicion and Control, pp. 1724–1729 (2004)
[12] Yang, F., Wilde, R.W.: Observers for linear systems with unknown inputs. IEEE Trans. Automatic Control 33, 677–681 (1988)
[13] Darouach, M., Zasadzinski, M., Xu, S.J.: Full-order observers for linear systems with unknown inputs. IEEE Trans. Automatic Control 39, 606–609 (1994)
[14] Syrmos, V.L.: Computational observer design techniques for linear systems with unknown inputs using the concept of transmission zeros. IEEE Transactions on Automatic Control 38(5), 790–794 (1993)
[15] Lin, S.F., Wang, A.P.: Unknown input observers for singular systems designed by eigenstructure assignment. Journal of the Franklin Institute 340(1), 43–61 (2003)
[16] Koenig, D.: Unknown input proportional T-S-integral observer design for descriptor systems: application to state and fault estimation. IEEE Transactions on Automatique Control 5(2), 213–217 (2005)
[17] Ha, Q.P., Trinh, H.: State and input simultaneous estimation for a class of nonlinear systems. Automatica 40(10), 1779–1785 (2004)
[18] Pikovsky, A., Rosemblum, M., Kurths, J.: Synchronization: A Universal Concept in Nonlinear Sciences. Cambridge University Press (2001) ISBN 0-521-53352-X
[19] Gonzalez-Miranda, J.M.: Synchronization and Control of Chaos. An introduction for scientists and engineers. Imperial College Press (2004) ISBN 1-86094-488-4
[20] Controlling chaos. In: Schuster, H.G. (ed.) Handbook of Chaos Control. Wiley-VCH, New York
[21] Sushchik, M.M., Rulkov, N.F., Tsimring, L.S., Abarbanel, H.D.I.: Generalized synchronization of chaos in directionally coupled chaotic systems. In: Proceedings of 1995 Intl. Symp. on Nonlinear Theory and Appl., vol. 2, pp. 949–952. IEEE (1995)
[22] Brown, R., Rulkov, N.F., Tracy, E.R.: Modeling and synchronization chaotic system from time-series data. Phys. Rev. E 49, 3784 (1994)
[23] Nolle, L., Goodyear, A., Hopgood, A.A., Picton, P.D., Braithwaite, N.: On Step Width Adaptation in Simulated Annealing for Continuous Parameter Optimisation. In: Reusch, B. (ed.) Fuzzy Days 2001. LNCS, vol. 2206, pp. 589–598. Springer, Heidelberg (2001)
[24] Nolle, L., Zelinka, I., Hopgood, A.A., Goodyear, A.: Comparison of an self organizing migration algorithm with simulated annealing and differential evolution for automated waveform tuning. Advances in Engineering Software 36(10), 645–653 (2005)
[25] Zelinka, I., Nolle, L.: Plasma reactor optimizing using differential evolution. In: Price, K.V., Lampinen, J., Storn, R. (eds.) Differential Evolution: A Practical Approach to Global Optimization, pp. 499–512. Springer, New York (2006)
[26] Zelinka, I.: Investigation on Evolutionary Deterministic Chaos Control. IFAC, Prague (2005)

[27] Ivan, Z.: Investigation on Evolutionary Deterministic Chaos Control – Extended Study. In: 19th International Conference on Simulation and Modeling (ECMS 2005), Riga, Latvia, June 1-4 (2005b)
[28] Zelinka, I., Senkerik, R., Navratil, E.: Investigation on Evolutionary Optimitazion of Chaos Control. Chaos, Solitons, Fractals (2007), doi:10.1016/j.chaos.2007.07.045
[29] Zelinka, I., Celikovsky, S., Richter, H., Chen, G.: Evolutionary Algorithms and Chaotic Systems. Springer, Germany (2010)
[30] Tanaka, K., Wang, H.O.: Fuzzy Control Systems Design and Analysis: A linear Matrix Inequality Approach. John Wiley & Sons, Inc. (2001)
[31] Chadli, M., Maquin, D., Ragot, J.: Stability analysis and design for continuous-time Takagi-Sugeno control systems. International Journal of Fuzzy Systems 7(3), 101–109 (2005)
[32] Johansson, M., Rantzer, A., Arzn, K.: Piecewise quadratic stability of fuzzy systems. IEEE Trans. on Fuzzy Systems 7(6), 713–722 (1999)
[33] Tanaka, K., Hori, T., Wang, H.O.: A multiple Lyapunov function approach to stabilization of fuzzy control systems. IEEE Transactions on Fuzzy Systems 11(4), 582–589 (2003)
[34] Chadli, M.: An LMI approach to design observer for unknown inputs Takagi-Sugeno fuzzy models. Asian Journal of Control 12(4), 524–530 (2010)
[35] Chilali, M., Gahinet, P.: H_∞ Design with pole placement constraints: an LMI approch. IEEE Transactions on Automatic Control 41(3), 358–367 (1996)
[36] Li, C., Liao, X., Wong, K.: Lag synchronization of hyperchaos with application to secure communications. Chaos, Solitons and Fractals 23, 183–193 (2005)
[37] Chen, M., Zhou, D., Shang, Y.: A new observer-based synchronization scheme for private communication. Chaos, Solitons and Fractals 24, 1025–1030 (2005)
[38] Boutayeb, M., Darouach, M., Rafaralahy, H.: Generalized State-Space Observers for Chaotic Synchronization and Secure Communication. IEEE Transactions on Circuits and Systems I: Fundamental Theory and Applications 49(3), 345–349 (2002)
[39] Alvares, G., Montoya, F., Romera, M., Pastor, G.: Breaking parameter modulated chaotic secure communication system. Chaos, Solitons & Fractals 21(4), 783–787 (2004)
[40] Akhenak, A., Chadli, M., Ragot, J., Maquin, D.: Unknown input multiple observer based approach: application to secure communication. In: 1st IFAC Conference on Analysis and Control of Chaotic Systems, Reims, France, June 28-30 (2006)
[41] Edwards, C., Spurgeon, S.K., Patton, R.J.: Sliding mode observers for fault detection and isolation. Automatica 36(4), 541–553 (2000)
[42] Vandenberghe, L., Boyd, S.: Semidefinite programming. SIAM Review 38(1), 49–95 (1996)
[43] Chadli, M., Karimi, H.R.: Robust Observer Design for Unknown Inputs Takagi-Sugeno Models. IEEE Trans. on Fuzzy Systems (2012), doi:10.1109/TFUZZ.2012.2197215

Entropy of Fractal Systems

Oldrich Zmeskal

Faculty of Chemistry, Brno University of Technology, Czech Republic

1 Historical Concept of Entropy

The **Kolmogorov *K*-entropy** [1, 2] is important characteristic which describes a degree of chaoticity of the systems. It gives the average rate of information loss about a position of the phase point on the attractor. It is well-known that $K_q = 0$ in an ordered system, $K_q \to \infty$ is infinite in a random system, and $0 < K_q < \infty$ in a chaotic system (deterministic chaos). The parameter q is order of entropy. Numerically, the Kolmogorov entropy can be estimated as the **Renyi entropy** S_q [3]. A special case of Renyi entropy for the $q = 1$ is in information theory the **Shanon entropy** S_1 [4]. By multiplying of Shanon entropy by Boltzmann constant k_B we will get the **thermodynamic entropy** S.

2 History of Fractal Geometry

The history of **fractals geometry** was founded in the 17th century when mathematician **Gottfried Leibniz** studied recursive self-similarity. He used the term "fractional exponents", which had not been used in geometry yet. After two centuries later in 1872, **Karl Weierstrass** presented the first definition of a function, what is today considered a fractal. From this year many mathematicians published examples subsets known as fractal structures [5].

English mathematician **Lewis Fry Richardson** is noted for his pioneering work on fractals and a method for solving a system of linear equations known as modified Richardson iteration [6]. This research was quoted by mathematician **Benoît Mandelbrot** in his 1967 paper "How Long Is the Coast of Britain"? [7]. French American mathematician **Benoît B. Mandelbrot** worked on a wide range of mathematical problems, but is best known as the father of fractal geometry. He coined the term fractal and described the Mandelbrot set. Mandelbrot also wrote books and gave popular lectures aimed at the general public [8].

3 Entropy and Fractal Dimension

Fractal structures are characterized by their fractal dimension. In truth, there exists an infinite family of fractal dimensions. By embedding the dataset in an E-dimensional space which cells have sides of size r, we can compute the probability with which data points fall into the i-th cell, and compute **generalized fractal**

I. Zelinka et al. (Eds.): Nostradamus: Mod. Meth. of Prediction, Modeling, AISC 192, pp. 25–26.
springerlink.com

dimension D_q. The **Rényi entropy** and **generalized fractal dimension** is connected by known relation $S_q = -D_q \ln r$.

We assume that this probability corresponds to the **density of fractal structure** $n(r) = Kr^{D-E}$ [9, 10]. By multiplying of the density of fractal structure by Boltzmann constant k_B we will get a density of heat capacity $c = k_B n(r)$. By application of Laplace (Poisson) equation to density of heat capacity we can calculate other thermodynamic quantities: the space distribution of the temperature at the fractal structure, the space and temperature dependence of density of energy (or of pressure), and the entropy $n = K \exp(\alpha S)$, where $a=(E-D)/D$ [11].

The described fractal theory is connected with the **El Naschie infinite theory** and the special cases give the **golden mean value** [12, 13].

Acknowledgement. This work was supported by the projects from the Ministry of Industry and Trade of the Czech Republic (Grant FR-TI1/144), and from "Centre for Materials Research at FCH BUT" No. CZ.1.05/2.1.00/01.0012 supported by ERDF.

References

[1] Grassberger, P., Procaccia, I.: Characterization of Strange Attractors. Physical Review Letters 50(5), 346–349 (1983), doi:10.1103/PhysRevLett.50.346
[2] Grassberger, P., Procaccia, I.: Estimation of the Kolmogorov entropy from a chaotic signal. Physical Review A 28(4), 2591–2593 (1983), doi:10.1103/PhysRevA.28.2591
[3] Rényi, A.: Probability theory. Elsevier (1970)
[4] Shannon, C.E.: A mathematical theory of communication. Bell Syst. Tech. J. 27, 379–423, 623–656 (1948)
[5] Trochet, H.: A History of Fractal Geometry. MacTutor History of Mathematics (2009)
[6] Richardson, L.F., Ashford, O.M., Charnock, H., Drazin, P.G., Hunt, J.C.R., Smoker, P., Sutherland, I.: The Collected Papers of Lewis Fry Richardson, Cambridge (1993) ISBN 978-0-521-38297-7; ISBN 978-0-521-38298-4
[7] Mandelbrot, B.B.: How Long Is the Coast of Britain? Statistical Self-Similarity and Fractional Dimension. Science, New Series 156, 636–638 (1967), doi:10.1126/science.156.3775.636
[8] Mandelbrot, B.B.: Fractal Geometry of Nature. W. H. Freeman and Co., New York (1983)
[9] Zmeskal, O., Nezadal, M., Buchnicek, M.: Fractal–Cantorian geometry, Hausdorff dimension and the fundamental laws of physics. Chaos, Solitons & Fractals 17, 113–119 (2003)
[10] Zmeskal, O., Nezadal, M., Buchnicek, M.: Field and potential of fractal–Cantorian structures and El Naschie's infinite theory. Chaos, Solitons & Fractals 19, 1013–1022 (2004)
[11] Zmeskal, O., Buchnicek, M., Vala, M.: Thermal properties of bodies in fractal and Cantorian physics. Chaos, Solitons & Fractals 25, 941–954 (2005)
[12] El Naschie, M.S.: Determining the temperature of the microwave background radiation from the topology and geometry of spacetime. Chaos, Solitons & Fractals 14, 1121–1126 (2002)
[13] El Naschie, M.S.: A review of E infinity theory and the mass spectrum of high energy particle physics. Chaos, Solitons & Fractals 19, 209–236 (2004)

Modeling Complexity: From Cellular Automata to Evolutionary Game Theory and Multiagent Systems with Coalitions

Juan Carlos Burguillo-Rial

Department of Telematics Engineering
University of Vigo
36310-Vigo (SPAIN)

1 Summary of the Presentation

Since its origins, Cellular Automata (CA), have been used to model many type of physical and computational phenomena. Interacting CAs in spatial lattices combined with evolutionary game theory have been very popular for modeling genetics or behavior in biological systems. Refining and extending the behavior of each automaton we can obtain a framework where multiple autonomous entities (agents) interact, i.e., a Multiagent System (MAS). Multiagent systems can be used to solve problems that are difficult to manage using monolithic approaches. The dynamic formation of coalitions is nowadays a well-known area of interest in MAS, as coalitions can help self-interested agents to successfully cooperate and coordinate in a mutually beneficial manner. In this presentation, we will explore the use of MAS, as refined cellular automata, combined with evolutionary game theory and coalitions. We shall consider several application scenarios (social sciences, economy, optimization, ...) and also different interaction topologies ranging from spatial to complex networks.

The first scenario [1] presents a framework for describing the spatial distribution and the global frequency of agents who play the spatial prisoner's dilemma with coalition formation. The agent interaction is described by a non-iterated game, where each agent only locally interacts with its neighbours. Every agent may behave as a defector or a cooperator when playing isolated, but they can join or lead coalitions (group of agents) where a leader decides the coalition strategy. Isolated agents' strategies or groups' strategies are public and therefore can be memetically imitated by neighbors. The agent strategy is selected between two possibilities: probabilistic Tit-for-Tat (pTFT) or learning automata (LA). Coalition dynamics are organized around two axes. On the one hand, agents get a percentage of compromise when cooperating with other agents. On the other hand, leaders impose taxes to the other agents belonging to its coalition. These two rules and their related parameters guide the coalition formation and the game evolution.

I. Zelinka et al. (Eds.): Nostradamus: Mod. Meth. of Prediction, Modeling, AISC 192, pp. 27–28.
springerlink.com

The second scenario [2] considers several strategies to compete in a spatial and extended version of the Iterated Prisoner's Dilemma (IPD). This scenario follows the model presented in the first one, but here, besides the classical strategies, cooperate (C) and defect (D), we consider two other strategies based on the property of resources: Possession (P), as the right to possess what one owns, and Trade (T), as the right to buy and sell ownership. This scenario also includes a set of simulation results showing how ownership and trade emerge from anarchy, as evolutionary stable strategies, to enable the peaceful resolution of property conflicts under certain environment conditions.

Finally, the last scenario [3] explores cellular genetic algorithms (cGAs) as a kind of genetic algorithms (GAs) with decentralized population in which interactions among individuals are restricted to the closest ones. The use of decentralized populations in GAs allows to keep the population diversity for longer, usually resulting in a better exploration of the search space and, therefore in a better performance of the algorithm. However, the use of decentralized populations supposes the need of several new parameters that have a major impact on the behavior of the algorithm. In the case of cGAs, these parameters are the population and neighborhood shapes. Hence, in this scenario we present a new adaptive technique based in Cellular Automata, Game Theory and Coalitions that allow to manage dynamic neighborhoods. As a result, the new adaptive cGAs (EACO) with coalitions outperform the compared cGA with fixed neighborhood for the selected benchmark of combinatorial optimization problems.

References

[1] Burguillo-Rial, J.C.: A Memetic Framework for Describing and Simulating Spatial Prisoner's Dilemma with Coalition Formation. In: Eighth International Conference on Autonomous Agents and Multiagent Systems, AAMAS (2009)

[2] Burguillo, J.C., Peleteiro, A.: Ownership and Trade in Spatial Evolutionary Memetic Games. In: Schaefer, R., Cotta, C., Kołodziej, J., Rudolph, G. (eds.) PPSN XI. LNCS, vol. 6238, pp. 455–464. Springer, Heidelberg (2010)

[3] Dorronsoro, B., Burguillo, J.C., Peleteiro, A., Bouvry, P.: Evolutionary Algorithms based on Game Theory and Cellular Automata with Coalitions. In: Handbook of Optimization. Springer Series on Intelligent Systems (2012)

On Evolutionary Synthesis of Chaotic Systems

Ivan Zelinka

VŠB-Technical University of Ostrava, Faculty of Electrical Engineering
and Computer Science, Department of Computer Science, 17. listopadu 15,
708 33 Ostrava-Poruba, Czech Republic
ivan.zelinka@vsb.cz

Abstract. This tutorial introduces the principles of chaos synthesis by means of selected evolutionary algorithms and propose a novel method for discrete as well as continuous chaotic systems synthesis. Method introduced in this tutorial is of the same nature like genetic programming or grammatical evolution algorithms and was applied along with three evolutionary algorithms: differential evolution, self-organizing migration and genetic algorithm. The aim of this tutorial is to demonstrate our results of syntheses artificial (no physical background behind) new and complex chaotic systems based on some simple building elements with a properly defined cost function. The investigation consists of two major case studies: the discrete chaotic system synthesis and continuous system synthesis. For all used algorithms, numerous simulations of chaos synthesis were repeated and then averaged to guarantee the reliability and robustness of the proposed method. The most significant results were carefully selected, visualized and commented in this keynote.

1 Used Evolutionary Algorithms, Possibilities and Techniques

This research used two evolutionary algorithms: Self-Organizing Migrating Algorithm (SOMA) [1] and Differential Evolution (DE) [2], [3] and genetic algorithm [4], [5].

Self-Organizing Migrating Algorithm is a stochastic optimization algorithm that is modeled on the basis of social behavior of cooperating individuals [1]. It was chosen because it has been proven that the algorithm has the ability to converge towards the global optimum [1] and due to the successful applications together with AP [4], [5].

DE is a population-based optimization method that works on real-number-coded individuals, [2], [3]. DE is quite robust, fast, and effective, with global optimization ability. It does not require the objective function to be differentiable, and it works well even with noisy and time-dependent objective functions. Setting of used algorithms is mentioned in the Table 1-4.

I. Zelinka et al. (Eds.): Nostradamus: Mod. Meth. of Prediction, Modeling, AISC 192, pp. 29–34.
springerlink.com

Table 1 Used algorithms and its abbreviation

Algorithms	Version	Abbreviation
SOMA	AllToOne	A
	AllToOneRandomly	B
	AllToAll	C
	AllToAllAdaptive	D
DE	DERand1Bin	E
	DERand2Bin	F
	DEBest2Bin	G
	DELocalToBest	H
	DEBest1JIter	I
	DERand1DIter	J
GA		K

Table 2 SOMA setting for 4 strategies: A, B, C, D

Algorithms	A	B	C	D
PathLength	3	3	3	3
Step	0,11	0,11	0,11	0,11
PRT	0,1	0,1	0,1	0,1
PopSize	200	200	200	200
Migrations	10	10	10	10
MinDiv	-0,1	-0,1	-0,1	-0,1
Individual length	50	50	50	50

Table 3 DE setting for 6 strategies: E, F, G, H, I, J

Algorithms	E	F	G	H	I	J
NP	200	200	200	200	200	200
F	0,9	0,9	0,9	0,9	0,9	0,9
CR	0,3	0,3	0,3	0,3	0,3	0,3
Generations	200	200	200	200	200	200
Indiv. length	50	50	50	50	50	50

Table 4 GA setting (cannonical version): K

Algorithms	K
Population size	200
Mutation	0,4
Generations	100
Individual length	50

Identification of various dynamical systems is vitally important in the case of practical applications as well as in theory. A rich set of various methods for dynamical system identification has been developed. In the case of chaotic dynamics, it is for example the well-known reconstruction of chaotic attractor based on research of Takens [22] who has shown that, after the transients have died out, one can reconstruct the trajectory on the attractor from the measurement of a single component. Because whole trajectory contains too much information, series of papers by [9], [10], [11] is introduced to show a set of averaged coordinate invariant numbers (generalized dimensions, entropies, and scaling indices) by which different strange attractors can be distinguished. Method presented here is based on evolutionary algorithms (EAs), see [12], [21], and allows reconstruction not only of chaotic attractors as a geometrical objects, but also their mathematical description. All those techniques belong to the class of genetic programming techniques; see [12], [13]. Generally, when it is used on data fitting, these techniques are called symbolic regression (SR).

The term SR represents a process, in which measured data is fitted by a suitable mathematical formula such as $x_2 + C$, $sin(x)+1/e^x$, etc., Mathematically, this process is quite well known and can be used when data of an unknown process is obtained. Historically SR has been in the preview of manual manipulation, however during the recent past, a large inroad has been made through the use of computers. Generally, there are two well-known methods, which can be used for SR by means of computers. The first one is called genetic programming or GP, [12, 13] and the other is grammatical evolution, [14, 15], or, as we have used by so called analytic programming, see [21].

The idea as to how to solve various problems using SR by means of EA was introduced by John Koza, who used genetic algorithms (GA) for GP. Genetic programming is basically a symbolic regression, which is done by the use of evolutionary algorithms, instead of a human brain. The ability to solve very difficult problems is now well established, and hence, GP today performs so well that it can be applied, e.g. to synthesize highly sophisticated electronic circuits [16].

In the last decade of the 20th century, C. Ryan developed a novel method for SR, called grammatical evolution (GE). Grammatical evolution can be regarded as an unfolding of GP due to some common principles, which are the same for both algorithms. One important characteristic of GE is that it can be implemented in any arbitrary computer language compared with GP, which is usually done (in its canonical form) in LISP. In contrast to other evolutionary algorithms, GE was used only with a few search strategies, for example with a binary representation of the populations in [17]. Another interesting investigation using symbolic regression was carried out by [18] working on Artificial Immune Systems or/and systems, which are not using tree structures like linear genetic programming.

Put simply, evolutionary algorithm simulates Darwinian evolution of individuals (solutions of given problem) on a computer and are used to estimate-optimize numerical values of defined cost function. Methods of GP are able to synthesize in an evolutionary way complex structures like electronic circuits, mathematical formulas etc. from basic set of symbolic (nonnumeric) elements.

Another alternative method of symbolic regression is in [21]. This method can be used as well as GP or GE for predictive model synthesis [8] or neural network synthesis [6], [7] for example.

In paper [20] and [21], analytic programming (AP) is applied, for the identification of selected chaotic system. Identification is not done on the "level" of strange attractor reconstruction, but it produces a symbolic-mathematical description of the identified system. Selected results are part of this tutorial.

2 Results

Some important and selected results are depicted in Fig. 1-3, Table 5 and fully referred in [20] and [21]. All used EAs have demonstrated good performance in model parameter identification and also in its synthesis.

Table 5 Selected results

Synthesized chaotic system	**Bifurcation and chaos observed in interval**
$A - x\left(-A + \frac{x\left(-\frac{A}{x} + x + Ax\right)}{A + x^2}\right)$	[0; 4]
$\frac{A\left(2A - 2x^2 - 3x(A - x + Ax)\right)}{-A + x - x^2}$	[0,1; 0,13] [0,8; 1,2]
$-x - \frac{1 - 2A + 2x + 2A^2 x}{1 - A + \frac{A^2 - x}{x} + x}$	[0,3; 0,5]
$\frac{x - A\left(A - x - 2x^2\right)}{-A - x + Ax^2 - A\left(-A + \frac{A^3}{x} + 2x\right)}$	[0,4; 0,5]
$\frac{2A(-2A + 2x)}{x + \frac{1 + A^2 + x}{x}}$	[0; 4]
$\frac{x}{(3A + 2x)\left(-1 - A - x + \frac{x(A + 2x)}{A^2 + x}\right)}$	[0,12; 0,23] [0,3; 0,36]

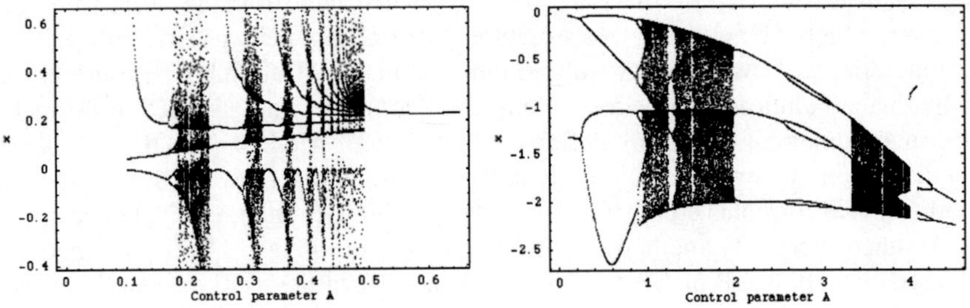

Fig. 1 Synthesized bifurcation diagrams

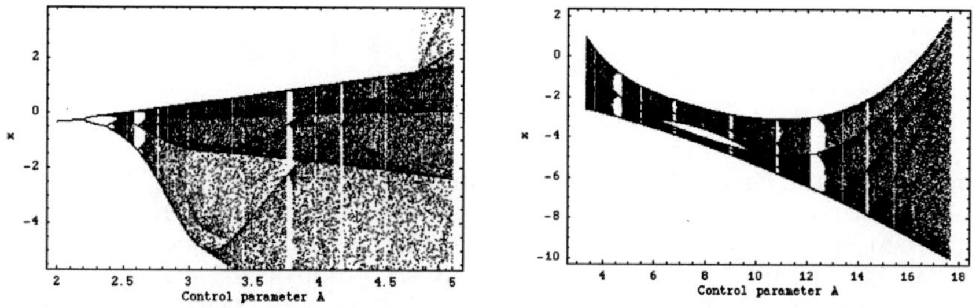

Fig. 2 Synthesized bifurcation diagrams

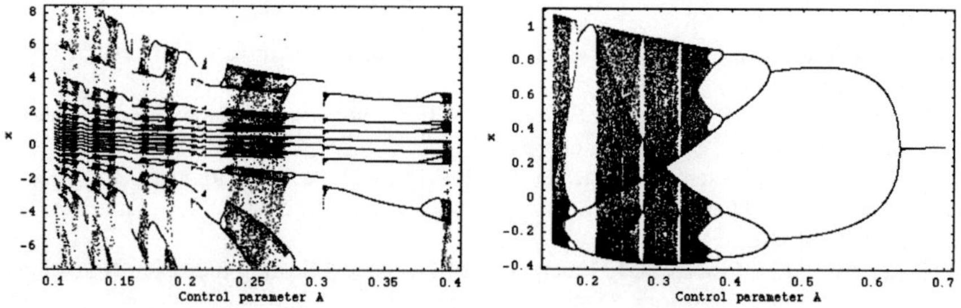

Fig. 3 Synthesized bifurcation diagrams

3 Conclusions

In this tutorial was described, how can be used various evolutionary algorithms for synthesis or identification of suitable complex models, that exhibit chaotic behavior. There were made two sets of simulations:

1. Synthesis of discrete chaotic systems, see also [20] and [21], partly reported here.
2. Synthesis of continuous chaotic systems, tested on Lorenz system, see [20].

As is visible and fully reported in [20], [21], all simulations were successful. Despite this fact, all simulations are temporal - they will be further improved and repeated in wider scale.

Acknowledgments. This work was supported by the Development of human resources in research and development of latest soft computing methods and their application in practice project, reg. no. CZ.1.07/2.3.00/20.0072 funded by Operational Programme Education for Competitiveness, co-financed by ESF and state budget of the Czech Republic.

References

[1] Zelinka, I.: SOMA – Self Organizing Migrating Algorithm. In: Babu, B.V., Onwubolu, G. (eds.) New Optimization Techniques in Engineering. Springer (2004) ISBN 3-540-20167X

[2] Price, K.: An Introduction to Differential Evolution. In: Corne, D., Dorigo, M., Glover, F. (eds.) New Ideas in Optimization, pp. 79–108. McGraw-Hill, London (1999)

[3] Price, K., Storn, R.: Differential evolution homepage (2001), http://www.icsi.berkeley.edu/~storn/code.html (accessed May 15, 2012)
[4] Holland, J.H.: Adaptation in Natural and Artificial Systems. Univ. Michigan Press, Ann Arbor (1975)
[5] Holland, J.H.: Genetic Algorithms. Scientific American, 44–50 (July 1992)
[6] Varacha, P., Jasek, R.: ANN Synthesis for an Agglomeration Heating Power Consumption Approximation. In: Recent Researches in Automatic Control, pp. 239–244. WSEAS Press, Montreux, ISBN 978-1-61804-004-6
[7] Varacha, P., Zelinka, I.: Distributed Self-Organizing Migrating Algorithm Application and Evolutionary Scanning. In: Proceedings of the 22nd European Conference on Modelling and Simulation ECMS, pp. 201–206 (2008) ISBN 0-9553018-5-8
[8] Soeterboek, A.R.M.: Predictive Control. Proefschrift, Technische Universiteit Delft, Rotterdam (1990)
[9] Grassberger, P., Procaccia, I.: Estimation of the Kolmogorov Entropy From a Chaotic Signal. Phys. Rev. 29 A, 2591 (1983b)
[10] Halsey, T.C., Jensen, M.H., Kadanoff, L.P., Procaccia, I., Schraiman, B.I.: Fractal Measures and Their Singularities: the Characterization of Strange Sets. Phys. Rev. 33 A, 1141 (1986)
[11] Eckmann, J.P., Procaccia, I.: Fluctuation of Dynamical Scaling Indices in Non-Linear Systems. Phys. Phys. Rev. 34 A, 659 (1986)
[12] Koza, J.R.: Genetic Programming II. MIT Press (1998) ISBN 0-262-11189-6
[13] Koza, J.R., Bennet, F.H., Andre, D., Keane, M.: Genetic Programming III. Morgan Kaufnamm Pub. (1999) ISBN 1-55860-543-6
[14] O'Neill, M., Ryan, C.: Grammatical Evolution. Evolutionary Automatic Programming in an Arbitrary Language. Kluwer Academic Publishers (2002) ISBN 1402074441
[15] Ryan, C., Collins, J.J., O'Neill, M.: Grammatical Evolution: Evolving Programs for an Arbitrary Language. In: Banzhaf, W., Poli, R., Schoenauer, M., Fogarty, T.C. (eds.) EuroGP 1998. LNCS, vol. 1391, p. 83. Springer, Heidelberg (1998)
[16] Koza, J.R., Keane, M.A., Streeter, M.J.: Evolving Inventions. Scientific American, 40–47 (February 2003) ISSN 0036-8733
[17] O'Sullivan, J., Ryan, C.: An Investigation into the Use of Different Search Strategies with Grammatical Evolution. In: Foster, J.A., Lutton, E., Miller, J., Ryan, C., Tettamanzi, A.G.B. (eds.) EuroGP 2002. LNCS, vol. 2278, pp. 268–277. Springer, Heidelberg (2002)
[18] Johnson, C.G.: Artificial Immune Systems Programming for Symbolic Regression. In: Ryan, C., Soule, T., Keijzer, M., Tsang, E.P.K., Poli, R., Costa, E. (eds.) EuroGP 2003. LNCS, vol. 2610, pp. 345–353. Springer, Heidelberg (2003)
[19] Zelinka, I., Chen, G., Celikovsky, S.: Chaos Synthesis by Means of Evolutionary Algorithms. International Journal of Bifurcation and Chaos 18(4), 911–942 (2008)
[20] Zelinka, I., Celikovsky, S., Richter, H., Chen, G.: Evolutionary Algorithms and Chaotic Systems, 550s p. Springer, Germany (2010)
[21] Zelinka, I., Davendra, D., Senkerik, R., Jasek, R., Oplatkova, Z.: Analytical Programming - a Novel Approach for Evolutionary Synthesis of Symbolic Structures. In: Eisuke Kita, Evolutionary Algorithms. InTech, ISBN 978-953-307-171-8
[22] Takens, F.: Detecting strange attractors in turbulence. In: Rand, D.A., Young, L.-S. (eds.) Dynamical Systems and Turbulence. Lecture Notes in Mathematics, vol. 898, pp. 366–381. Springer

On the Evolutionary Optimization of Chaos Control – A Brief Survey

Roman Senkerik

Tomas Bata University in Zlin, Faculty of Applied Informatics,
Nam T.G. Masaryka 5555, 760 01 Zlin, Czech Republic
senkerik@fai.utb.cz

Abstract. This work represents the brief introduction into the issues of evolutionary optimization of discrete chaotic systems. This work introduces and compares evolutionary approach representing tuning of parameters for an existing control method either with the standard cost function using the numerical desired state as the one of the input or blackbox type cost function, as well as meta-evolutionary approach representing synthesis of a whole control law by means of Analytic Programming (AP). The main part of this work is focused on the proper development of the cost function used in evolutionary process. As an example of discrete chaotic system, one-dimensional Logistic equation was used. For the experiments following soft computing tools were utilized: Symbolic regression tool Analytic Programming and evolutionary algorithms Self-Organizing Migrating Algorithm (SOMA) and Differential Evolution (DE).

1 Introduction

During the recent years, usage of new intelligent systems in engineering, technology, modeling, computing and simulations has attracted the attention of researchers worldwide. The most current methods based on soft computing are mostly: neural networks, evolutionary algorithms, fuzzy logic, and genetic programming. Presently, evolutionary algorithms are known as a powerful set of tools for almost any difficult and complex optimization problem.

The interest about the interconnection between evolutionary techniques and control of chaotic systems is spread daily. The first steps were done in [1] - [3], where the control law was based on the Pyragas method, which is Extended delay feedback control (ETDAS) [4]. These papers were concerned with tuning several parameters inside the control technique for chaotic system. Compared to this simple approach, this presented brief survey also shows a possibility for generating the whole control law (not only to optimize several parameters) for the purpose of stabilization of a chaotic systems on desired Unstable Periodic Orbits (UPO). The synthesis of control law is inspired by the Pyragas's delayed feedback control TDAS and ETDAS [5], [6]. These two methods are very advantageous for evolutionary computation, due to the amount of accessible control parameters, which can be easily tuned by means of evolutionary algorithms (EA).

I. Zelinka et al. (Eds.): Nostradamus: Mod. Meth. of Prediction, Modeling, AISC 192, pp. 35–48.
springerlink.com

Together with simple EA utilization, analytic programming (AP) is used in this work. AP is a superstructure of EAs and is used for synthesis of analytic solution according to the required behaviour [7]. The new control law for chaotic system can be viewed as a symbolic structure, which can be synthesized according to the requirements for the stabilization of the chaotic system.

This work is a cumulation of previous work [3], [8], [9] – [11] and expansion of initial studies [12], [13]. Furthermore, it briefly introduces and compares three possible approaches:

- Simple evolutionary approach with the cost function utilizing the position of desired UPO.
- Blackbox evolutionary approach utilizing special cost functions, thus without knowledge about exact UPO position in the chaotic attractor, is introduced. This means that EA is used to find the best control parameter set up based only on the demanded type of chaotic system behavior and not based on the position of UPO.
- AP application for synthesis of a whole control law instead of parameters tuning for existing and commonly used control method to stabilize chaotic systems.

In general, this work will briefly introduce the background of the research, mainly the development of the cost function and show obtained results.

2 Problem Design

The brief description of used chaotic system, original feedback chaos control method ETDAS and the process of development of used cost functions are given here. The ETDAS control technique was used in this research either as the main control method, or as the inspiration for synthesizing a new feedback control law by means of AP.

2.1 Selected Chaotic System

The chosen example of chaotic system was the one-dimensional Logistic equation as in form (1):

$$x_{n+1} = rx_n\left(1 - x_n\right). \tag{1}$$

The Logistic equation (Logistic map) is a one-dimensional discrete-time example of how a complex chaotic behavior can arise from very simple non-linear dynamical equation. [14]. This chaotic system was introduced and popularized by the biologist Robert May [15]. It was originally introduced as a demographic model of a typical predator – prey relationship. The chaotic behavior can be observed by varying the parameter *r*. When r = 3.57, this is the beginning of chaos, at the end of the period-doubling behavior. When $r > 3.57$, the system exhibits chaotic behavior. The example of this behavior can be clearly seen from the bifurcation diagram in Figure 1.

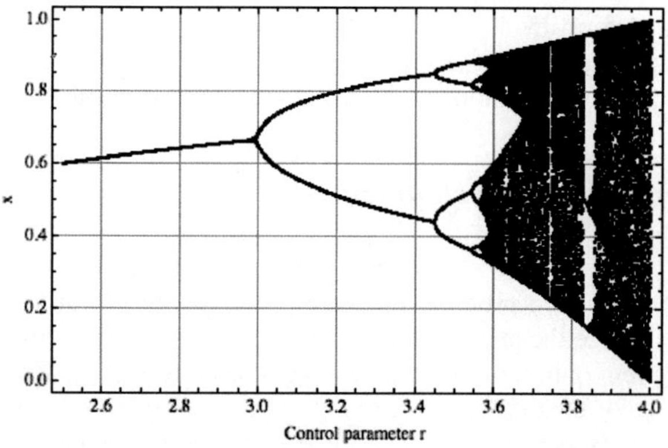

Fig. 1 Bifurcation diagram of Logistic equation

2.2 ETDAS Control Method

This work is focused on explanation of application of AP for synthesis of a whole control law (meta-evolutionary approach) as well as tuning of parameters for ETDAS method control laws (simple and blackbox evolutionary approach) to stabilize desired Unstable Periodic Orbits (UPO). In this work, desired UPOs were p-1 (stable state) and p-2 (two-periodic orbit, which represents oscillation between two values).

1. Within the research concentrated on synthesis of control law, an inspiration for preparation of sets of basic functions and operators for AP was also ETDAS control method. The original control method – ETDAS has form (2).

$$
\begin{aligned}
F(t) &= K\left[(1-R)S(t-\tau_d)-x(t)\right] \\
S(t) &= x(t)+RS(t-\tau_d)
\end{aligned}
\tag{2}
$$

Where: K and R are adjustable constants, F is the perturbation; S is given by a delay equation utilizing previous states of the system and τ_d is a time delay.

The original control method – ETDAS in the discrete form suitable for Logistic equation has the form (3).

$$
\begin{aligned}
x_{n+1} &= rx_n(1-x_n)+F_n \\
F_n &= K\left[(1-R)S_{n-m}-x_n\right] \\
S_n &= x_n + RS_{n-m}
\end{aligned}
\tag{3}
$$

Where: m is the period of m-periodic orbit to be stabilized. The perturbation F_n in equations (3) may have arbitrarily large value, which can cause diverging of the system outside the output interval of Logistic equation {0, 1}. Therefore, F_n should have a value between $-F_{\max}$, $F_{\max}$. The suitable $F_{\max}$ value was either directly obtained form evolutionary optimization process or in the case of meta-evolutionary approach it was set up based on experiences with the numerous experiments dealing with evolutionary chaos control optimization.

2.3 Cost Functions Development

The proposal of the basic cost function (CF) is in general based on the simplest CF, which could be used problem-free only for the stabilization of p-1 orbit. The idea was to minimize the area created by the difference between the required state and the real system output on the whole simulation interval – τ_i. This CF design is very convenient for the evolutionary searching process due to the relatively favorable CF surface. (see Fig. 2.). Nevertheless this simple approach has one big disadvantage, which is the including of initial chaotic transient behavior of not stabilized system into the cost function value. As a result of this, the very tiny change of control method setting for extremely sensitive chaotic system causing very small change of CF value, can be suppressed by the above-mentioned including of initial chaotic transient behavior. Moreover, another cost function had to be used for stabilizing of the higher periodic orbit. In this case, it is not possible to use the simple rule of minimizing the area created by the difference between the required and actual state on the whole simulation interval – τ_i, due to simple and serious reason, which is the degrading of the possible best solution by a phase shift of periodic orbit.

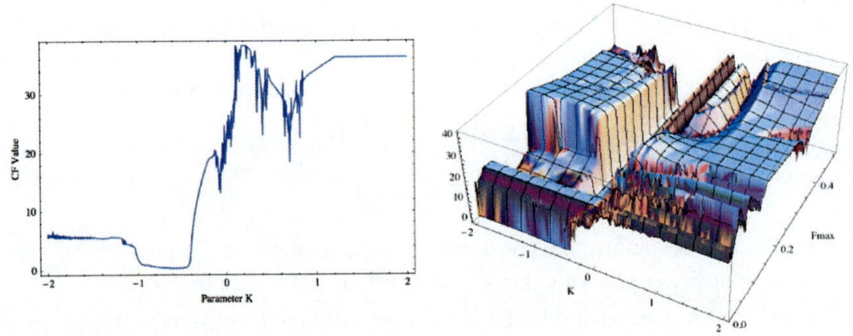

Fig. 2 Dependence of CF value parameter K (left), and on parameter K and F_{max} (right); Simple cost function, p-1 orbit

Other cost functions (CF_1 and CF_2) had to be used for the stabilizing of the chaotic system in "evolutionary blackbox mode", ie. without exact numerical value of target state, which means the exact numerical position of desired UPO. These cost functions also include initial chaotic transient behavior of not stabilized system into the cost function value. But these cost functions are based on searching for periodic orbits in chaotic attractor and there is no direct connection between the area created by the difference between the required state and the real system output during initial chaotic transient and during the rest of the simulation interval, where the system is stabilized. Thus the optimal setting for control method given by evolutionary process is not too influenced by initial chaotic behavior (exceptions and more about this issue is described in the results and conclusion sections). On the other hand, the issue of searching for periodic orbits causes more chaotic, erratic and discrete type CF surface (see Fig. 2.), thus less convenient for evolutionary algorithms.

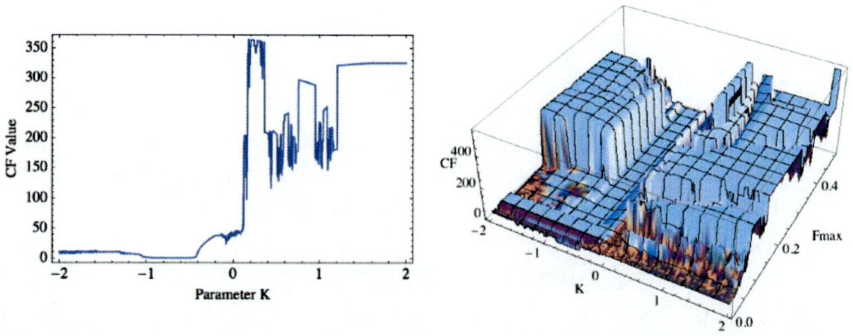

Fig. 3 Dependence of CF value parameter K (left), and on parameter K and F_{max} (right); BlackBox type cost function, p-1 orbit

In the case of blackbox approach, obviously it is not possible to use the simple rule of minimizing the area created by the difference between the required and actual state on the whole simulation interval – τ or its arbitrary part.

This approach is based on searching for periodic orbits in chaotic attractor and stabilizing the system on these periodic orbits by means of applying the optimal feedback perturbation F_n. It means that this new CF does not take any numerical target state into consideration, but the selected target behavior of system. Therefore, the new CF is based on the searching for optimal feedback perturbation F_n securing the stabilization on any type of selected UPO. The slight disadvantage of this approach is that for each UPO (i.e. different behavior) a different CF is needed.

The proposal of CF_1 used for the p-1 orbit stabilization is based on the following simple rule. In discrete systems, the iteration $y(n)$ and $y(n+1)$ of output value must be the same. The idea was to minimize the area created by the difference between the n and $n+1$ output iteration on the whole simulation interval – τ, thus at the same time this proposal of CF should secure fast targeting into the close neighborhood of p-1 orbit and its stabilization. The CF_1 has form (4).

$$CF_1 = p + \sum_{t=0}^{\tau} \left| y(n+1) - y(n) \right| \tag{4}$$

where: p = *penalization*

The next proposal of CF_2 used in the case of p-2 orbit is based on the following simple rule. The iteration *y(n)* and *y(n+2)* must have the same value. But this rule is also valid for the case of – p-1 orbit, where in discrete systems, the iteration *y(n)* and *y(n+1)* of output value must be the same. Thus another condition had to be added. It says that in the case of p-2 orbit there must be some difference between the *n* and *n+1* output iteration. Considering the fact of minimizing the CF the value this condition had to be rewritten into this suitable form (5):

$$\frac{1}{\left|y(n+1)-y(n)\right|+c}\,, \tag{5}$$

where: *c* – small constant 1.10^{-16} which was added to prevent the evolutionary optimization from crashing, since upon finding the suboptimal solution stabilized at p-1 orbit it returns the division by zero. The CF_2 has the form (6).

$$CF_2 = p1+\sum_{t=0}^{\tau}\left|y(n+2)-y(n)\right|+\frac{1}{\left|y(n+1)-y(n)\right|+c}\,, \tag{6}$$

where: p1 = penalization.

In the proposed CFs (4) and (6) there had to be included penalization, which should avoid the finding of solutions, where the stabilization on saturation boundary values {0, 1} or oscillation between them (i.e. artificial p-2 orbit) occurs. This penalization was calculated as the sum of the number of iterations, where the system output reaches the saturation boundary value.

All other used cost functions are purely based on searching for the desired stabilized periodic orbit and thereafter calculation of the difference between desired and found actual periodic orbit on the short time interval - τ_s (20 iterations – p-1 orbit and 40 iterations – p-2 orbit) from the point, where the first minimal value of difference between desired and actual system output is found. Such a design of CF should secure the successful stabilization of either p-1 orbit (stable state) or higher periodic orbit anywise phase shifted. The CF_{Meta} used for meta-evolutionary approach, which is very time demanding, has the form (7).

$$CF_{Meta} = pen_1 + \sum_{t=\tau 1}^{\tau 2}\left|TS_t - AS_t\right| \tag{7}$$

where:
TS - target state
AS - actual state
τ_1 - the first min value of difference between TS and AS
τ_2 – the end of optimization interval ($\tau_1 + \tau_s$),
$pen_1 = 0$ if $\tau_i - \tau_2 \geq \tau_s$
$pen_1 = 10*(\tau_i - \tau_2)$ if $\tau_i - \tau_2 < \tau_s$ (i.e. late stabilization).

Within the simple evolutionary approach, which is less time demanding, advanced CF securing the very fast stabilization was used. It was necessary to modify the definition of basic CF in order to decrease the average number of iteration required for the successful stabilization and avoidance of any associated problem. The easiest but the most problematic way is that the whole CF value is multiplied by the number of iterations (*NI*) of the first found minimal value of difference between desired and actual system output (i.e. the beginning of fully stabilized UPO). To avoid errors associated with CF returning value 0 and other problems, the small constant (*SC*) is added to CF value before penalization (multiplying by *NI*). The *SC* value (9) is computed with the aid of power of non-penalized basic part of CF (8), thus it is always secured that the penalization is at similar level as the non-penalized CF value.

$$\mathrm{ExpCF} = \log_{10}\left(\sum_{t=\tau 1}^{\tau 2} \left|TS_t - AS_t\right| + 10^{-15}\right) \tag{8}$$

$$SC = 10^{\mathrm{ExpCF}} \tag{9}$$

The CF used for evolutionary approach is given in (10):

$$CF_{Adv} = \sum_{1}^{n}\left(\left(NI \cdot SC\right) + penalization1 + \sum_{t=\tau 1}^{\tau 2}\left|TS_t - AS_t\right|\right), \tag{10}$$

where $x_{initial}$ is from the range 0.05 – 0.95 and uses step 0.1 (i.e. n = 10).

Here the number of steps for stabilization (*NI*) multiplies only the small constant (*SC*). Finally, to avoid the problems with fast stabilization only for limited range of initial conditions, the final CF value is computed as a sum of n repeated simulations for different initial conditions. Consequently, the EA should find the robust solutions securing the fast targeting into desired behavior of system for almost any initial conditions.

The issue of pure searching for periodic orbits causes very chaotic, erratic and discrete type CF surfaces (see Fig. 4.)

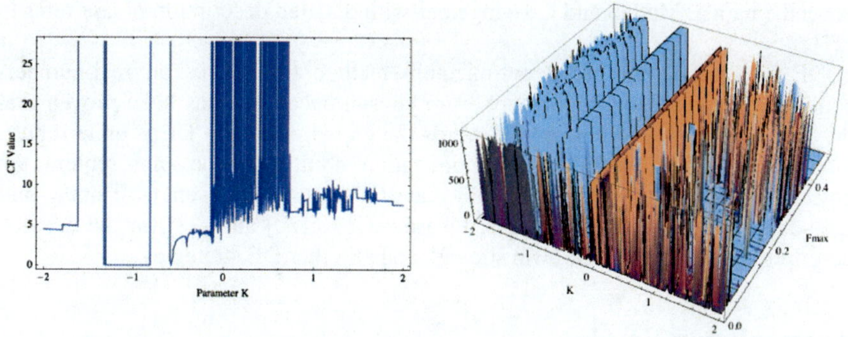

Fig. 4 Dependence of CF value parameter K (left), and on parameter K and $F_{\max}$ (right); Advanced cost function, for simple evolutionary process, p-1 orbit

3 Used Soft – Computing Tools

This chapter briefly introduces used symbolic regression tool, which is Analytic Programming, and two evolutionary algorithms.

3.1 Analytic Programming

Basic principles of the AP were developed in 2001 [16], [17]. Until that time only Genetic Programming (GP) and Grammatical Evolution (GE) had existed. GP uses Genetic Algorithms (GA) while AP can be used with any EA, independently on individual representation. To avoid any confusion, based on the nomenclature according to the used algorithm, the name - Analytic Programming was chosen, since AP represents synthesis of analytical solution by means of EA. Various applications of AP are described in [7], [18].

AP exists in 3 versions – basic without constant estimation, AP_{nf} – estimation by means of nonlinear fitting package in *Wolfram Mathematica* environment and AP_{meta} – constant estimation by means of another evolutionary algorithms; meta implies meta-evolution.

3.2 Evolutionary Algorithms

This work used two evolutionary algorithms: Self-Organizing Migrating Algorithm (SOMA) [19] and Differential Evolution (DE) [20].

SOMA is a stochastic optimization algorithm that is modeled on the social behavior of cooperating individuals. Numerous applications either of canonical [21] – [23] or special version [24] of SOMA together with successful applications with AP [25], [26] have proven that these heuristics are suitable for solving a difficult class of problems. The detailed principle is described in [19]. For the source codes in Mathematica, Matlab and C++ together with detailed description please refer to [27].

DE is a population-based optimization method that works on real-number-coded individuals. Both algorithms were chosen because it has been proven that they have the ability to converge towards the global optimum. DE is quite robust, fast, and effective, with global optimization ability. It does not require the objective function to be differentiable, and it works well even with noisy and time-dependent objective functions. Please refer to [20] and [28] for the detailed description of used DERand1Bin strategy and all other DE strategies.

4 Simulation Results

In the case of tuning of parameters for ETDAS method within both simple evolutionary and blackbox approaches, only SOMA algorithm was used.

In the case of a new control law synthesis, AP_{meta} version was used. Meta means usage of one evolutionary algorithm for main AP process and the second algorithm for coefficient estimation. SOMA algorithm was used for the main AP process and DE was used in the second evolutionary process.

Settings of EA parameters for all approaches were based on performed numerous experiments with chaotic systems and evolutionary algorithms.

The results for the first simple evolutionary approach shown in Table 1 represent the best founded solution of parameters set up for ETDAS control method for both case studies (p-1 and p-2 orbit stabilization) together with the cost function value comprising 10 runs of chaotic system with initial conditions in the range 0.05 – 0.95 with step 0.1, and average CF value per 1 run.

Table 2 contains the best founded solutions of parameters set up for ETDAS control method in the case of blackbox evolutionary approach and also for both case studies (p-1 and p-2 orbit stabilization) together with the corresponding best cost function value.

The optimization results in Table 3 represent the best examples of synthesized control laws for the p-1 orbit stabilization as well as for p-2 orbit stabilization. Description of the selected simulation results covers output from AP representing the synthesized control law with simplification after estimation of constants by means of second algorithm DE and corresponding CF value.

From previous experiments and numerical analysis of this system, it follows that unperturbed Logistic equation has this p-1 orbit: $x_F = 0.7368$ and this p-2 orbit:

$$x_1 = -0.3737, x_2 = 0.8894.$$

The ranges of all estimated parameters within the both evolutionary approaches were these: $-2 \le K \le 2$, $0 \le F_{\max} \le 0.5$ and $0 \le R \le 0.99$,

where $F_{\max}$ is a limitation of feedback perturbation, securing the avoidance of diverging of the chaotic system outside the interval {0, 1}.

Basic set of elementary functions for AP:

- Simple operators: +, -, /, *, ^
- Constants and output data: $data_{n-1}$ to $data_n$, K (for p-1 orbit)
- Constants and output data: $data_{n-9}$ to $data_n$, K (for p-2 orbit)

Due to the recursive attributes of delay equation S utilizing previous states of the system in discrete ETDAS (3), the data set for AP had to be expanded and cover longer system output history, thus to imitate inspiring control method for the successful synthesis of control law securing the stabilization of higher periodic orbits.

The novelty of the meta-evolutionary approach represents the synthesis of feedback control law F_n (perturbation) (11) for the Logistic equation inspired by original ETDAS control method.

$$x_{n+1} = rx_n(1 - x_n) + F_n \tag{11}$$

4.1 Basic Evolutionary – Not Blackbox – Approach

From Table 1 and Figure 5, it follows, that ETDAS control method with optimized parameters given by means of SOMA algorithm is able to secure robust, fast and precise (for p-1 orbit) stabilization of chaotic system on desired behavior.

Table 1 Simulation results for control of Logistic equation with simple evolutionary approach

UPO	K	Fmax	R	CF Value	CF Value per 1 run
p-1	-0.9336	0.4154	0.4994	$2.54\ 10^{-14}$	$2.54\ 10^{-15}$
p-2	0.4093	0.2824	0.1949	0.6392	0.0639

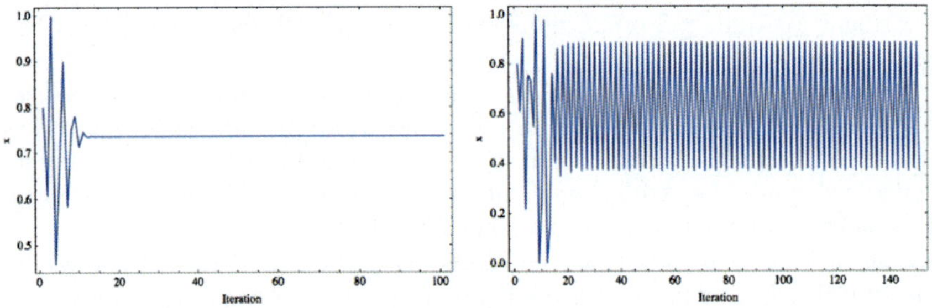

Fig. 5 Simulation results for the optimized ETDAS method settings – simple evolutionary approach: p-1 orbit (left) and p-2 orbit (right)

4.2 Evolutionary Blackbox Approach

Based on obtained results presented in Table 2 and depicted in Figure 6, it may be stated that the ETDAS control parameters estimated by means of SOMA algorithm and blackbox type CF ensured fast reaching of a desired stable state, without any knowledge about its exact position.

Obtained simulations results depicted in Figure 6 (right) and CF value for p-2 orbit in Table 2 may evoke the impression, that the relatively high CF value is not corresponding with the system fully stabilized on desired p-2 UPO. This was caused by the including of initial chaotic transient behaviour into the CF value and by the phase-shift (p-2 orbit iteration order) of the output values towards to the predefined notation in the CF (6).

Table 2 Simulation results for control of Logistic equation with blackbox evolutionary approach

UPO	K	Fmax	R	CF Value
p-1	-0.6703	0.1077	0.0402	0.6724
p-2	0.6243	0.0529	0.4157	150.4922

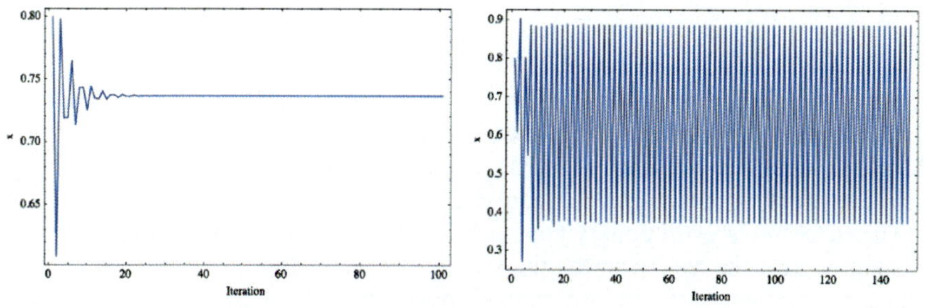

Fig. 6 Simulation results for the optimized ETDAS method settings – blackbox evolutionary approach: p-1 orbit (left) and p-2 orbit (right)

4.3 Meta-Evolutionary Approach – Synthesis of a New Control Law

Simulations depicted in Figure 7 lend weight to the argument, that AP is able to synthesize a new control laws securing very quick and very precise stabilization. Within the both case studies, new synthesized control laws were able to stabilize the system at required UPO more precisely compared with the CF values in Table 1 – simple evolutionary approach. The interesting fact was that for the p-1 orbit AP has found the notation (12) of original simple TDAS [6] control method suitable for lower periodic orbits.

$$F(t) = K[x(t-\tau_d)-x(t)] \tag{12}$$

Table 3 Results for control of Logistic equation with meta-evolutionary approach

UPO	Control law with coefficients	CF Value
p-1	$F_n = -0.527311(x_{n-1} - x_n)$	$6.6613.10^{-16}$
p-2	$F_n = -0.0169(-x_{n-6} - 23.4428)x_{n-5}(x_{n-2} - x_n)$	0.0206

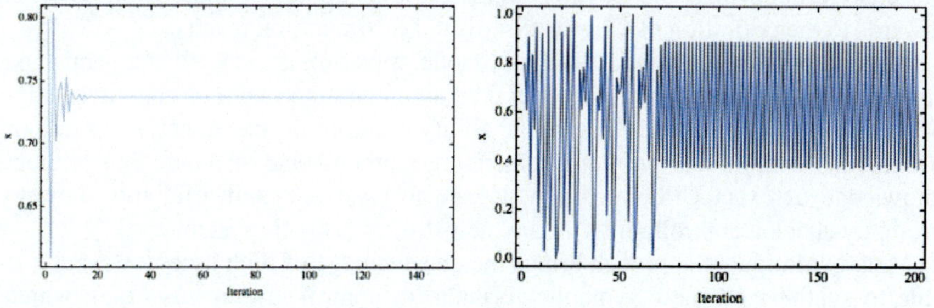

Fig. 7 Simulation results for the best new synthesized control laws – meta-evolutionary approach, p-1 orbit: (left) and p-2 orbit (right)

5 Conclusion

This work introduces and compares three possible approaches for the optimization of stabilization of Logistic equation, which was selected as an example of discrete chaotic system. The main part of this work is focused on the issue of developing the proper cost function for evolutionary searching process.

The first simple evolutionary approach represented tuning of parameters for an existing control method, whereas the second blackbox approach represented also tuning of parameters for control method, however without the knowledge of desired UPO position, when only the definition of type of chaotic system behavior was used as a cost function input. The third meta-evolutionary approach represented synthesis of a whole control law by means of Analytic Programming (AP). The control method used within the first two approaches was an inspiration for the creation of data sets for AP.

Obtained results show that synthesized control laws have given better results than the original control method, which served as an inspiration. This fact reinforce the argument that AP is able to solve this difficult problems and to produce a new synthesized control law in a symbolic way securing desired behavior of chaotic system. Precise and fast stabilization lends weight to the argument, that AP is a powerful symbolic regression tool, which is able to strictly and precisely follow the rules given by cost function and synthesize any symbolic formula. In the case of this research, it means to synthesize the feedback controller for chaotic system.

The question of energy costs and more precise stabilization will be included into future research together with the development of better cost functions, different AP data set, and performing of numerous simulations to obtain more results and produce better statistics, thus to confirm the robustness of this approach. Presented data and statistical comparison can be summarized as follows:

Simple evolutionary approach is easy to implement, very fast and gives satisfactory results. But the quality of results are restricted by the limitations of the mathematical formulas, control laws, models etc., for which the parameters are tuned by EA.

The Blackbox approach brings the advantage of avoidance of mathematical analysis of chaotic systems, nevertheless in this case an unasked influence of initial chaotic behaviour and possible phase shift of higher periodic orbit solution towards to the definition of CF was discovered in simulation results.

Nevertheless the proposed Blackbox mode approach is very advantageous and simple to implement in the case of an unknown chaotic system or chaotic oscillations in any system, because of its ability to control the chaotic system or oscillations without any previous demanding mathematical analysis, thus without knowledge of exact UPOs position. It can be used as a powerful tool, how to promptly check a controllability of any new discrete chaotic system.

Meta-evolutionary approach brings the disadvantage of high time-costs, but it is able to synthesize a new symbolic formula, (a control law in this case), which gives even better results than the best ones obtained by the first simple evolutionary approach. Also in this case an interesting phenomenon was discovered. AP synthesized some examples of control laws, which followed the rule given by cost function and secured the stabilization on the part of this short time interval, and than the system freely escaped to either chaotic behaviour or to different artificial controlled orbit. It is very interesting, that these control laws are able to stabilize the chaotic system on optional artificial periodic orbits. Most of common control method was developed for stabilization only on real UPO with low energy costs, thus, when the system enters the UPO, there is no perturbation.

Finally from the time-demands point of view, to obtain a one solution for both evolutionary approaches, 5400 CF evaluations were required (approx. 1 minute), whereas for meta-evolutionary approach, more than 32 millions CF evaluations were required (approx. 72 hours).

Acknowledgments. This work was supported by European Regional Development Fund under the project CEBIA-Tech No. CZ.1.05/2.1.00/03.0089.

References

1. Zelinka, I., Senkerik, R., Navratil, E.: Investigation on evolutionary optimization of chaos control. Chaos, Solitons & Fractals 40(1), 111–129 (2009)
2. Senkerik, R., Zelinka, I., Davendra, D., Oplatkova, Z.: Evolutionary Design of Chaos Control in 1D. In: Zelinka, I., Celikovsky, S., Richter, H., Chen, G. (eds.) Evolutionary Algorithms and Chaotic Systems. SCI, vol. 267, pp. 165–190. Springer, Heidelberg (2010)
3. Senkerik, R., Zelinka, I., Davendra, D., Oplatkova, Z.: Utilization of SOMA and differential evolution for robust stabilization of chaotic Logistic equation. Computers & Mathematics with Applications 60(4), 1026–1037 (2010)
4. Pyragas, K.: Control of chaos via extended delay feedback. Physics Letters A 206, 323–330 (1995)
5. Just, W.: Principles of Time Delayed Feedback Control. In: Schuster, H.G. (ed.) Handbook of Chaos Control. Wiley-Vch (1999)
6. Pyragas, K.: Continuous control of chaos by self-controlling feedback. Physics Letters A 170, 421–428 (1992)
7. Zelinka, I., Davendra, D., Senkerik, R., Jasek, R., Oplatkova, Z.: Analytical Programming - a Novel Approach for Evolutionary Synthesis of Symbolic Structures. In: Kita, E. (ed.) Evolutionary Algorithms. InTech (2011)
8. Senkerik, R., Zelinka, I., Davendra, D., Oplatkova, Z.: Evolutionary Optimization Of Henon Map Control – A Blackbox Approach. International Journal of Operational Research 13(2), 129–146 (2012)
9. Senkerik, R., Oplatkova, Z., Zelinka, I., Davendra, D.: Synthesis of feedback controller for three selected chaotic systems by means of evolutionary techniques: Analytic programming. Mathematical and Computer Modelling (2010), doi:10.1016/j.mcm.2011.05.030
10. Senkerik, R., Oplatkova, Z., Zelinka, I., Davendra, D., Jasek, R.: Application of Analytic Programming for Evolutionary Synthesis of Control Law - Introduction of Two Approaches. In: Byrski, A., Carvalho, Z.O.M., Dorohinicki, M.K. (eds.) Advances in Intelligent Modelling and Simulation: Simulation Tools and Applications. Springer Series. SCI, pp. 253–268 (2012) ISBN: 978-3-642-28887-6

11. Senkerik, R., Oplatkova, Z., Zelinka, I., Davendra, D., Jasek, R.: Application of Evolutionary Techniques for Optimization of Chaos Control – Introduction of Three Approaches. In: Zelinka, I., Snasel, V., Abraham, A. (eds.) Handbook of Optimization. Springer Series "Inteligent Systems" (in press), ISBN 978-3-642-30503-0
12. Senkerik, R., Oplatkova, Z., Zelinka, I., Davendra, D., Jasek, R.: Evolutionary Synthesis of Control Law for Higher Periodic Orbits of Chaotic Logistic Equation. In: Proceedings of 25th European Conference on Modelling and Simulation, pp. 452–458 (2011)
13. Senkerik, R., Oplatkova, Z., Zelinka, I., Davendra, D., Jasek, R.: Evolutionary Synthesis of Controller for Stabilization of Synthesized Chaotic System Oscillations. In: Proceedings of 17th International Conference on Soft Computing – MENDEL 2011, pp. 73–79 (2011) ISBN 978-80-214-4302-0
14. Hilborn, R.C.: Chaos and Nonlinear Dynamics: An Introduction for Scientists and Engineers. Oxford University Press (2000)
15. May, R.M.: Stability and Complexity in Model Ecosystems. Princeton University Press (2001) ISBN: 0-691-08861-6
16. Zelinka, I., Oplatkova, Z., Nolle, L.: Boolean Symmetry Function Synthesis by Means of Arbitrary Evolutionary Algorithms-Comparative Study. International Journal of Simulation Systems, Science and Technology 6(9), 44–56 (2005) ISSN: 1473-8031
17. Lampinen, J., Zelinka, I.: New Ideas in Optimization – Mechanical Engineering Design Optimization by Differential Devolution, vol. 1, 20 p. McGraw-hill, London (1999) ISBN 007-709506-5
18. Oplatková, Z., Zelinka, I.: Investigation on Evolutionary Synthesis of Movement Commands. Modelling and Simulation in Engineering (2009) ISSN: 1687-559
19. Zelinka, I.: SOMA – Self Organizing Migrating Algorithm. In: Babu, B.V., Onwubolu, G. (eds.) New Optimization Techniques in Engineering, ch. 7, 33. Springer (2004)
20. Price, K., Storn, R.M., Lampinen, J.A.: Differential Evolution: A Practical Approach to Global Optimization. Natural Computing Series. Springer (1995)
21. Coelho, L.D.: Self-organizing migration algorithm applied to machining allocation of clutch assembly. Mathematics and Computers in Simulation 80(2), 427–435 (2009)
22. Coelho, L.D.: Self-Organizing Migrating Strategies Applied to Reliability-Redundancy Optimization of Systems. IEEE Transactions on Reliability 58(3), 501–510 (2009)
23. Coelho, L.D., Mariani, V.C.: An efficient cultural self-organizing migrating strategy for economic dispatch optimization with valve-point effect. Energy Conversion and Management 51(12), 2580–2587 (2010)
24. Davendra, D., Zelinka, I., Senkerik, R.: Chaos driven evolutionary algorithms for the task of PID control. Computers & Mathematics with Applications 60(4), 1088–1104 (2010)
25. Varacha, P., Jasek, R.: ANN Synthesis for an Agglomeration Heating Power Consumption Approximation. In: Recent Researches in Automatic Control, pp. 239–244. WSEAS Press, Montreux, ISBN 978-1-61804-004-6
26. Varacha, P., Zelinka, I.: Distributed Self-Organizing Migrating Algorithm Application and Evolutionary Scanning. In: Proceedings of the 22nd European Conference on Modelling and Simulation, ECMS, pp. 201–206 (2008) ISBN 0-9553018-5-8
27. Zelinka, I.: SOMA homepage, http://www.fai.utb.cz/people/zelinka/soma/ (accessed June 15, 2012)
28. Price, K., Storn, R.M.: Differential evolution homepage, `http://www.icsi.berkeley.edu/~storn/code.html` (accessed June 15, 2012)

The Use of Soft Computing Methods for Forecasting in Business, Their Applications in Practice

Petr Dostál

Brno University of Technology, Faculty of Business and Management,
Institute of Informatics, Kolejní 2906/4, 612 00 Brno, Czech Republic

Abstract. Forecasting methods have had successful applications in business. Nowadays the new theories of soft computing are used for these purposes. The applications in business have specific features in comparison with others. The forecasting plays very important roles especially in business because it helps to reduce costs that can lead to higher profits and to success in the competitive fight. There are various forecasting methods used: classical ones and methods using soft computing. There are especially the methods such as fuzzy logic, neural networks, evolutionary algorithms, and the theory of chaos.

1 Introduction

There are various ways of using the methods for forecasting such as fuzzy logic, neural networks, evolutionary algorithms and others. The forecasting methods are used for forecasting of future trends and cycles of time series. The program MATLAB® with Fuzzy Logic Toolbox, Neural Network, and Global Optimization is used [14, 25]. The fields of applications of forecasting methods in business cover a wide area of applications [1, 7, 8, 9, 10, 16, 17, 18].

2 Fuzzy Logic

Let us mention an example of the use of fuzzy logic for forecasting of a time series. At first it is necessary to say that the time series is a sequence of values that are dependent on time t. The value at time $t = 1$ is denoted x_1, at time $t = 2$ is denoted x_2, and so on, and the value in time $t = N$ is denoted x_N, when N signifies the number of values in the time series. The time series can be expressed as a vector of values $x = (x_1, x_2, \ldots, x_N)$. For needs of forecasting we specify that the value x_N will be the last known value of the time series and it will correspond to the present. The value $x_{N+1}^{\bullet}$ will be the first future value forecasted, the value $x_{N+2}^{\bullet}$ will be the second value forecasted, etc. (The symbol $^{\bullet}$ is be used to denote

I. Zelinka et al. (Eds.): Nostradamus: Mod. Meth. of Prediction, Modeling, AISC 192, pp. 49–60.
springerlink.com

forecasted values.) The interval between measurement is very often constant, then $\Delta = t_2 - t_1 = t_3 - t_2 = \ldots = t_N - t_{N-1} = \text{const}$. This interval in the economy (in contrast to technical sciences) has values in the range of minutes, hours, days, weeks, months, years, and their fractions. In this respect we speak about time series with very high (minutes), high (hours), medium (days), low (weeks), and very low (year) frequencies.

The following verbal notes were chosen for the solution of forecasting by means of the FuzzyTech program of Inform GmbH. There are defined input variables: Delta_1 = $x_N - x_{N-1}$, Delta_2 = $x_{N-1} - x_{N-2}$, Delta_3 = $x_{N-2} - x_{N-3}$, Delta_4 = $x_{N-3} - x_{N-4}$ (the signs of these differences express the trend of the time series). The build-up model for forecasting has four input variables Delta_1, Delta_2, Delta_3, Delta_4, one rule box, and one output variable Prediction. See the fuzzy model in Fig. 1.

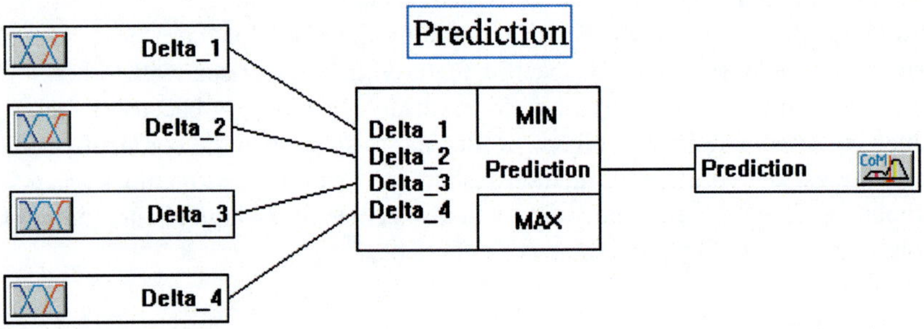

Fig. 1 Fuzzy logic – Model

The input variables have five attributes defined by values Delta, specified by their signs and the size of difference of neighbouring values (high positive, positive, zero, negative, high negative difference). As membership functions the shapes Λ, S, and Z are used. The output variable Prediction has five attributes that evaluate the future course of the time series (high increase, increase, stagnation, decrease, high decrease), specifying the situation at time $N+1$ (the value of forecasting $x_{N+1}^{\bullet}$). The membership functions are spline curves of Λ, Π, S, and Z shapes.

The membership functions of input variables Delta_1, Delta_2, Delta_3, Delta_4 must be set up. Also the fuzzy rule box must be set up on the basis of knowledge, preferably by the experts who understand the problem. The setup of a fuzzy rule box depends on the type of solved case. For example a suitable rule can be similar to the following:

When inputs Delta_1 and Delta_2 and Delta_3 and Delta_4 are high negative, it means that the time series is decreasing and a large increase of the time series forecasting is expected in future.

This situation can be verbally described in capital markets: after a great and long decrease of share values they tend to start a fast increase with 90% probability. The rule can be described by this form:

<When> Delta_1 << 0 <And> Delta_2 << 0 <And> Delta_3 << 0 <And> Delta_4<< 0 <Then> Prediction = High increase <With> $s = 0.90$.

The rule for the opposite case can be verbally described in a capital market: after a great and long increase of share values they tend to start a fast decrease. The rule has the form

<When> Delta_1 >> 0 <And> Delta_2 >> 0 <And> Delta_3 >> 0 <And> Delta_4 >> 0 <Then> Prediction = High decrease <With> $s = 0.90$.

It is necessary to set up other rules that are combinations of these two described extreme variants. Figure 2 presents setup attributes and membership functions for the output variable Prediction.

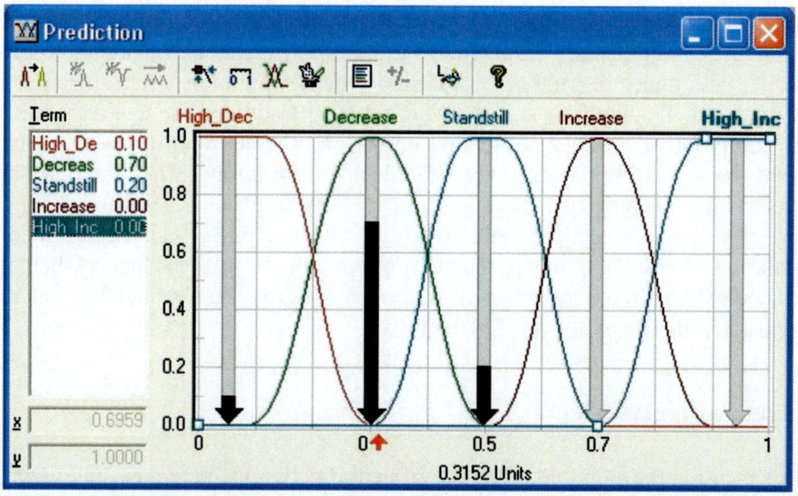

Fig. 2 Membership functions of output variable Prediction

For another case an unsatisfactory model could be set up and it is necessary to choose another number of variables, to define variables in another way, to choose other attributes and membership functions. The model must be tuned to give us good forecasting.

Let us mention another example that solves the problem of decision making in capital markets: whether to trade in the stock market or not. The model for the FuzzyTech program is presented in Fig. 3.

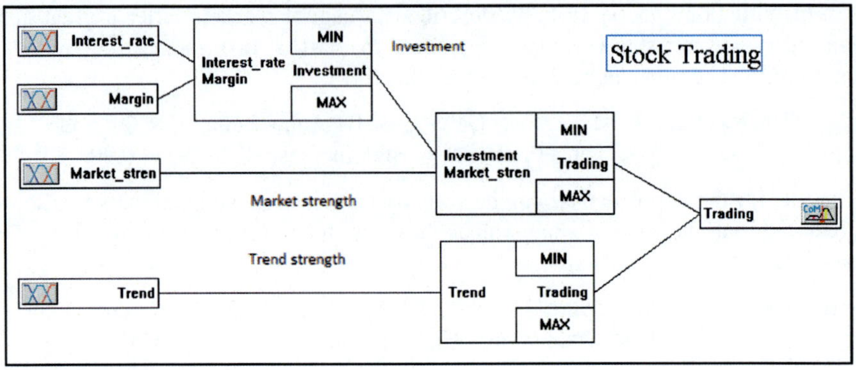

Fig. 3 The diagram of model – stock market

The input variables and their attributes are as follows: Margin (insignificant, significant), Interest rates (low, medium, high), Strength of market (low, medium, high), and Trend the course of time series (deterministic, stochastic).

The rules and attributes are as follows: the box Investment (unsuitable, neutral, suitable) determines the rule of desirability of depositing money in a stock market; the first block Trading (yes, no) evaluates whether trading in the market is suitable from the point of profitability of investment and the strength of the market; the second block Trading (yes, no) gives the decision for trading when the time series is stochastic, meaning that there is no possibility to make a good forecasting of future development of time series (the rate of randomness can be searched by calculation of Hurst exponent). The output variable Trading evaluates whether to trade with share, index, commodity, or currency ratio. The membership functions were used in the shapes of Λ, Π, S, and Z.

3 Neural Networks

Let us mention the example of the use of neural networks for forecasting of values of time series that are created by development of currency ratios of euro EUR and dollar USD. The calculation was done by means of the NeuroForecaster program.

The input of neural networks for forecasting is a matrix of values whose rows represent the single time intervals and columns present the values of single time series. For example, the values used in the stock market are presented by open, close, a minimum value, a maximum price, and a volume of trades. For the forecasting of a time series it is necessary to set up parameters. See Fig. 4.

It is necessary to determine the topology of neural networks and to choose the input and output variables, the type of transfer function, and the number of layers of neural networks. Then it is necessary to determine the data that will be used for learning and for testing. After successfully finishing the process of learning and testing, it is possible to start the process Forecast. After finishing the process of calculation we obtain the results in a graph that presents real values and values of forecasting. See Fig. 5.

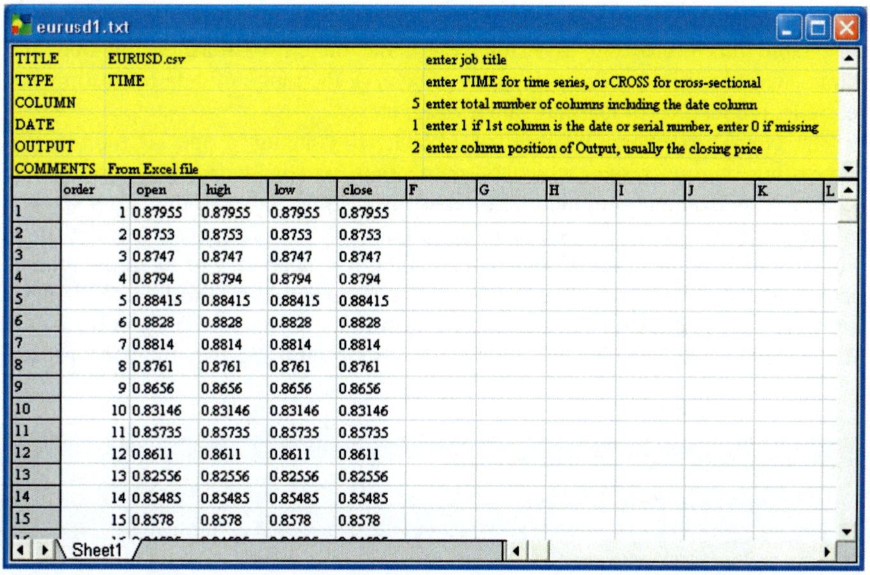

eurusd1.txt

TITLE	EURUSD.csv	enter job title
TYPE	TIME	enter TIME for time series, or CROSS for cross-sectional
COLUMN	5	enter total number of columns including the date column
DATE	1	enter 1 if 1st column is the date or serial number, enter 0 if missing
OUTPUT	2	enter column position of Output, usually the closing price
COMMENTS	From Excel file	

	order	open	high	low	close	F	G	H	I	J	K	L
1	1	0.87955	0.87955	0.87955	0.87955							
2	2	0.8753	0.8753	0.8753	0.8753							
3	3	0.8747	0.8747	0.8747	0.8747							
4	4	0.8794	0.8794	0.8794	0.8794							
5	5	0.88415	0.88415	0.88415	0.88415							
6	6	0.8828	0.8828	0.8828	0.8828							
7	7	0.8814	0.8814	0.8814	0.8814							
8	8	0.8761	0.8761	0.8761	0.8761							
9	9	0.8656	0.8656	0.8656	0.8656							
10	10	0.83146	0.83146	0.83146	0.83146							
11	11	0.85735	0.85735	0.85735	0.85735							
12	12	0.8611	0.8611	0.8611	0.8611							
13	13	0.82556	0.82556	0.82556	0.82556							
14	14	0.85485	0.85485	0.85485	0.85485							
15	15	0.8578	0.8578	0.8578	0.8578							

Sheet1

Fig. 4 Input matrix

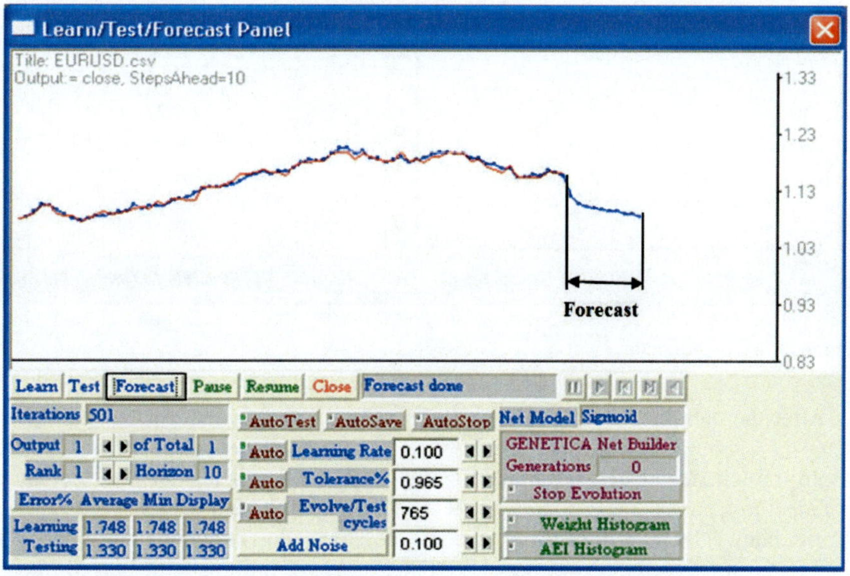

Fig. 5 The graph of real and forecasted values

It is possible to predict any time series. The quality of forecasting can be influenced by the choice of a start of the time series (a short time series can be as unsuitable for forecasting as a long one), the choice of transfer function and parameters of neural network (number of layers), the range of data for testing and learning, the method of learning and testing, etc.

The Neural Network Toolbox of MATLAB contains a special toolbox for forecasting of time series, which is quite often used in the business and public sectors. This example presents the forecasting of time series from history created by the number of sold products recorded in defined file. The menu is called from workspace using the command *ntstool*. At first, it is necessary to setup the type of neural network. Here we chose the *Nonlinear Autoregressive Model* (*NAR*). Then it is necessary to click on button *Next* and download the data from the defined file. When the data are downloaded, the window is closed using the command *Finish*. See Fig. 6.

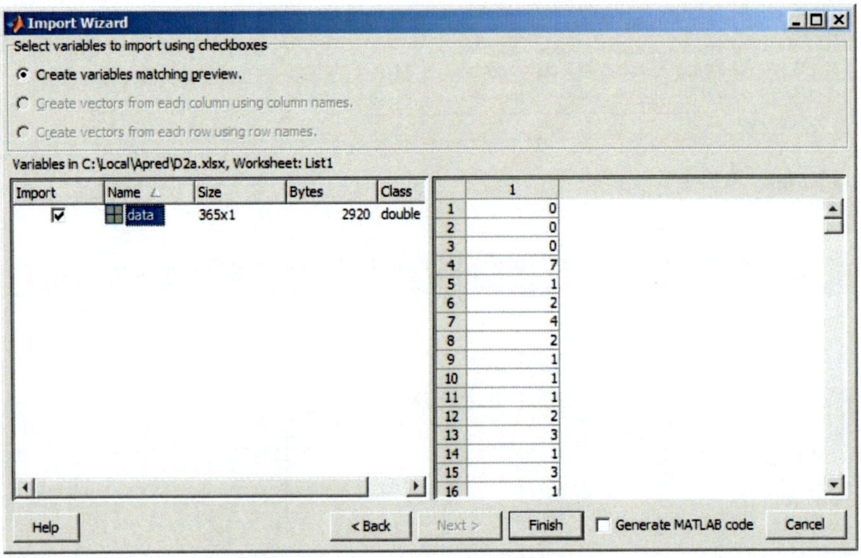

Fig. 6 Download of data

After the data have been selected, the command *Next* opens a window for the setup of *Training*, *Validation,* and *Testing*. The default value can be used or it could be changed. The command *Next* enables the setup of the network architecture represented by *Number of Hidden Neurons* and *Number of Delays*. The default value can be used or it could be changed. The commands *Next* and *Train* start training the neural network. The process of training is displayed in the window and after the process of training it is possible to evaluate the graph to determine whether the correct training was done, for example performance *Performance*. It is possible to evaluate the mean square error *MSE* and fit of forecasting. The process of training may be repeated several times using the command *Retrain*. When the results of training correspond to the required demands, the output has to be saved as a variable, for example *output*. Saved data sets could be used for evaluation, for example to display them at workplace or transferring data into MS Excel. The graph is presented in Fig. 7.

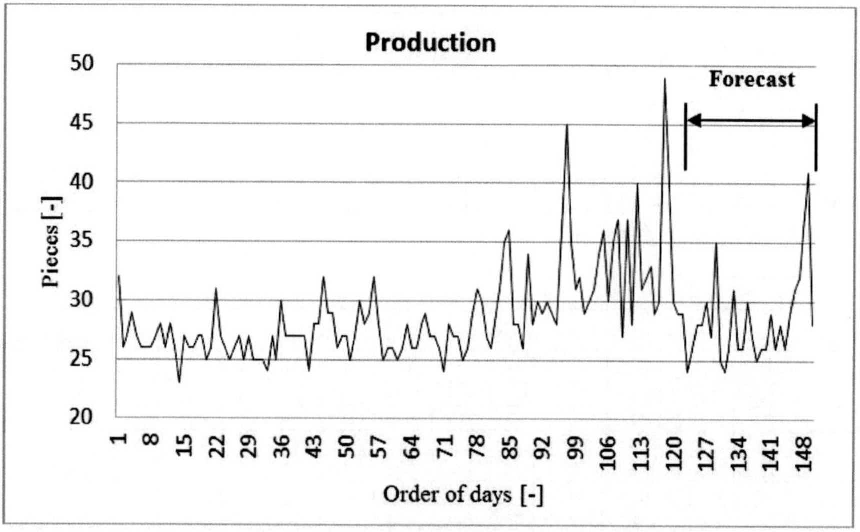

Fig. 7 Historical and forecasted data of time series

The graph presents history of 120 days of selling products and it's forecasting for next 30 days to help to set up a plan of production.

4 Genetic Algorithm

The genetic algorithm can be used for forecasting. An example of the forecasting of the tendency of Microsoft Corp. shares (MSFT) by means of the GeneHunter program is presented. The inputs are represented by the values of the time series *MSFT* that are presented by four independent time series with *Open*, *High*, *Low*, and *Close* prices. The principle of the method is that a rule for trading is suggested and this rule is optimized by genetic algorithm to maximum profit. The tuned rule is used for the forecasting whether the tendency of development of searched time series will be increasing or decreasing. This result can be used for decision making whether to buy the share or not. One of the possible rules can be in the form

<When> $(O,H,L,C)_i > (O,H,L,C)_j$ <And> $(O,H,L,C)_k > (O,H,L,C)_l$ <Then> shares increase.

The symbols used in the rule are as follows: O = Open, H = High, L = Low, C = Close values of the time series. The subscripts i, j, k, l determine the number of backward steps from the last value to be taken for evaluation of the emphasised cells. See Table 1.

Table 1 Forecasting of the future trend of *MSFT* share

Date	*Open*	*High*	*Low*	*Close*	*MSFT*											
8.11.2010	26.68	28.87	26.58	26.81			I	II	III	IV						
9.11.2010	26.81	27.11	26.71	26.95			0	2	2	0	<---Row code (0..3)					
10.11.2010	27.01	27.80	26.81	26.94			2	1	28	14	<--- Days back(0..40)					
11.11.2010	26.68	26.72	26.28	26.68			1	0			<--- Sign (0..1)					
12.11.2010	26.47	26.52	26.10	26.27												
15.11.2010	26.33	26.50	26.17	26.20	<When> O 2 > H 1 <And> C 21 <= O 14 <Then> share increases											
16.11.2010	26.04	26.40	25.65	25.81												
17.11.2010	25.90	25.91	25.55	25.57	27.77	>	27.61	And	27.98	<=	28.12	1	1	1		
18.11.2010	25.71	26.80	25.61	25.84												
......	...	...	....	...												
3.1.2011	28.05	28.18	27.92	27.98												
4.1.2011	27.94	28.17	27.85	28.90	The day evaluations of the rule									Bef.	Change	Back
5.1.2011	27.90	28.01	27.77	28.00	I	P	II		III	P	IV	1	2	P		
6.1.2011	28.40	28.85	27.86	28.82	27.94	>	27.77	And	25.17	<=	27.76	1	1	1	0.39	18
7.1.2011	28.64	28.74	28.25	28.60	27.90	>	27.86	And	24.93	<=	27.92	1	1	1	0.15	17
10.1.2011	28.26	28.40	28.40	28.22	28.40	>	28.25	And	25.00	<=	27.95	1	1	1	0.10	16
11.1.2011	28.20	28.25	28.50	28.11	28.64	>	28.40	And	25.56	<=	27.85	1	1	1	0.20	15
12.1.2011	28.12	28.59	28.70	28.55	28.26	>	28.50	And	26.20	<=	28.01	0	1	0	-0.60	14
13.1.2011	28.33	28.39	28.10	28.19	28.20	>	28.70	And	26.78	<=	27.97	0	1	0	-0.19	13
14.1.2011	28.08	28.38	27.91	28.30	28.12	>	28.10	And	26.76	<=	28.12	1	1	1	0.23	12
18.1.2011	28.16	28.74	28.14	28.66	28.33	>	27.91	And	26.85	<=	27.97	1	1	1	0.13	11
19.1.2011	28.46	28.68	28.27	28.47	28.08	>	28.14	And	26.80	<=	27.94	0	1	0	-0.14	10
20.1.2011	28.50	28.55	28.13	28.35	28.16	>	28.27	And	27.10	<=	27.92	0	1	0	0.07	9
21.1.2011	28.40	28.43	28.20	28.20	28.46	>	28.13	And	27.11	<=	27.08	1	0	0	-0.21	8
24.1.2011	28.02	28.56	27.99	28.38	28.50	>	28.20	And	27.17	<=	28.05	1	1	1	0.13	7
25.1.2011	28.14	28.45	28.12	28.45	28.40	>	27.99	And	27.26	<=	27.94	1	1	1	0.38	6
26.1.2011	28.51	28.99	28.50	28.78	28.02	>	28.12	And	27.53	<=	27.90	0	1	0	-0.01	5
27.1.2011	28.75	29.46	28.49	28.87	28.14	>	28.50	And	27.66	<=	28.40	0	1	0	-1.04	4
28.1.2011	28.90	28.93	27.45	27.75	28.51	>	28.49	And	27.75	<=	28.64	1	1	1	-0.03	3
31.1.2011	27.77	27.90	27.42	27.73	28.75	>	27.45	And	27.68	<=	28.26	1	1	1	0.19	2
1.2.2011	27.80	27.61	27.61	27.99	28.90	>	27.42	And	27.76	<=	28.20	1	1	1	0.27	1
2.2.2011	27.93	28.11	27.88	27.94	27.94	>	27.77	And	25.17	<=	27.76	1	1	1	0.39	0

In this case the fitness function is defined to obtain the maximum profit in searched period. The changed variables are as follows: the values of suffixes *i, j, k, l*, the sign of inequality in the rule $>$, $\leq$ and symbols *O*, *H*, *L*, *C* that mark which time series is represented in the rule.

After the process of calculation the optimized rule that presents the maximum profit is obtained. This rule is in the form

<When> Open 2 > High 1 <And> Close 21 $\leq$ Open 14 <Then> share increases.

This rule has a verbal interpretation: When the value of share price *Open* on Day 2 (back) is greater than the value of share price *High* on Day 1 and also the value of share price *Close* on Day 28 is less than or equal to value of share price *Open* on

Day 14, then the share MSFT increases. (When the condition is not fulfilled, the opposite is true.) When we substitute the real values into the optimized rule we obtain the condition in the form

<When> 27.77 > 27.61<And> 27.98 $\leq$ 28.12 <Then> share increases.

As the condition is fulfilled the rule gives us advice for the decision making in the stock market that the share price will increase in the following day.

Table 1 presents the method of calculation of the fitness function that is searched for the maximum profit. The fitness function is defined as the sum of profits in the searched period. The algorithm is as follows: when an increase is predicted, a Buy signal is issued at the beginning of the following day and a Sell signal at the end of that market day. The correct (wrong) decision of trend means a profit (loss) during the market day. During the process of optimization the rule is changed which leads to a different profit. The best profit marks the best rule that could be used for the determination of decrease or increase of the share on a following day.

Neither of the mentioned methods are limited nor are the rules (it is possible to think up other rule), nor the number of inputs (the input can be any index, shares, commodities, currency ratio, etc.), nor the length of the searched history.

5 Theory of Chaos

The complicated behaviour in economy could be illustrated by the equation (in literature called a logistic function)

$$x_{i+1} = r \,.\, x_i \,(\, 1 - x_i \,),$$

where r is a constant and x_i is a variable. The equation enables generation of behaviours that are considered to be point, periodic, and chaotic. Figure 8 presents the graph generated by this equation with initial values $x_0 = 0.85$ and $r = 3.7$.

The vertical axis represents the values x_{i+1} and horizontal axis represents the values i. The curve in the graph represents chaotic behaviour, in that it has a hidden order that is described by the above-mentioned logistic equation, even if this reality is hardly recognizable. The presented curve has a fractal character because the doubling is repeated with an accurate rate.

The complicated behaviour in economy could be illustrated by the equation in the form

$$P_{t+1} = \frac{P_t}{1 + a\sigma_t} + (V_t - P_t),$$

where P_{t+1} simulates the market price of the share, σ_t is the volatility of yield of share, a is a coefficient of market aversion against risk, V_t is fundamental "inner" value of share in time t, $V_t - P_t$ is the response of market on the change of inner value. The initial values (P_0, σ_0, V_0) could be different in the same way as the coefficients of market aversion against risk. If the value $\sigma = 0$ and market is neutral to risk $a = 0$, then $P_{t+1} = V_t$. Figure 9 presents graphically the course of P_{t+1} dependence on time with aversion to risk $a = 0.25$.

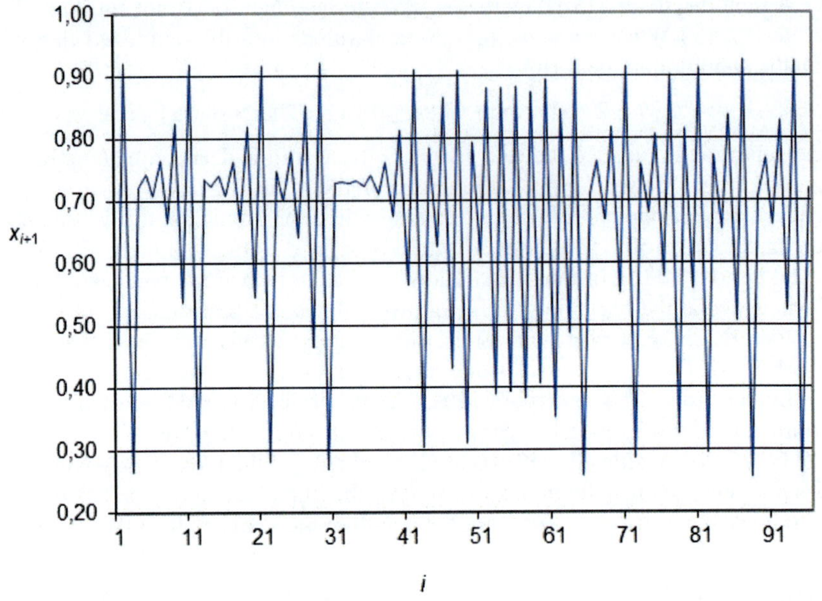

Fig. 8 The graph of logistic function

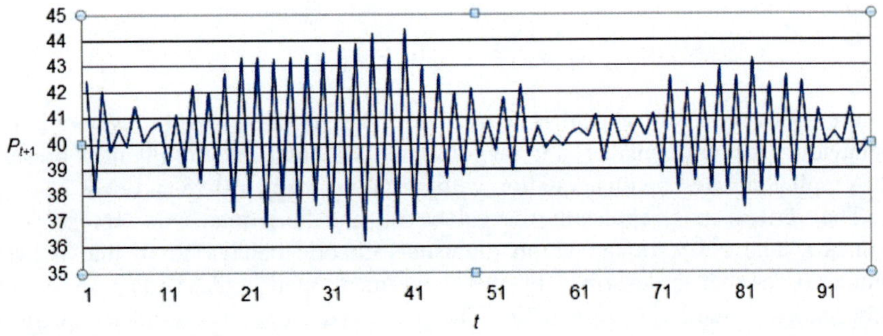

Fig. 9 The graph of function P_{t+1}

The course of the curve reminds one of real values of time series of shares, commodities, currency ratios, and values of indices on the stock market.

6 Conclusion

The business applications have specific features in comparison with others. The processes are focused on obtaining income or profit, or decreasing expenses. Therefore, such applications, both successful and unsuccessful, are not published very often because of secrecy in the highly competitive environments among firms and institutions. The soft computing methods can help in decentralization of decision-making processes to become standardized, reproduced, and documented. The forecasting methods play very important roles in companies because they help to reduce costs that can lead to higher profit and they can help to compete successfully, or decrease expenses in institutions. The forecasting processes in business are very complicated because they include political, social, psychological, economic, financial, and other factors. Some variables are difficult to measure; they may be characterized by imprecision, uncertainty, vagueness, semi-truth, approximations, and so forth.

The use of these computing methods can lead to higher quality of forecasting that can be used for decision-making in business, economy, financial areas, or the other sectors.

References

1. Aliev, A., Aliev, R.: Soft Computing and Its Applications, 444 p. World Scientific Pub., UK (2002)
2. Altroc, C.: Fuzzy Logic & Neurofuzzy—Applications in Business & Finance, 375 p. Prentice Hall, USA (1996) ISBN 0-13-591512-0
3. Azoff, E.M.: Neural Network Time Series Forecasting of Financial Markets, 196 p. John Wiley, USA (1994) ISBN 0-471-94356-8
4. Barnsley, M.F.: Fractals Everywhere, 531 p. Academic Press Professional, USA (1993) ISBN 0-12-079061-0
5. Bose, K., Liang, P.: Neural Networks. In: Fundamentals with Graphs, Algorithms and Applications, 478 p. McGraw-Hill, USA (1996) ISBN 0-07-114064-6
6. Davis, L.: Handbook of Genetic Algorithms, 385 p. Int. Thomson Com. Press, USA (1991) ISBN 1-850-32825-0
7. Dostál, P.: Advanced Decision Making in Business and Public Services, 168 p. CERM, Brno (2011) ISBN 978-80-7204-747-5
8. Dostál, P.: Advanced Economic Analyses, 80 p. VUT – FP – Brno (2008) ISBN 978-80-214-3564-3
9. Ehlers, F.J.: Cybernetic Analysis for Stock and Futures, 256 p. John Wiley, USA (2004) ISBN 0-471-46307-8
10. Fransws, P.H.: Time Series Models for Business and Economic Forecasting, 213 p. Cambridge University Press, UK (2001) ISBN 0-521-58641-0
11. Gately, E.: Neural Networks for Financial Forecasting, 169 p. John Wiley, USA (1996) ISBN 0-471-11212-7
12. Gleick, J.: Chaos, 330 p. Ando Publishing (1996) ISBN 80-86047-04-0
13. Hagan, T., Demuth, B.: Neural Network Design, 2nd edn., 702 p. PWS Publishing, USA (1996) ISBN 0-534-94332-2
14. Hanselman, D., Littlefield, B.: Mastering MATLAB 7, 852 p. Prentice Hall, USA (2005) ISBN 0-13-185714-2
15. Chen, G.: Controlling Chaos and Bifurcations in Engineering Systems, 648 p. CRC Press, China (2000) ISBN 978-3-44098-7

16. Chen, P., Jain, L., Tai, C.: Computational Economics: A Perspective from Computational Intelligence. Idea Group Publishing (2005) ISBN 1-591-406-498
17. Chen, S., Wang, P., Wen, T.: Computational Intelligence in Economics and Finance. Springer (2004) ISBN 978-3-540-72820-7
18. Chen, S., Wang, P., Wen, T.: Computational Intelligence in Economics and Finance, vol. II. Springer (2007) ISBN 978-3-540-72820-7
19. Kazabov, K., Kozma, R.: Neuro-Fuzzy—Techniques for Intelligent Information Systems, 427 p. Physica-Verlag, Germany (1998) ISBN 3-7908-1187-4
20. Klir, G.J., Yuan, B.: Fuzzy Sets and Fuzzy Logic, Theory and Applications, 279 p. Prentice Hall, New Jersey (1995) ISBN 0-13-101171-5
21. Li, Z., Halong, W.A., Chen, G.: Integration of Fuzzy Logic and Chaos Theory. Springer (2006) ISBN 3-540-26899-5
22. Peters, E.E.: Fractal Market Analysis—Applying Chaos Theory to Investment & Economics, 315 p. John Wiley, USA (1994) ISBN 0-471-58524-6
23. Peters, E.E.: Chaos and Order in the Capital Markets: A New View of Cycles, Prices, 274 p. Wiley Finance Edition, USA (1996) ISBN 0-471-13838-6
24. Trippi, R.R.: Chaos & Nonlinear Dynamics in the Financial Markets, 505 p. Irwin Professional Publishing, USA (1995) ISBN 1-55738-857-1
25. The MathWorks, MATLAB – User's Guide, The MathWorks (2010)

Adaptive Predictive Control of Time-Delay Systems

Vladimír Bobál, Marek Kubalčík, Petr Dostál, and Jakub Matějíček

Tomas Bata University in Zlín, Faculty of Applied Informatics, Nam T.G. Masaryka 5555, 760 01 Zlin, Czech Republic
{bobal,kubalcik,dostalp}@fai.utb.cz, jakub.matejicek@seznam.cz

Abstract. Design of an optimal controller for higher-order systems often leads to complex control algorithms. One of the possibilities of control of such processes is their approximation by a lower-order model with a time-delay (dead time). These time-delay processes can be effectively handled by model-based predictive control (MPC) method. The paper deals with design of an algorithm for adaptive predictive control of higher-order processes which are approximated by a second-order model of the process with a time-delay. The controller was tested and verified by simulation on experimental data from a laboratory model of a heat exchanger.

1 Introduction

Some technological processes in industry are characterized by high-order dynamic behaviour or large time constants and time-delays. Time-delay in a process increases the difficulty of controlling it. However using the approximation of higher-order process by lower-order model with time-delay provides simplification of the control algorithms. Processes with time-delay are difficult to control using standard feedback controllers. One of the possible approaches to control processes with time delay is predictive control [1], [2]. The aim of the paper is implementation of a predictive controller for control of high-order processes which are approximated by second-order model with time delay. The designed controller was tested and verified by control of a model of a laboratory heat exchanger.

The paper is organized in the following way: problems of implementation of predictive control based on the minimization of the quadratic criterion (MPC) are described in Section 2 and the computation of the predictor in Section 3. The experimental identification of the time-delay systems is briefly introduced in Section 4. The use of the predictive control algorithm on a control of laboratory heat exchanger in simulation conditions is demonstrated in Section 5. Section 6 concludes the paper.

I. Zelinka et al. (Eds.): Nostradamus: Mod. Meth. of Prediction, Modeling, AISC 192, pp. 61–72.
springerlink.com

2 MPC Based on Minimization of Quadratic Criterion

The standard cost function used in MPC contains quadratic terms of control error and control increments on a finite horizon into the future [1], [3]

$$J=\sum_{i=N_1}^{N_2}\left[\hat{y}(k+i)-w(k+i)\right]^2+\sum_{i=1}^{N_u}\left[\lambda(i)\Delta u(k+i-1)\right]^2 \tag{2.1}$$

where $\hat{y}(k+i)$ is the process output of i steps in the future predicted on the base of information available upon the time k, $w(k+i)$ is the sequence of the reference signal and $\Delta u(k+i-1)$ is the sequence of the future control increments that have to be calculated. Parameters N_1, N_2 and N_u are called minimum, maximum and control horizon. The parameter $\lambda(i)$ is a sequence which affects future behaviour of the controlled process, generally, it is chosen in the form of constants or exponential weights. The output of the model (predictor) is computed as the sum of the forced response $\boldsymbol{y}_n$ and the free response $\boldsymbol{y}_n$

$$\hat{\boldsymbol{y}}=\boldsymbol{y}_n+\boldsymbol{y}_0 \tag{2.2}$$

It is possible to compute the forced response as the multiplication of the matrix $\boldsymbol{G}$ (Jacobian Matrix of the model) and the vector of future control increments $\Delta\boldsymbol{u}$, which is generally a priori unknown

$$\boldsymbol{y}_n=\boldsymbol{G}\Delta\boldsymbol{u} \tag{2.3}$$

where

$$\boldsymbol{G}=\begin{bmatrix} g_1 & 0 & 0 & \cdots & 0 \\ g_2 & g_1 & 0 & \cdots & 0 \\ h_3 & h_2 & g_1 & \cdots & 0 \\ \vdots & \vdots & \vdots & \ddots & \vdots \\ g_{N_2} & g_{N_2-1} & g_{N_2-2} & \cdots & g_{N_2-N_u+1} \end{bmatrix} \tag{2.4}$$

is matrix containing step responses.

It follows from equations (2.3) and (2.4) that the predictor in a vector form is given by

$$\hat{\boldsymbol{y}}=\boldsymbol{G}\Delta\boldsymbol{u}+\boldsymbol{y}_0 \tag{2.5}$$

Minimization of the cost function (2.1) now becomes a direct problem of linear algebra. The solution in an unconstrained case can be found by setting partial derivative of J with respect to $\Delta\boldsymbol{u}$ to zero and yields

$$\Delta\boldsymbol{u}=-\left(\boldsymbol{G}^T\boldsymbol{G}+\lambda\boldsymbol{I}\right)^{-1}\boldsymbol{G}^T\left(\boldsymbol{y}_0-\boldsymbol{w}\right)=-\boldsymbol{G}_H^{-1}\boldsymbol{g} \tag{2.6}$$

where G_H and g are the Hesse-Matrix and the gradient. Denoting the first row of the matrix $\left(G^T G + \lambda I\right)^{-1} G^T$ by K, the actual control law can be calculated as

$$u(k) = K(w - y_0) + u(k-1) \tag{2.7}$$

3 Computation of Predictor

An important task is computation of predictions for arbitrary prediction and control horizons. Dynamics of most of processes requires horizons of length where it is not possible to compute predictions in a simple straightforward way. Recursive expressions for computation of the free response and the matrix G in each sampling period had to be derived. There are several different ways of deriving the prediction equations for transfer function models. Some papers make use of Diophantine equations to form the prediction equations [4]. In [5] matrix methods are used to compute predictions. We derived a method for recursive computation of both the free response and the matrix of the dynamics.

Computation of the predictor for the time-delay system can be obtained by modification of the predictor for the corresponding system without a time-delay. At first we will consider the second order system without time-delay and then we will modify the computation of predictions for the time-delay system.

3.1 Second Order System without Time-Delay

The model is described by the transfer function

$$G\left(z^{-1}\right) = \frac{b_1 z^{-1} + b_2 z^{-2}}{1 + a_1 z^{-1} + a_2 z^{-2}} = \frac{B\left(z^{-1}\right)}{A\left(z^{-1}\right)} \tag{3.1}$$

$$A\left(z^{-1}\right) = 1 + a_1 z^{-1} + a_2 z^{-2};\ B\left(z^{-1}\right) = b_1 z^{-1} + b_2 z^{-2} \tag{3.2}$$

The model can be also written in the form

$$A\left(z^{-1}\right) y(k) = B\left(z^{-1}\right) u(k) \tag{3.3}$$

A widely used model in general model predictive control is the CARIMA model which we can obtain from the nominal model (3.1) by adding a disturbance model

$$A\left(z^{-1}\right) y(k) = B\left(z^{-1}\right) u(k) + \frac{C\left(z^{-1}\right)}{\Delta} n_c(k) \tag{3.4}$$

where $n_c(k)$ is a non-measurable random disturbance that is assumed to have zero mean value and constant covariance and the operator $\Delta = 1 - z^{-1}$. Inverted Δ is then an integrator. The polynomial $C\left(z^{-1}\right)$ will be further considered as $C\left(z^{-1}\right) = 1$. The CARIMA description of the system is then in the form

$$\Delta A\left(z^{-1}\right)y(k)=B\left(z^{-1}\right)\Delta u(k-1)+n_c(k) \tag{3.5}$$

The difference equation of the CARIMA model without the unknown term $n_c(k)$ can be expressed as:

$$\begin{aligned}y(k)=&(1-a_1)y(k-1)+(a_1-a_2)y(k-2)+a_2y(k-3)+\\&+b_1\Delta u(k-1)+b_2\Delta u(k-2)\end{aligned} \tag{3.6}$$

It was necessary to compute three step ahead predictions in straightforward way by establishing of lower predictions to higher predictions. The model order defines that computation of one step ahead prediction is based on three past values of the system output. The three step ahead predictions are as follows

$$\begin{aligned}\hat{y}(k+1)=&(1-a_1)y(k)+(a_1-a_2)y(k-1)+a_2y(k-2)+\\&+b_1\Delta u(k)+b_2\Delta u(k-1)\\\hat{y}(k+2)=&(1-a_1)y(k+1)+(a_1-a_2)y(k)+a_2y(k-1)+\\&+b_1\Delta u(k+1)+b_2\Delta u(k)\\\hat{y}(k+3)=&(1-a_1)y(k+2)+(a_1-a_2)y(k+1)+a_2y(k)+\\&+b_1\Delta u(k+2)+b_2\Delta u(k+1)\end{aligned} \tag{3.7}$$

1. The predictions after modification can be written in a matrix form

$$\begin{aligned}&\begin{bmatrix}\hat{y}(k+1)\\\hat{y}(k+2)\\\hat{y}(k+3)\end{bmatrix}=\begin{bmatrix}g_1&0\\g_2&g_1\\g_3&g_2\end{bmatrix}\begin{bmatrix}\Delta u(k)\\\Delta u(k+1)\end{bmatrix}+\begin{bmatrix}p_{11}&p_{12}&p_{13}&p_{14}\\p_{21}&p_{22}&p_{23}&p_{24}\\p_{31}&p_{32}&p_{33}&p_{34}\end{bmatrix}\begin{bmatrix}y(k)\\y(k-1)\\y(k-2)\\\Delta u(k-1)\end{bmatrix}=\\&=\begin{bmatrix}b_1&0\\b_1(1-a_1)+b_2&b_1\\(a_1-a_2)b_1+(1-a_1)^2b_1+(1-a_1)b_2&b_1(1-a_1)+b_2\end{bmatrix}\begin{bmatrix}\Delta u(k)\\\Delta u(k+1)\end{bmatrix}+\\&+\begin{bmatrix}(1-a_1)&(a_1-a_2)&a_2&b_2\\(1-a_1)^2+(a_1-a_2)&(1-a_1)(a_1-a_2)+a_2&a_2(1-a_1)&b_2(1-a_1)\\(1-a_1)^3+2(1-a_1)(a_1-a_2)+a_2&(1-a_1)^2(a_1-a_2)+a_2(1-a_1)+(a_1-a_2)^2&a_2(1-a_1)^2+(a_1-a_2)a_2&b_2(1-a_1)^2+(a_1-a_2)b_2\end{bmatrix}\begin{bmatrix}y(k)\\y(k-1)\\y(k-2)\\\Delta u(k-1)\end{bmatrix}\end{aligned} \tag{3.8}$$

It is possible to divide computation of the predictions to recursion of the free response and recursion of the matrix of the dynamics. Based on the three previous predictions it is repeatedly computed the next row of the free response matrix in the following way:

$$\begin{aligned} p_{41} &= (1-a_1)\,p_{31} + (a_1-a_2)\,p_{21} + a_2 p_{11} \\ p_{42} &= (1-a_1)\,p_{32} + (a_1-a_2)\,p_{22} + a_2 p_{12} \\ p_{43} &= (1-a_1)\,p_{33} + (a_1-a_2)\,p_{23} + a_2 p_{13} \\ p_{44} &= (1-a_1)\,p_{34} + (a_1-a_2)\,p_{24} + a_2 p_{14} \end{aligned} \tag{3.9}$$

The first row of the matrix is omitted in the next step and further prediction is computed based on the three last rows including the one computed in the previous step. This procedure is cyclically repeated. It is possible to compute an arbitrary number of rows of the matrix.

The recursion of the dynamics matrix is similar. The next element of the first column is repeatedly computed in the same way as in the previous case and the remaining columns are shifted to form a lower triangular matrix. This procedure is performed repeatedly until the prediction horizon is achieved. If the control horizon is lower than the prediction horizon a number of columns in the matrix is reduced. Computation of the new element is performed as follows:

$$g_4 = (1-a_1)\,g_3 + (a_1-a_2)\,g_2 + a_2 g_1 \tag{3.10}$$

3.2 *Second Order System with Time-Delay*

The nominal model with two steps time-delay is considered as

$$G(z^{-1}) = \frac{B(z^{-1})}{A(z^{-1})} z^{-2} = \frac{b_1 z^{-1} + b_2 z^{-2}}{1 + a_1 z^{-1} + a_2 z^{-2}} z^{-2} \tag{3.11}$$

The CARIMA model for time-delay system takes the form

$$\Delta A(z^{-1}) y(k) = z^{-d} B(z^{-1}) \Delta u(k-1) + n_c(k) \tag{3.12}$$

where d is the dead time. In our case d is equal to 2. In order to compute the control action it is necessary to determine the predictions from d+1 (2+1 in our case) to $d+N_2$ ($2+N_2$).

The predictor (3.8) is then modified to

$$\begin{aligned} \begin{bmatrix} \hat{y}(k+3) \\ \hat{y}(k+4) \\ \hat{y}(k+5) \end{bmatrix} &= \begin{bmatrix} p_{31} & p_{32} & p_{33} \\ p_{41} & p_{42} & p_{43} \\ p_{51} & p_{52} & p_{53} \end{bmatrix} \begin{bmatrix} y(k) \\ y(k-1) \\ y(k-2) \end{bmatrix} + \\ &+ \begin{bmatrix} g_1 & 0 \\ g_2 & g_1 \\ g_3 & g_2 \end{bmatrix} \begin{bmatrix} \Delta u(k) \\ \Delta u(k+1) \end{bmatrix} + \begin{bmatrix} g_2 & g_3 & p_{34} \\ g_3 & g_4 & p_{44} \\ g_4 & g_5 & p_{54} \end{bmatrix} \begin{bmatrix} \Delta u(k-1) \\ \Delta u(k-2) \\ \Delta u(k-3) \end{bmatrix} \end{aligned} \tag{3.13}$$

Recursive computation of the matrices is analogical to the recursive computation described in the previous section.

4 System Identification of Time-Delay Processes

The time-delay processes can by identified either using off-line methods or recursive (on-line) methods. The off-line method is used for the process model identification for the simulation verification of the predictive control. The recursive identification algorithm is a part or the adaptive predictive algorithm.

4.1 Off-Line Process Identification

In this paper, the time-delay is assumed to be approximately known or possibly to be obtained separately from an off-line identification using the least squares method (LSM)

$$\hat{\boldsymbol{\Theta}} = \left(\boldsymbol{F}^T\boldsymbol{F}\right)^{-1}\boldsymbol{F}^T\boldsymbol{y} \tag{4.1}$$

where the matrix $\boldsymbol{F}$ has dimension (N-n-d, $2n$), the vector $\boldsymbol{y}$ (N-n-d) and the vector of parameter model estimates $\hat{\boldsymbol{\Theta}}$ ($2n$). N is a number of samples of measured input and output data, n is the model order. Equation (4.1) serves for an one-off calculation of the vector of parameter estimates $\hat{\boldsymbol{\Theta}}$ using N samples of measured data. The individual vectors and matrix in equation (4.1) have the form [6]

$$\boldsymbol{y}^T = \left[y(n+d+1) \quad y(n+d+2) \quad \cdots \quad y(N)\right] \tag{4.2}$$

$$\hat{\boldsymbol{\Theta}}^T = \left[\hat{a}_1 \quad \hat{a}_2 \quad \cdots \quad \hat{a}_n \quad \hat{b}_1 \quad \hat{b}_2 \quad \cdots \quad \hat{b}_n\right] \tag{4.3}$$

$$\boldsymbol{F} = \left[\begin{matrix} -y(n+d) & -y(n+d-1) & \cdots & -y(d+1) \\ -y(n+d+1) & -y(n+d) & \cdots & -y(d+2) \\ \vdots & \vdots & \cdots & \vdots \\ -y(N-1) & -y(N-2) & \cdots & -y(N-n) \end{matrix}\right. \\ \left.\begin{matrix} u(n) & u(n-1) & \cdots & u(1) \\ u(n+1) & u(n) & \cdots & u(2) \\ \vdots & \vdots & \cdots & \vdots \\ u(N-d-1) & u(N-d-2) & \cdots & u(N-d-n) \end{matrix}\right] \tag{4.4}$$

For identification of time-delay systems it is also available the optimization MATLAB function „fminsearch“. This function is generally referred to as an unconstrained nonlinear optimization [7].

4.2 Recursive Identification Algorithm

The regression (ARX) model of the following form

$$y(k) = \boldsymbol{\Theta}^T(k)\boldsymbol{\Phi}(k) + n_c(k) \tag{4.5}$$

is used in the identification part of the designed controller algorithms, where

$$\boldsymbol{\Theta}^T(k) = [a_1 \quad a_2 \quad b_1 \quad b_2] \tag{4.6}$$

is the vector of model parameters and

$$\boldsymbol{\Phi}^T(k-1) = [-y(k-1) \quad -y(k-2) \quad u(k-d-1) \quad u(k-d-2)] \tag{4.7}$$

is the regression vector. The non-measurable random component $n_c(k)$ is assumed to have zero mean value $E[n_c(k)] = 0$ and constant covariance (dispersion) $R = E[n_c^2(k)]$.

The adaptive predictive controller uses the recursive identification algorithm based on the Recursive Least Squares Method (RLSM) extended to include the technique of directional (adaptive) forgetting. Numerical stability is improved by means of the LD decomposition [8], [9]. This method is based on the idea of changing the influence of input-output data pairs to the current estimates. The weights are assigned according to amount of information carried by the data.

5 Experimental Example

The use of the predictive control algorithm is demonstrated on a control of laboratory heat exchanger in simulation conditions. The laboratory heat exchanger [10] is based on the principle of transferring heat from a source through a piping system using a heat transferring media to a heat-consuming appliance.

5.1 Laboratory Heat Exchanger Description

A scheme of the laboratory heat exchanger is depicted in Fig. 4. The heat transferring fluid (e. g. water) is transported using a continuously controllable DC pump (6) into a flow heater (1) with max. power of 750 W. The temperature of a fluid at the heater output T_1 is measured by a platinum thermometer. Warmed liquid then goes through a 15 meters long insulated coiled pipeline (2) which causes the significant delay in the system. The air-water heat exchanger (3) with two cooling fans (4, 5) represents a heat-consuming appliance. The speed of the first fan can be continuously adjusted, whereas the second one is of on/off type. Input and output temperatures of the cooler are measured again by platinum thermometers as T_2, resp. T_3. The laboratory heat exchanger is connected to a standard PC via technological multifunction I/O card. For all monitoring and control functions the MATLAB/SIMULINK environment with Real Time Toolbox.

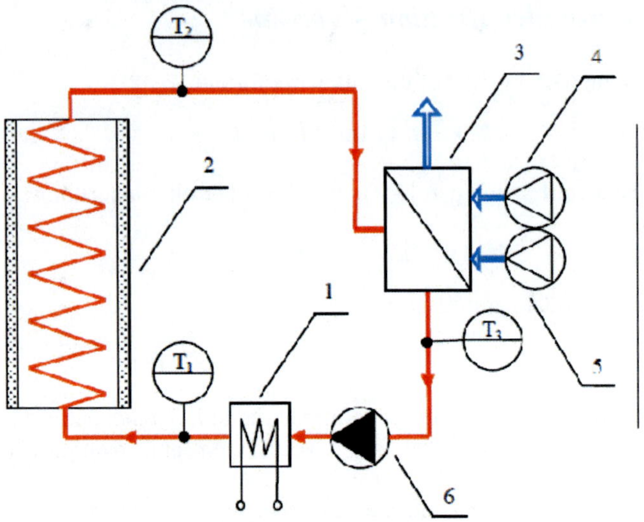

Fig. 1 Scheme of laboratory heat exchanger

A scheme of the laboratory heat exchanger is depicted in Fig. 1. The heat transferring fluid (e. g. water) is transported using a continuously controllable DC pump (6) into a flow heater (1) with max. power of 750 W. The temperature of a fluid at the heater output T_1 is measured by a platinum thermometer. Warmed liquid then goes through a 15 meters long insulated coiled pipeline (2) which causes the significant delay in the system. The air-water heat exchanger (3) with two cooling fans (4, 5) represents a heat-consuming appliance. The speed of the first fan can be continuously adjusted, whereas the second one is of on/off type. Input and output temperatures of the cooler are measured again by platinum thermometers as T_2, resp. T_3. The laboratory heat exchanger is connected to a standard PC via technological multifunction I/O card. For all monitoring and control functions the MATLAB/SIMULINK environment with Real Time Toolbox.

5.2 *Identification of Laboratory Heat Exchanger*

The dynamic model of the laboratory heat exchanger was obtained from processed input (the power of a flow heater P [W]) and output (the temperature of a T_2 [deg] of the cooler) data. The input signal $u(k)$ was generated using a pseudo-random binary signal (prbs), random gaussian signal (rgs) and a signal that is a sum of sinusoids (sine). The MATLAB code

```
u=idinput(N,'type',[0 B],[Umin, Umax])
```

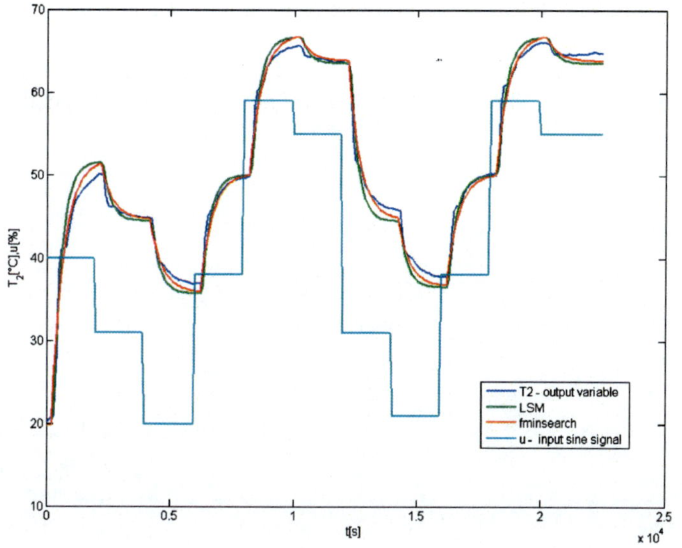

Fig. 2 Results of experimental off-line identification

generates a type signal of the length N, where [0 B] determines the frequency passband. Umin, Umax defines the minimum and maximum values of u. The signal level is such that Umin is the mean value of the signal, minus one standard deviation, while Umax is the mean value plus one standard deviation. The laboratory heat exchanger was identified using three input signals (prbs, rgs and sine) and two identification methods (LSM and „fminsearch"). The comparison of the dynamic behaviour of the real process which was excitated by sine input signal and identified using LSM and „fminsearch" is depicted in Fig. 2.

Using the input sine signal and the MATLAB function „fminsearch", the followed discrete transfer function for sampling period $T_0 = 100$ s was identified

$$G(z^{-1}) = \frac{0.1494z^{-1} + 0.028z^{-2}}{1 - 0.6376z^{-1} - 0.1407z^{-2}} z^{-2} \qquad (5.1)$$

5.3 Simulation Results

A typical control scheme used is depicted in Fig. 3. This scheme is used for systems with time-delay of two sample steps. The process was simulated by model (5.1), the process output was excitated by the nonmeasurable white noise and the step disturbance 10 deg in the time 7 000 s. The simulation of control process

(5.1) was realized for different value of the weighting factor. The individual control courses are depicted in Fig. 4 and Fig. 5. From Fig. 4 and 5 it is obvious that the control quality is very good, including of intervals, during the transient response. The predictive control algorithm is well robust also in the case of the step disturbance excitation. The control courses are not significantly dependent on the weighting factor λ.

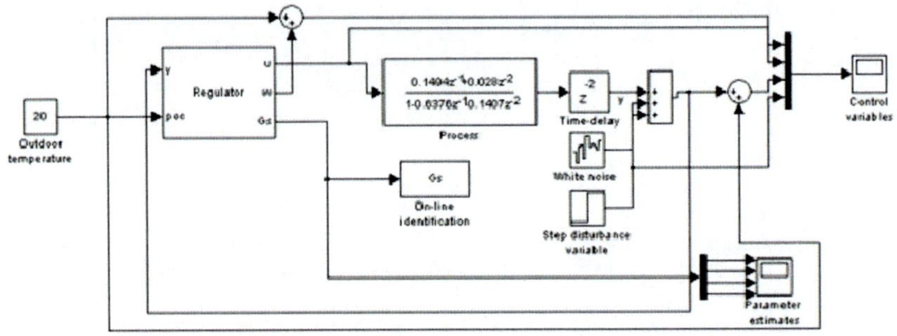

Fig. 3 Simulink control scheme

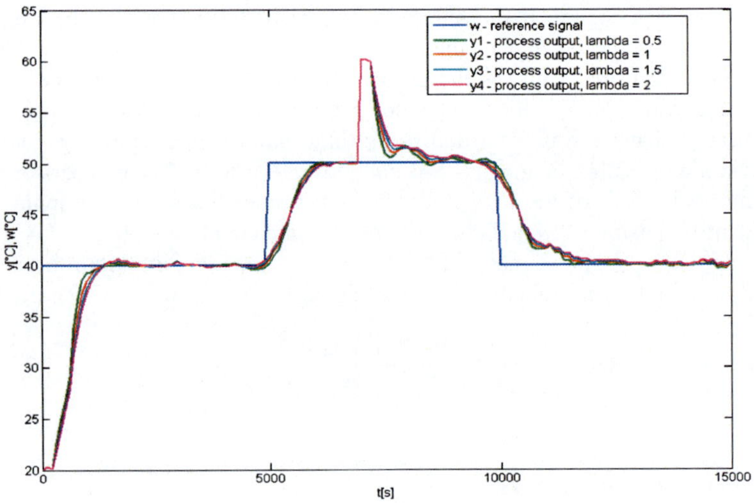

Fig. 4 Course of process outputs $y(k)$

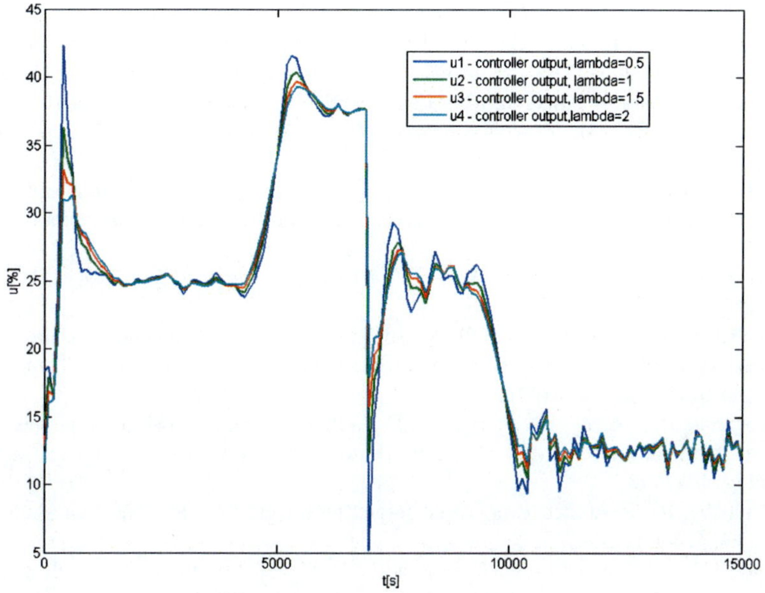

Fig. 5 Course of process inputs $u(k)$

6 Conclusion

1. The contribution presents adaptive model-based predictive control applied to time-delay processes. A linear model with constant coefficients used in pure model predictive control cannot describe the control system in all its modes. Therefore, an on-line identification was incorporated into the controller to obtain self-tuning capabilities. MPC based on quadratic values criterion was derived and tested. The laboratory heat exchanger system was identified by an experimental on-line method and its discrete model was used for verification of proposed predictive controller. The simulation experiments confirmed that predictive approach is able to cope with the given control problem. The designed predictive controller was successfully verified not only by simulation but also in real-time laboratory conditions for control of the heat exchanger.

Acknowledgment. This article was created with support of Operational Programme Research and Development for Innovations co-funded by the European Regional Development Fund (ERDF), national budget of Czech Republic within the framework of the Centre of Polymer Systems project (reg. number: CZ.1.05/2.1.00/03.0111).

References

1. Camacho, E.F., Bordons, C.: Model Predictive Control. Springer, London (2004)
2. Bobál, V., Chalupa Petr, P., Kubalčík, M., Dostál, P.: Self-tuning Predictive Control of Non-linear Servo-motor. Journal of Electrical Engineering 61, 365–372 (2010)
3. Mikleš, J., Fikar, M.: Process Modelling, Optimisation and Control. Springer, Berlín (2008)
4. Kwon, W.H., Choj, H., Byun, D.G., Noh, S.: Recursive solution of generalized predictive control and its equivalence to receding horizon tracking control. Automatica 28, 1235–1238 (1992)
5. Rossiter, J.A.: Model Based Predictive Control: a Practical Approach. CRC Press (2003)
6. Bobál, V., Chalupa, P., Dostál, P., Kubalčík, M.: Design and simulation verification of self-tuning Smith Predictors. International Journal of Mathematics and Computers in Simulation 5, 342–351 (2011)
7. Lagarias, J.C., Reeds, J.A., Wright, M.H., Wright, P.E.: Convergence properties of the Nelder-Mead Simplex Method in Low Dimensions. SIAM Journal of Optimization 9, 112–147 (1998)
8. Kulhavý, R.: Restricted exponential forgetting in real time identification. Automatica 23, 586–600 (1987)
9. Bobál, V., Böhm, J., Fessl, J., Macháček, J.: Digital Self-tuning Controllers: Algorithms, Implementation and Applications. Springer, London (2005)
10. Pekař, L., Prokop, R., Dostálek, P.: An anisochronic model of a laboratory heating system. In: Proceedings of 13th WSEAS International Conference on Systems, Rhodes, Greece, pp. 165–172 (2009)

Specific Behaviour of GPA-ES Evolutionary System Observed in Deterministic Chaos Regression

Tomas Brandejsky[1] and Ivan Zelinka[2]

[1] Faculty of Transportation Sciences, CTU in Prague
[2] Faculty of Electrical Engineering and Computer Science, TU of Ostrava

Abstract. The paper deals with symbolic regression of deterministic chaos systems using GPA-ES system. Lorenz attractor [1], Rősler attractor [2], Rabinovich-Fabrikant equations [3] and Van der Pol oscillator [4] are used as examples of deterministic chaos systems to demonstrate significant differences in efficiency of symbolic regression of systems described by equations of similar complexity. Within the paper, the source of this behaviour is identified in presence of structures which are hard to be discovered during evolutionary process due to low probability of their occurrence in initial population and by low chance to produce them by standard evolutionary operators given by low probability to form them in single step and low fitness function magnitudes of inter-steps when GPA tries to form them in more steps. This low magnitude of fitness function for particular solutions tends to elimination of them thus increases number of needed evolutionary steps. The results are significant for prediction of unknown system states on the base of its model symbolic regression on the base of its previous states.

1 Introduction

Symbolic regression of chaotic systems is nowadays frequently studied due to its influence not only into the area of modeling [5] but also of control and communication [6, 7] and security [8, 9]. It brings many unexpected surprises. Especially it is hard to estimate complexity of the task and not only in the case of regression of natural data of really unknown model, but also in the case of backward regression of artificial data prepared by known deterministic chaos model. For example, it is easy to reconstruct Lorenz attractor equations, a little bit more difficult is Rőssler attractor, but Rabinovich-Fabrikant equations or Van der Pol oscillator one requires many times large populations and numbers of evolutionary iterations non looking that complexity of the equations is not much higher. The paper tries to study this phenomenon, which is significant for real data modelling and prediction.

2 Hybrid GPA-ES Evolutionary System

Modeling of complex systems, as chaotic ones are, requires big structural sensitivity of the algorithm and elimination of random influences, especially in parameters

I. Zelinka et al. (Eds.): Nostradamus: Mod. Meth. of Prediction, Modeling, AISC 192, pp. 73–81.
springerlink.com

estimation. These requirements tend to use of Genetic Programming Algorithms extended in comparison to original Koza's one [10] by specialized Evolutionary Strategy algorithm used in each cycle of GPA work for each individual of its population for optimization of parameters of model developed by this GPA. Only this computationally expensive (it is possible to translate as slow) way gives certainty, that GPA algorithm will compare models with perfectly fitted parameters, it means without random influences. Other words, hybrid two-layer algorithms are more efficient than original GPAs not looking that there is large computational effort consumed by created structures parameter estimation. Thus it is better applicable to chaotic system analysis [11], where many evolutionary systems are also successfully applied [12].

The structure of GPA-ES algorithm was published e.g. in [13] and it is outlined at Fig. 1. The algorithm is able to reconstruct equations describing above mentioned systems of deterministic chaos, as it is published in [14, 15]. Convergence of GPA-ES algorithm depends on many parameters, the most significant are listed below:

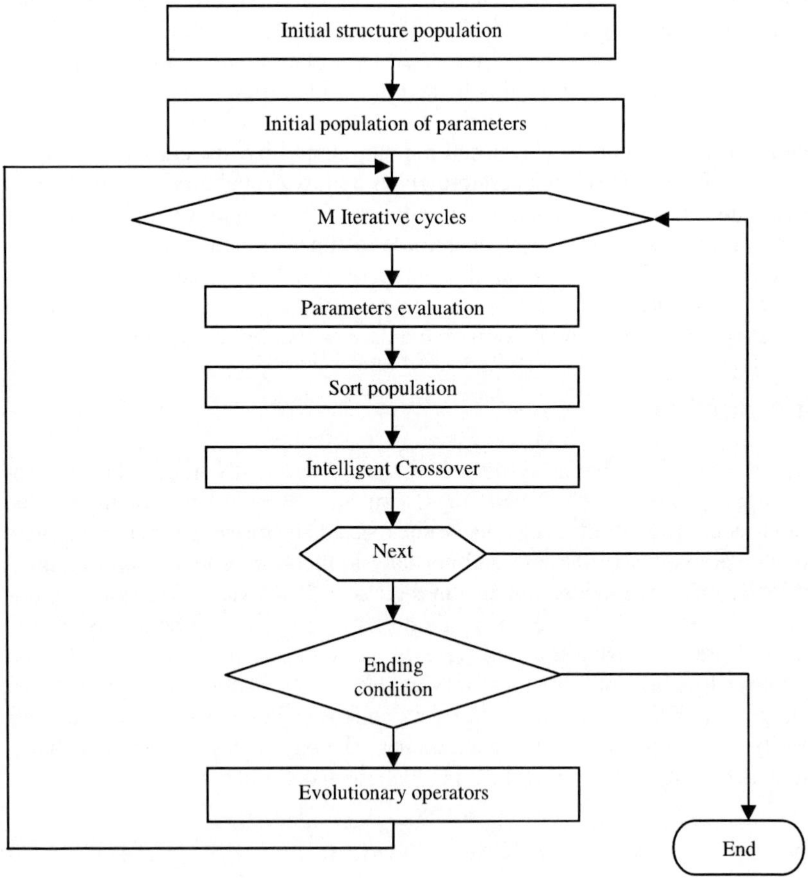

Fig. 1 GPA-ES algorithm structure

n	number of GPA individuals
m	number of ES individuals
l	complexity of structures created by GPA (number of operators and terminals)
k	average number of constants in GPA genes, where
p	number of GPA populations
q	number of ES population

In the following text, the influence of parameters l and k will be discussed, because these parameters influences efficiency (computational complexity) of the algorithm in nonlinear way that it is much more complicated, than it was described in the work [8].

3 GPA and ES Population Size Influence

GPA-ES algorithm work is strongly influenced by size of GPA population and ES populations and numbers of their iterations (evolutionary cycles). ES algorithm used to optimize parameters of each structure developed by GPA consumes the largest part of computational time. It works with fixed number of iterations. Their number is tied with number of iterations. When the population is large, number of iterations must be larger too to harness all available information in the population, because such operations as crossover operate only on the pair of individuals, so it is useful to set number of ES iterations (number of ES populations) at least to magnitude given by equation (1) to obtain chance that all combinations will be tested:

$$q \geq \log_2(m) \tag{1}$$

Small populations create stronger evolutionary pressure than the lager ones. On the opposite side, large populations tend to produce of more stable results. It means that small populations converge to optima faster, but there is larger risk that they get stuck in local optima far from searched global one. Thus, in the case of ES part of the algorithm, larger populations are preferred to decrease probability of good structure loses due to wrongly identified parameters.

To obtain more information about relationships between convergence speed and GPA and ES population sizes in symbolic regression of deterministic chaos system of Lorenz attractor (2), the dependencies presented by Fig. 2-4 were obtained.

$$\begin{aligned} x'[t] &= \sigma\ (y[t]-x[t]), \\ y'[t] &= x[t]\ (\rho - z[t])-y[t], \\ z'[t] &= x[t]\ y[t]\ -\beta\ z[t] \end{aligned} \tag{2}$$

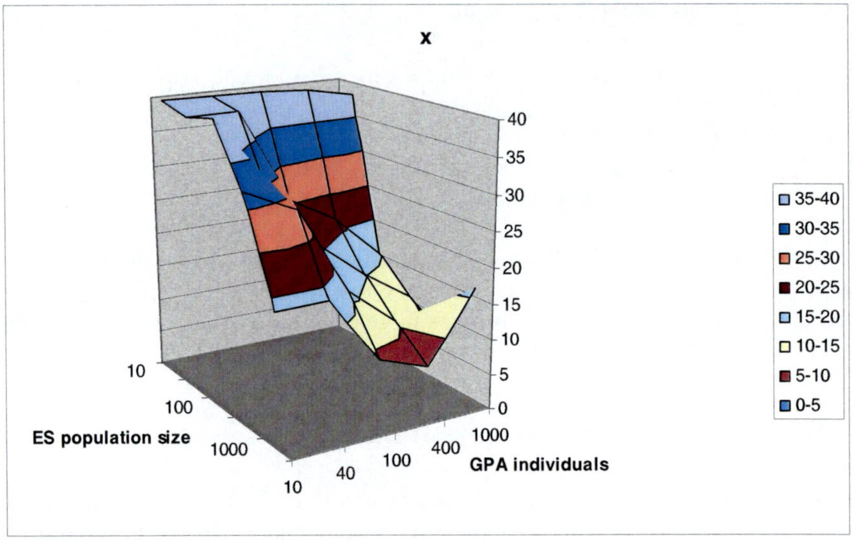

Fig. 2 Average precision of outputs of symbolic regression of variable x of Lorenz attractor provided by GPA-ES system

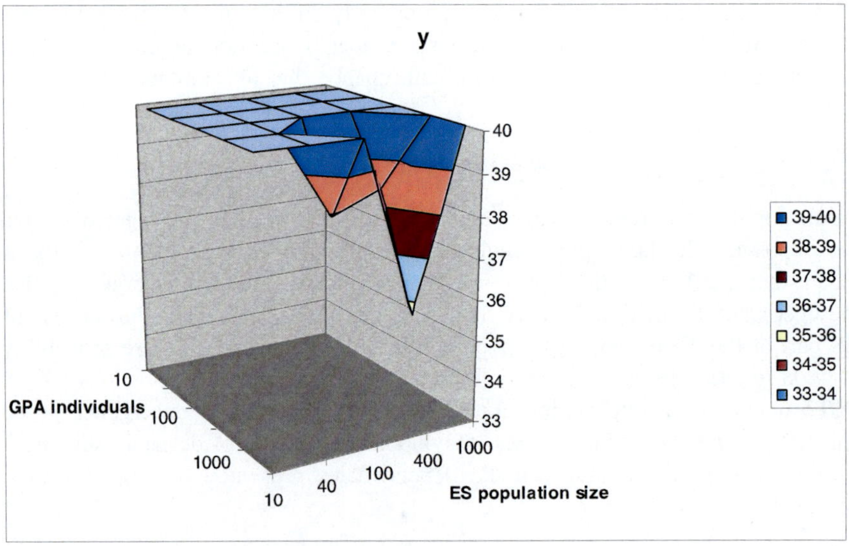

Fig. 3 Average precision of outputs of symbolic regression of variable y of Lorenz attractor provided by GPA-ES system

The loose of efficiency, visible especially on the Fig. 4 but also on the rest, is given by the loose of computational efficiency in the large populations combined with small limit of iterations near the limit given by (1). Magnitude of m was 1000, q was 40 iterations and equation (1) gives in this case q=10. This limit is satisfied, but (1) does not respect, that not all evolutionary are efficient – not all products of the operation are better than their parents. If they are not better, they are not included into population, so they cannot influence the next population and thus the better estimation of q is given by (3):

$$q \geq \alpha \log_2(m), \tag{3}$$

where α denotes probability of crossover operation success (probability that it is able to produce better individual than its parents are). First test points that it is good to estimate $\alpha = c$, where c is current computation cycle number.

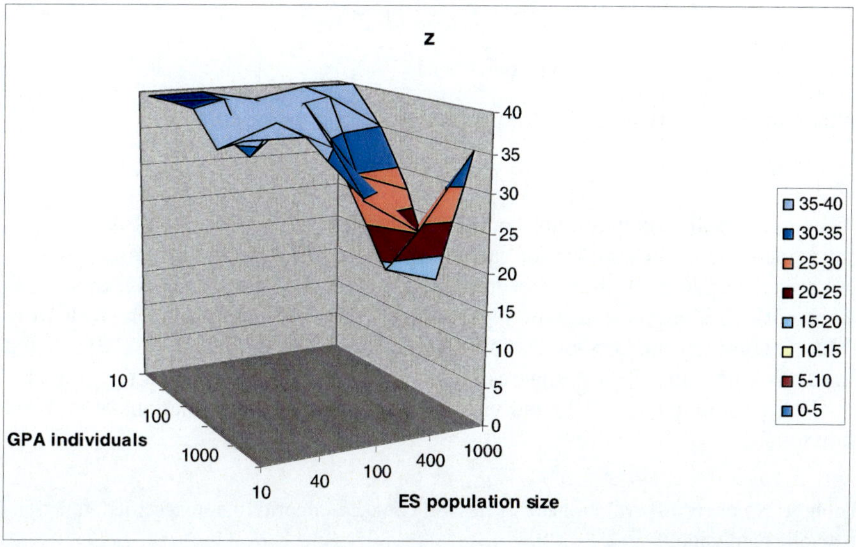

Fig. 4 Average precision of outputs of symbolic regression of variable z of Lorenz attractor provided by GPA-ES system

4 Typical Structures in Equations Describing Deterministic Chaos Systems

Above mentioned well known systems of deterministic chaos defined on continuous time domain and real value space are described as (2) for Lorenz attractor, (4) for Rősler attractor, (5) for Van der Pol oscilator and (6) Rabinovich-Fabricant system.

$$\begin{aligned} x' &= -y - z \\ y' &= x + ay \\ z' &= b + z(x - c) \end{aligned} \tag{4}$$

$$x'_1 = \mu * (x_1 - 1/3 * x_1^3 - x_2)$$

$$x'_2 = \frac{x_1}{\mu} \tag{5}$$

$$\begin{aligned} x' &= y(z - 1 + x^2) + \gamma x \\ y' &= x(3z + 1 - x^2) + \gamma y \\ z' &= -2z(\alpha + xy) \end{aligned} \tag{6}$$

In these equations, typical structures listed in (7) are identified:

$$\left[x^2, x^3, xy, x^2y, x - y, xyz\right] \tag{7}$$

To compare real computational complexity of their regression by GPA-ES algorithm, the table of 36 samples for each above listed function and for variables x, y, and z was prepared. With respect to presence of such functions as (7), this list is short, but sufficient for symbolic regression by the used algorithm. The following Table 2 presents numbers of evolutionary cycles needed to regress above listed functions with sum of error squares smaller than 10^{-6}. For presented test, the size of GPA population was 100 individuals and size of ES populations was 1000 individuals.

Table 1 Numbers of evolutionary cycles of GPA-ES algorithm for specified functions symbolic regression

Function	Number of needed iteration cycles	Number of fitness function evaluations
x^2	39	3,9e10^6
x^3	147	1,47e10^7
xy	3	3e10^5
x^2y	41	4,1e10^6
$x-y$	3	3e10^5
xyz	8	8es10^5

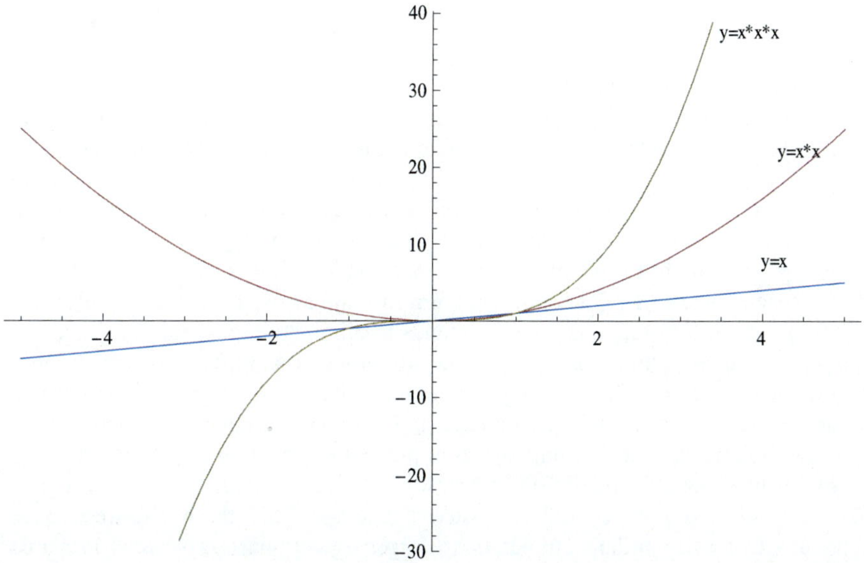

Fig. 5 Discussed functions y=x, y=x^2 and y=x^3 illustrating problem of movement from original imprecise approximation by monotonously increasing function y=x into better form y=x^3 across function y=x^2, which is not monotonous

Table 1 points that some pairs of equivalent complexity structures might tend to different computational times (and numbers of evolutionary cycles) as in the case of pairs xyz and x^3 or xy and x^2, where the computation in one case might be more than 10 times faster than in another case. Small probability of occurrence of structures and in initial population might be computed by Markov chains [16]. Evolutionary operators as mutation and crossover ones tend to analogical results in the case of mutation. Crossover operator gives higher probability to broke structure close to final one than to update it to searched structure. There is also another problem – there is larger similarity between functions y=x and y=x^2 than between y=x and y=x^3 or between y=x^2 and y=x^3 as it is illustrated by Figure 5. This fact decreases probability of movement from y=x to y=x^3 because this change requires typically two mutations and the first one must produce function y=x^2, which gives worse fitness than original y=x in the case of reaching final y=x^3.

5 Prediction

GPA-ES algorithm demonstrated that it is able to produce models of complex dynamic systems comparable to original algebraic model of data, or these models are overcomplicated the way that it is possible to automatically simplify them to original model by symbolic mathematic tool like Mathematica. Typically, on the place of single constant, say 2.0, GPA-ES algorithm uses expression like (1.3+0.7). Because obtained models are fully comparable to model of original data, see e.g. [14] and [15], they might serve for future system state prediction.

6 Conclusions

The paper explains unexpected behavior of GPA-ES algorithm (and standard GPAs), when symbolic regression of some structures and functions might require more effort than another one of the same complexity. The source of this behavior was identified in presence of structures which are hard to be discovered during evolutionary process due to low probability of their occurrence in initial population and by low chance to produce them by standard evolutionary operators given by low probability to form them in single step and low fitness function magnitudes of inter-steps when GPA tries to form them in more steps. This low magnitude of fitness function for particular solutions tends to elimination of them thus increases number of needed evolutionary steps. The solution is to modify crossover and mutation operators in the GPA-ES algorithm or to modify evolutionary process to be similar to simulated annealing (increase and decrease "temperature" and thus to work with variable length of population to prevent elimination of partial structures in some situations). Modification of crossover operator should be focused to preservation of sub-trees complexity variety because [17] these sub-trees have tendency to reach similar complexity in larger evolutionary processes. Prepared future applications are now focused to areas of transportation and mechanical systems control.

Acknowledgement. The work is supported by research project of MŠMT ČR No 6840770043 „Improvement of methods of design and employment of transportation networks from optimization viewpoint".

References

[1] Lorenz, E.N.: Deterministic nonperiodic flow. J. Atmos. Sci. 20(2), 130–141 (1963)

[2] Rössler, O.E.: An Equation for Continuous Chaos. Physics Letters 57A(5), 397–398 (1976)

[3] Van der Pol, B.: On relaxation-oscillations. The London, Edinburgh and Dublin Phil. Mag. & J. of Sci. 2(7), 978–992 (1927)

[4] Rabinovich, M.I., Fabrikant, A.L.: Stochastic Self-Modulation of Waves in Nonequilibrium Media. Sov. Phys. JETP 50, 311 (1979)

[5] Zelinka, I., Senkerik, R., Navratil, E.: Investigation on Evolutionary Optimization of Chaos Control. Chaos, Solitons & Fractals 40(1), 111–129 (2009)

[6] Kommenda, M., Kronberger, G., Feilmayr, C., Schickmair, L., Affenzeller, M., Winkler, S.M., Wagner, S.: Application of Symbolic Regression on Blast Furnace and Temper Mill Datasets. In: Moreno-Díaz, R., Pichler, F., Quesada-Arencibia, A. (eds.) EUROCAST 2011, Part I. LNCS, vol. 6927, pp. 400–407. Springer, Heidelberg (2012)

[7] Cuomo, K.M., Oppenheim, A.V., Strogatz, S.H.: Synchronization of Lorenz-based chaotic circuits with application to communication. IEEE Trans. Circuits Syst. I 40, 626–633 (1993)

[8] Wu, C.W., Chua, L.O.: A simple way to synchronize chaotic system with application to secure communication system. Int. J. Bifurc. Chaos 3(6), 1619–1627 (1993)

[9] Yang, T., Wu, C.W., Chua, L.O.: Cryptography based on chaotic systems. IEEE Trans. Circuits Syst. I 44, 469–472 (1997)
[10] Koza, J.R., Bennett III, F.H., Andre, D., Keane, M.A.: Genetic Programming III – Darwinian Invention and Problem Solving. Morgan Kaufmann Publishers, San Francisco (1999)
[11] Abarbanel, H.D.I.: Analysis of observed chaotic data. Springer Inc., New York (1996)
[12] Zelinka, I., Lampinen, J.: Evolutionary Identification of Predictive Models. In: Fye, C. (ed.) International Symposium on Engineering of Intelligent Systems, Paisley, Scotland, UK. ICSC Academic Press International Computer Science Conventions, Canada, Switzerland, ISBN 3-906454-21-5
[13] Brandejsky, T.: Symbolic regression of deterministic chaos. In: Proceedings of 17th International Conference on Soft Computing (MENDEL 2011), pp. 90–93 (2011)
[14] Brandejsky, T.: Multi-layered evolutionary system suitable to symbolic model regression. In: Recent Researches in Applied Informatics, vol. 1, pp. 222–225. WSEAS Press, Athens (2011)
[15] Brandejsky, T.: Nonlinear system identification by GPA-ES. In: Petráš, I., Podlubny, I., Kostúr, K., Kačúr, J., Mojžišová, A. (eds.) Proceedings of the 2012 13th International Carpathian Control Conference (ICCC). CD (2012) IEEE Catalog Number: CFP1242L-CDR, ISBN: 978-1-4577-1866-3
[16] Markov, A.A.: Extension of the limit theorems of probability theory to a sum of variables connected in a chain; reprinted in Appendix B of: Howard, R.: Dynamic Probabilistic Systems, volume 1: Markov Chains. John Wiley and Sons (1971)
[17] Langdon, W.B., Poli, R.: Foundations of Genetic Programming. Springer, Heidelberg (1998) ISBN: 978-3540424512

Chaos Synchronization Based on Unknown Inputs Takagi-Sugeno Fuzzy Observer

Mohammed Chadli[1] and Ivan Zelinka[2]

[1] University of Picardie Jules Verne, Laboratory of "modélisation, Information et Systèmes", UPJV-MIS, 7 Rue Moulin neuf, 80000, Amiens, France
mohammed.chadli@u-picardie.fr

[2] VSB - Technical University of Ostrava, Faculty of Electrical Engineering and Computer Science, 17. listopadu 15, 708 33 Ostrava-Poruba, Czech Republic
ivan.zelinka@vsb.cz

Abstract. This note deals with the chaos synchronization problem using unknown inputs Takagi-Sugeno fuzzy observer. The design of observers for Takagi-Sugeno (T-S) fuzzy models subject to unknown inputs is first considered. Based on Linear Matrix Inequalities (LMI) terms and Lyapunov method, sufficient design conditions are given. The pole placement in an LMI region is also considered to improve the observer performances. The proposed approach can be also used in a chaotic cryptosystem procedure where the plaintext (message) is encrypted using chaotic signals at the drive system side. The resulting ciphertext is embedded to the state of the drive system and is sent via public channel to the response system. The plaintext is retrieved via the designed unknown input observer. An example is given to illustrate the effectiveness of the derived results.

Keywords: Fuzzy model, unknown inputs, state estimation, Lyapunov method, linear matrix inequalities (LMI).

1 Introduction

Synchronization is dynamical process during which is one system (synchronized, slaved) remoted by another (synchronizing, master) so that synchronized system is in certain manner following behavior of the master system. The word "synchronization" come from Greek word "synchronos" ($\sigma\upsilon\nu\chi\rho o\nu o\varsigma$) in which $\sigma\upsilon\nu$ (syn) means the same (common,...) and $\chi\rho o\nu o\varsigma$ (chronos) means the "time". Synchronization can be dividen into following classes (1), (2) or (3):

- **Identical synchronization.** This synchronization that may occur when two identical chaotic oscillators are mutually coupled (unidirectional or bidirectional coupling), or when one of them drives the other, which is the case of numerical study A (Lorenz-Lorenz), reported in this chapter. Basically if $\{x_1, x_2, ...x_n\}$ is set of state (dynamical) variables of master system as well as $\{x_1', x_2', ...x_n'\}$ of

I. Zelinka et al. (Eds.): Nostradamus: Mod. Meth. of Prediction, Modeling, AISC 192, pp. 83–92.
springerlink.com

slave system, then both systems are synchronized if under certain initial conditions and $t \to \infty$ is true that $|x_1 - x_1^{'}| \to 0$. This say nathong more than for time large enough is dynamics of both systems in a good approximation. This kind of synchronization is usually called identical synchronization.

- **Generalized synchronization** differ from the previous case by fact that coupled chaotic oscillators are different and that the dynamical state of one of the oscillators is completely determined by the state of the other. This is a case of numerical study B (Rössler-Lorenz), reported in this chapter.
- **Phase synchronization** is another case of synchronization which occurs when the oscillators coupled are not totally identical and the amplitudes of the oscillator remain unsynchronized, while only oscilator phases evolve in synchrony. There is a geometrical interpretation of this case of synchronization.It is possible to find a so called plane in phase space in which the projection of the trajectories of the oscillator follows a rotation around a well-defined center. The phase is defined by the angle $\varphi(t)$, described by the segment which is joining the center of rotation and the projection of the trajectory point onto the plane.
- **Anticipated and lag synchronization.** Lets say that we have synchronizing system with state variables $\{x_1, x_2, ...x_n\}$ and synchronized system with state variables $\{x_1^{'}, x_2^{'}, ...x_n^{'}\}$. Anticipated and lag synchronization occurs when is true that $x_1^{'}(t) = x_1(t+\tau)$. This relation, in fact, say, that the dynamics of one of the systems follows, or anticipates, the dynamics of the other and whose dynamics is described by delay differential equations.
- **Amplitude envelope synchronization** is kind of synchronization which may appear between two weakly coupled chaotic oscillators. Comparing with another cases of synchronization, there is no correlation between phases or amplitudes. One can observe a periodic envelope that has the same frequency in the two systems. Magnitude of that envelope has the same order than the difference between the average frequencies of oscillation of both systems. It is important to note that phase synchronization can develops from aplitude one, when the strength of the coupling force between two amplitude envelope synchronized oscillators increase in time.

A rich amount of literature, working with synchronization, exist. Wecan recommend as a representative literature (1), (2) or (3), all three books are well written and highly readable. Another research works are (4), (5) (synchronization based on time series analysis), (6) (there is studied robustness of synchronizad systems), and many others. As a very good starting reference can be used above mentioned books (1), (2) or (3).

Evolutionary algorithms (EA) are also capable to synchronize simple chaotic systems, without knowledge of internal system structure. The ability of EAs to successfully work with problem kind of black box have been proven many times; see for example real-time control of plasma reactor (7), (8), (9) or coupled map lattices (CML), non real-time control by evolutionary algorithms (10), (11), (12) or in (13) where is described an application of EAs on chaotic dynamic control, synthesis, synchronization and identification.

The main goal of this paper concerns the chaos synchronization problem by using unknown inputs Takagi-Sugeno fuzzy observer. Indeed, for Takagi-Sugeno (T-S) fuzzy models (15), there have been several studies concerning stability and state estimation (16; 17; 18; 19; 20) and particularly for T-S models subject to unknown inputs (see (21) and references therein). Sufficient convergence conditions have been given Linear Matrix Inequalities (LMI) formulation.

This paper is organized as follows. In section 2, the considered structure of unknown inputs T-S fuzzy model is given. In section 3, the main results are given. Indeed, a structure of unknown input T-S observer is proposed and then synthesis conditions are given in LMI terms. To improve the performances of the proposed observer, the pole assignment in a LMI region is also studied. Finally, numerical example is given in section 4 to show the effectiveness of the given results.

Notation. Throughout this paper, the notation $X > Y$ where X and Y are symmetric matrices, means that $X - Y$ is positive definite, $\mathbf{R}^n$ and $\mathbf{R}^{n\times m}$ denote, respectively, the n dimensional Euclidean space and the set of all $n \times m$ real matrices. Superscript "T" denotes matrix transposition, $\otimes$ is the Kronecker product, $\mathbf{I}$ is the identity matrix with compatible dimensions, the symbol $(*)$ denotes the transpose elements in the symmetric positions, $I_M = \{1, 2, \cdots, M\}$.

2 Unknown Input T-S Fuzzy Model Representation

In this work, we consider the case of continuous-time fuzzy models with unknown inputs defined as

$$\begin{cases} \dot{x}(t) = \sum_{i=1}^{M} \mu_i(\xi(t))(A_i x(t) + B_i u(t) + R_i v(t)) \\ y(t) = Cx(t) + Fv(t) \end{cases} \tag{1}$$

with

$$\mu_i(\xi(t)) \geq 0, \sum_{i=1}^{M} \mu_i(\xi(t)) = 1 \tag{2}$$

M being the number of sub-models, $x(t) \in \mathbf{R}^n$ the state vector, $u(t) \in \mathbf{R}^m$ the input vector, $v(t) \in \mathbf{R}^q$, the unknown input and $y \in \mathbf{R}^p$ the measured outputs. $A_i \in \mathbf{R}^{n\times n}$, $B_i \in \mathbf{R}^{n\times m}$, $D_i \in \mathbf{R}^n$ and $C \in \mathbf{R}^{p\times n}$ define the ith local model. Matrices $R_i \in \mathbf{R}^{n\times q}$ and $F \in \mathbf{R}^{p\times q}$ represent the influence of the unknown inputs. We assume that $q < p$ and, without loss of generality, that

Assumption 1: $rank(F) = q$ and $rank(R_i) = q$, i.e. F and R_i are full column ranks.

Assumption 2: $rank(C) = p$, i.e. C is full row rank.

The *activation functions* $\mu_i(.)$ depend on the so-called decision vector $\xi(t)$ assumed to depend on measurable variables.

In this paper, we are concerned by the reconstruction of state variable $x(t)$ of unknown inputs T-S model (1) using only the available information, namely known input $u(t)$ and measured output $y(t)$. Before closing this section, let us recall the following lemma that will be used in the rest of the paper.

3 Unknown Input T-S Fuzzy Observer Design

In order to estimate the state of the unknown input T-S fuzzy model (1), the considered unknown input T-S observer structure has the following form

$$\begin{cases} \dot{z}(t) = \sum_{i=1}^{M} \mu_i(\xi) \Big(N_i z(t) + G_{i1} u(t) + L_i y(t) \Big) \\ \hat{x}(t) = z(t) - E y(t) \end{cases} \tag{3}$$

The considered observer only uses known variables ($u(t)$ and $y(t)$) and the same activation functions $\mu_i(.)$ as used for the T-S model (1). The unknown inputs $v(t)$ are considered non available. The variables $N_i \in \mathbf{R}^{n\times n}$, $G_{i1} \in \mathbf{R}^{n\times m}$, $G_{i2} \in \mathbf{R}^{n}$, $L_i \in \mathbf{R}^{n\times p}$ and $E \in \mathbf{R}^{n\times p}$ are the observer gains to be determined in order to estimate the state of the unknown input T-S model (1). For that purpose, let us define the state estimation error:

$$\tilde{x}(t) = x(t) - \hat{x}(t) \tag{4}$$

The following subsections give LMI conditions satisfying $\tilde{x}(t) \to 0$ when $t \to \infty$. The general case of T-S model subject to unknown inputs which simultaneously influence states and outputs of the system is studied. Poles placement in LMI region to improve the performance of the designed T-S observer finish this section.

3.1 LMI Synthesis Conditions

The following result gives sufficient LMI conditions guaranteeing the global asymptotic convergence of state estimation error (4).

Theorem 1. *The state estimation error between T-S observer (3) and unknown input T-S model (1) converges globally asymptotically towards zero, if there exists matrices $X > 0$, S and W_i such that the following conditions hold $\forall\, i \in I_M$:*

$$A_i^T X + X A_i + A_i^T C^T S^T + S C A_i - W_i C - C^T W_i^T < 0 \tag{5a}$$

$$(X + SC) R_i = W_i F \tag{5b}$$

$$SF = 0 \tag{5c}$$

Then T-S observer (3) is completely defined by:

$$E = X^{-1} S \tag{6a}$$

$$G_{i1} = (\mathbf{I} + X^{-1} SC) B_i \tag{6b}$$

$$N_i = (\mathbf{I} + X^{-1} SC) A_i - X^{-1} W_i C \tag{6c}$$

$$L_i = X^{-1} W_i - N_i E \tag{6d}$$

Remark 1. Classical numerical tools may be used to solve LMI problem (5a) on variables $X > 0$, S, W_i and scalars ε_i with the linear equality constraints (5b). The developed results can easily be solved using numerical tools such as the LMITOOL (29). A numerical example is given in section 4 to illustrate the derived stability conditions.

Remark 2. The proposed result deals with the case of common output matrix $C_i = C, \forall i \in I_M$. The case of different multiple output matrices leads to non convex constraints not easy to resolve with existing numerical tools and not considered in this paper.

3.2 Unknown Inputs Estimation

A lot of works have been considered for the unknown input estimation problem (see for example (21; 28)). The method proposed in this chapter is based on the hypothesis of the good estimation of the state variables (21). Indeed, when the state estimation error is equal to zero; by replacing x by $\hat{x}$ in the equation (1) we obtain the following approximation:

$$\hat{y} = C\hat{x} + F\hat{v} \tag{7}$$

Since the assumption 1 holds, i.e. the matrix F is of full column rank, an estimation of unknown inputs can be carried out in a simpler way by

$$\hat{v} = F^{+}(y - \hat{y}) \tag{8}$$

where F^{+} is any generalized inverse of F such that $FF^{+}F = F$.

3.3 Performance Improvement

In this part, we investigate how to improve the performances of the proposed observer (3) for T-S fuzzy model (1). In order to achieve a desired transient performance, a pole placement should be considered. For many problems, exact pole assignment may not be necessary; it suffices to locate the pole in a sub-region of the complex left half plane (22). This section discusses a pole assignment in LMI regions $S(\alpha, \beta)$.

Since a prescribed LMI region will be added as supplementary constraint to these of theorem 1, it suffices to locate the poles of matrix $\sum_{i=1}^{M} \mu_i(\xi(t))N_i$ in prescribed LMI regions. Indeed, estimation error is D-stable if there exists a matrix $X > 0$ such that (22):

$$\alpha \otimes X + \beta \otimes (N_i X) + \beta^T \otimes (N_i X)^T < 0 \tag{9}$$

We obtain the following result.

Corollary 1: If there exit matrices $X > 0$, S and W_i such that the following conditions hold $\forall\, i \in I_M$:

$$\alpha \otimes X + \beta \otimes (XA_i + SCA_i - W_iC) +$$
$$\beta^T \otimes (XA_i + SCA_i - W_iC)^T < 0 \tag{10a}$$
$$(X + SC)R_i = W_iF \tag{10b}$$
$$SF = 0 \tag{10c}$$

Then, T-S observer (3) is globally asymptotically convergent with the performance defined by complex region $S(\alpha, \beta)$. The observer gains are defined by (6).

For example, to ensure a given performance of the state estimation error, we define region $S_r(\alpha, \beta)$ as the intersection between a circle, of center $(0,0)$ and of radius β, and the left half plane limited by a vertical straight line of x-coordinate equal to $(-\alpha)$, $\alpha > 0$. The corresponding LMI formulation of the corollary 2 is given by the following corollary.

Corollary 2: If there exit matrices $X > 0$, S and W_i such that the following LMI conditions hold $\forall\, i \in I_M$:

$$\begin{bmatrix} -\beta X & XA_i + SCA_i - W_iC \\ (*) & -\beta X \end{bmatrix} < 0 \tag{11a}$$
$$XA_i + SCA_i - W_iC + (XA_i + SCA_i - W_iC)^T + 2\alpha X < 0 \tag{11b}$$
$$(X + SC)R_i = W_iF \tag{11c}$$
$$SF = 0 \tag{11d}$$

Then T-S observer (3) defined by (6) is globally asymptotically convergent with the performance defined by complex region $S_r(\alpha, \beta)$.

4 Simulation Examples

Results developed in section 3 can be applied to reconstruct states of chaotic system and also for a secure communication system. Indeed, the problem we are faced with consists of transmitting some coded message with a signal broadcasted by a communication channel. At the receiver side, the hidden signal is recovered by a decoding system. The increasing need of secure communications leads to the development of many techniques which make difficult the detecting of transmitted message (se for example (23), (24), (25), (26), (27)). In this section, our goal to show how the designed observer could be used in chaotic system reconstruction and in a secure communication scheme. For this purpose we use the nonlinear Lorenz model as chaotic systems represented by his equivalent chaotic T-S model:

$$\begin{cases} \dot{x} = \sum_{i=1}^{2} \mu_i(y_1) \Big(A_i x + R_i v(t) \Big) \\ y = Cx + Fv(t) \end{cases} \tag{12}$$

with:

$$A_1 = \begin{bmatrix} -10 & 10 & 0 \\ 28 & -1 & -30 \\ 0 & 30 & -8/3 \end{bmatrix}, A_2 = \begin{bmatrix} -10 & 10 & 0 \\ 28 & -1 & 30 \\ 0 & -30 & -8/3 \end{bmatrix}$$

$$B_1 = \begin{bmatrix} 0 \\ 0 \\ 0 \end{bmatrix} B_2 = \begin{bmatrix} 0 \\ 0 \\ 0 \end{bmatrix} C = \begin{bmatrix} 1 & 0 & 0 \\ 0 & 1 & 0 \end{bmatrix} F = \begin{bmatrix} 1 \\ 1 \end{bmatrix}$$

$$\mu_1(x_1(t)) = \frac{1}{2}\left(1 + \frac{x_1(t)}{d}\right) \ \mu_2(x_1(t)) = \frac{1}{2}\left(1 - \frac{x_1(t)}{d}\right)$$

Note that the obtained T-S model exactly represents the nonlinear Lorenz model under $x_1(t) \in [-d, d]$ with $d = 30$.

The simulation of the T-S model with the initial value $x_0 = (1\ 1\ 1)^\top$ shows the chaotic behavior of the example plotted in the phase plan of the system (see figure 1).

The message to be encoded constitutes the so-called unknown input of the T-S model which plays the role of the encoder. The output of this model is transmitted using a public channel. On the receiving side, an unknown input T-S observer serves as a decoder in order to re-build the message. Clearly, the choice of LTI local models, their number, as well as the nature of the function $\mu_i(\xi(t)))$ are key elements for an external person to be able to decode the embodied crypted message from only the signal $y(t)$. The goal of the proposed example is only to show the feasibility of the proposed design in chaotic system reconstruction and in secure communication procedure.

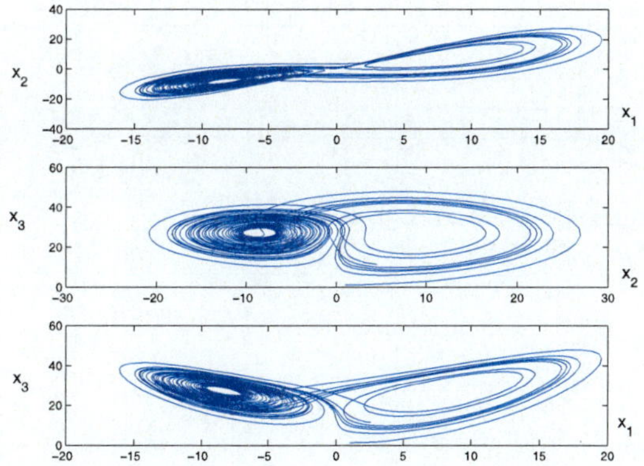

Fig. 1 Phase plan of the chaotic T-S model (4) with v=0

Indeed, the unknown input can represent the hidden message to be transmitted. Thus the transmitted signal y is embedded with the hidden message v.

The considered T-S observer for this application is

$$\begin{cases} \dot{z} = \sum_{i=1}^{2} \mu_i(y_1)\left(N_i z + L_i y\right) \\ \hat{x} = z - Ey \end{cases} \tag{13}$$

The resolution of conditions (5) with $B_1 = B_2 = (0,0,0)^\top$ lead to the following result:

$$X = \begin{bmatrix} 1.750 & 1.650 & -0.003 \\ 1.650 & 1.750 & -0.003 \\ -0.003 & -0.003 & 0.195 \end{bmatrix}, E = \begin{bmatrix} -3.05 & 3.05 \\ 3.99 & -3.99 \\ -0.004 & 0.004 \end{bmatrix}$$

$$N_1 = \begin{bmatrix} 33.66 & 47.89 & -91.72 \\ -36.18 & -45.78 & 89.96 \\ 61.12 & -32 & -2.79 \end{bmatrix}, N_2 = \begin{bmatrix} 35.06 & 49.54 & 91.72 \\ -37.82 & -47.14 & -89.96 \\ -62.08 & 31.20 & -2.53 \end{bmatrix}$$

$$L_1 = \begin{bmatrix} -16.44 & 17.44 \\ -14.94 & 15.94 \\ 253.9 & -252.9 \end{bmatrix} \quad L_2 = \begin{bmatrix} -19.38 & 17.32 \\ -13.67 & 17.67 \\ -252.38 & 253.37 \end{bmatrix}$$

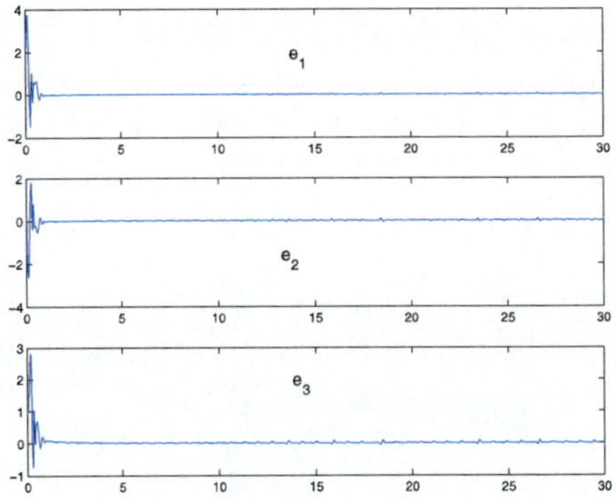

Fig. 2 Estimation errors $e_i = x_i - \hat{x}_i$, $i \in \{1,2,3\}$

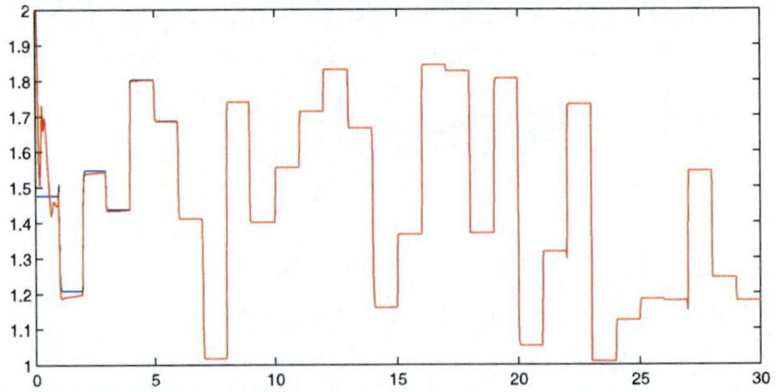

Fig. 3 Hidden message v and its estimate

Figure 2 represent the state estimation error with the initial conditions $x_0 = [1\ 1\ 1]^\top$ and $\hat{x}_0 = [0\ 0\ 0]^\top$. It shows the good reconstruction of chaotic system state. Figure 3 displays the hidden transmitted message and its estimate. Excepted around the time origin, the unknown input (transmitted message) is perfectly estimated.

5 Conclusion

In this paper we have shown how to design an observer for T-S fuzzy models with unknown inputs affecting both the state and the output of the system simultaneously. Uncertainties on state matrices are also considered. Sufficient conditions to design the proposed T-S observer are given. These conditions are formulated in LMI terms with additional equality constraints that are easy to compute with classical numerical tools. To improve the performances of the proposed unknown inputs observer, the pole assignment in LMI regions is also addressed. Simulation results have shown the effectiveness of the proposed results.

References

[1] Pikovsky, A., Rosemblum, M., Kurths, J.: Synchronization: A Universal Concept in Nonlinear Sciences. Cambridge University Press (2001) ISBN 0-521-53352-X

[2] Gonzalez-Miranda, J.M.: Synchronization and Control of Chaos. An introduction for scientists and engineers. Imperial College Press (2004) ISBN 1-86094-488-4

[3] Controlling chaos. In: Schuster, H.G. (ed.) Handbook of Chaos Control. Wiley-VCH, New York

[4] Sushchik, M.M., Rulkov, N.F., Tsimring, L.S., Abarbanel, H.D.I.: Generalized synchronization of chaos in directionally coupled chaotic systems. In: Proceedings of 1995 Intl. Symp. on Nonlinear Theory and Appl., vol. 2, pp. 949–952. IEEE (1995)

[5] Brown, R., Rulkov, N.F., Tracy, E.R.: Modeling and synchronization chaotic system from time-series data. Phys. Rev. E 49, 3784 (1994)

[6] Rulkov, N.F., Sushchik, M.M.: Robustness of synchronized chaotic oscillations. International Journal of Bifurcation and Chaos 7, 625 (1997)

[7] Nolle, L., Goodyear, A., Hopgood, A.A., Picton, P.D., Braithwaite, N.: On Step Width Adaptation in Simulated Annealing for Continuous Parameter Optimisation. In: Reusch, B. (ed.) Fuzzy Days 2001. LNCS, vol. 2206, pp. 589–598. Springer, Heidelberg (2001)

[8] Nolle, L., Zelinka, I., Hopgood, A.A., Goodyear, A.: Comparison of an self organizing migration algorithm with simulated annealing and differential evolution for automated waveform tuning. Advances in Engineering Software 36(10), 645–653 (2005)

[9] Zelinka, I., Nolle, L.: Plasma reactor optimizing using differential evolution. In: Price, K.V., Lampinen, J., Storn, R. (eds.) Differential Evolution: A Practical Approach to Global Optimization, pp. 499–512. Springer, New York (2006)

[10] Zelinka, I.: Investigation on Evolutionary Deterministic Chaos Control. IFAC, Prague (2005)

[11] Ivan, Z.: Investigation on Evolutionary Deterministic Chaos Control – Extended Study. In: 19th International Conference on Simulation and Modeling (ECMS 2005), Riga, Latvia, June 1-4 (2005b)

[12] Zelinka, I., Senkerik, R., Navratil, E.: Investigation on Evolutionary Optimitazion of Chaos Control. Chaos, Solitons, Fractals (2007), doi:10.1016/j.chaos.2007.07.045
[13] Zelinka, I., Celikovsky, S., Richter, H., Chen, G.: Evolutionary Algorithms and Chaotic Systems. Springer, Germany (2010)
[14] Boyd, S., et al.: Linear matrix inequalities in systems and control theory. SIAM, Philadelphia (1994)
[15] Takagi, T., Sugeno, M.: Fuzzy identification of systems and its application to modelling and control. IEEE Trans. on Systems, Man, Cybernetics 15(1), 116–132 (1985)
[16] Tanaka, K., Wang, H.O.: Fuzzy Control Systems Design and Analysis: A linear Matrix Inequality Approach. John Wiley & Sons, Inc. (2001)
[17] Chadli, M., Maquin, D., Ragot, J.: Stability analysis and design for continuous-time Takagi-Sugeno control systems. International Journal of Fuzzy Systems 7(3), 101–109 (2005)
[18] Johansson, M., Rantzer, A., Arzén, K.: Piecewise quadratic stability of fuzzy systems. IEEE Trans. on Fuzzy Systems 7(6), 713–722 (1999)
[19] Xiaodiong, L., Qingling, Z.: New approach to H_∞ controller designs based on observers for T-S fuzzy systems via LMI. Automatica 39, 1571–1582 (2003)
[20] Tanaka, K., Hori, T., Wang, H.O.: A multiple Lyapunov function approach to stabilization of fuzzy control systems. IEEE Transactions on Fuzzy Systems 11(4), 582–589 (2003)
[21] Chadli, M.: An LMI approach to design observer for unknown inputs Takagi-Sugeno fuzzy models. Asian Journal of Control 12(4), 524–530 (2010)
[22] Chilali, M., Gahinet, P.: ÒH_∞ Design with pole placement constraints: an LMI approch. IEEE Transactions on Automatic Control 41(3), 358–367 (1996)
[23] Li, C., Liao, X., Wong, K.: Lag synchronization of hyperchaos with application to secure communications. Chaos, Solitons and Fractals 23, 183–193 (2005)
[24] Chen, M., Zhou, D., Shang, Y.: A new observer-based synchronization scheme for private communication. Chaos, Solitons and Fractals 24, 1025–1030 (2005)
[25] Boutayeb, M., Darouach, M., Rafaralahy, H.: Generalized State-Space Observers for Chaotic Synchronization and Secure Communication. IEEE Transactions on Circuits and Systems I: Fundamental Theory and Applications 49(3), 345–349 (2002)
[26] Alvares, G., Montoya, F., Romera, M., Pastor, G.: Breaking parameter modulated chaotic secure communication system. Chaos, Solitons & Fractals 21(4), 783–787 (2004)
[27] Akhenak, A., Chadli, M., Ragot, J., Maquin, D.: Unknown input multiple observer based approach: application to secure communication. In: 1st IFAC Conference on Analysis and Control of Chaotic Systems, Reims, France, June 28-30 (2006)
[28] Edwards, C., Spurgeon, S.K., Patton, R.J.: Sliding mode observers for fault detection and isolation. Automatica 36(4), 541–553 (2000)
[29] Vandenberghe, L., Boyd, S.: Semidefinite programming. SIAM Review 38(1), 49–95 (1996)

Modeling and Model Predictive Control of Nonlinear Hydraulic System

Petr Chalupa and Jakub Novák

Department of Process Control, Faculty of Applied Informatics, Tomas Bata University in Zlin, nám. T.G. Masaryka 5555, 760 01 Zlín, Czech Republic
chalupa@fai.utb.cz

Abstract. This paper deals with modeling and control of a hydraulic three tank system. A process of creating a computer model in MATLAB / Simulink environment is described and optimal PID and model predictive controllers are proposed. Modeling starts with creation of an initial mathematical model based on first principles approach. Further, the initial model is modified to obtain better correspondence with real-time system and parameters of the modified system are identified from measurements. The real time system contains nonlinearities which cannot be neglected and therefore are identified and included in the final mathematical model. Resulting model is used for control design. As the real-time system has long time constants, usage of Simulink model dramatically speeds up design process. Optimal PID and MPC controllers are proposed and compared. Described techniques are not limited to one particular modeling problem but can be used as an illustrative example for modeling of many technological processes.

1 Introduction

Most of current control algorithms is based on a model of a controlled plant [1]. A plant model can be also used to investigate properties and behavior of the modeled plant without a risk of damage of violating technological constraints of the real plant. Two basic branches of modeling are used in practice: the black box approach and the first principles modeling.

The black box approach to the modeling [2], [3] is based on analysis of input and output signals of the plant and knowledge of physical principle of modeled plant is not required. On the other hand, model obtained by black box approach is generally valid only for signals it was calculated from.

The first principle modeling provides general model which is in optimal case valid for whole range of plant inputs and states. The model is created by analyzing the modeled plant and combining physical laws [4]. On the other hand, there is usually a lot of unknown constants and relations when performing analysis of a plant. Thus, first-principle modeling is suitable especially for simple controlled plants with small number of parameters or for obtaining basic information about controlled plant (rage of gain, rank of suitable sample time, etc.).

I. Zelinka et al. (Eds.): Nostradamus: Mod. Meth. of Prediction, Modeling, AISC 192, pp. 93–102.
springerlink.com

Combination of both methods is used in the paper. Basic relations between plant inputs and outputs are derived using mathematical physical analysis an obtained model is further improved on the basis of measurements. Obtained relations are used to design a Simulink environment with characteristics as close as possible to the real time system DTS 200 – Three tank System [5]. The major reason for creating the model of this laboratory equipment are big time constants of the plant and thus time consuming experiments. The model can dramatically decrease time needed for controller development because only promising control strategies are applied to the real plant and verified.

The paper is organized as follows. Section 2 presents the modeled system – Amira DTS200. Derivation of initial ideal using first principles modeling and enhancement of this model based on real-time experiments is carried out in Section 3. Section 4 presents the resulting Simulink model in detail and Section 5 copes with PID and MPC control of the plant.

2 The DTS200 System

The Amira DTS200 system consists of three interconnected cylindrical tanks, two pumps, six valves, pipes, water reservoir in the bottom, measurement of liquid levels and other elements. The pumps pump water from the bottom reservoir to the top of the left and right tanks. Valve positions are controlled and measured by electrical signals, which allow precise positioning.

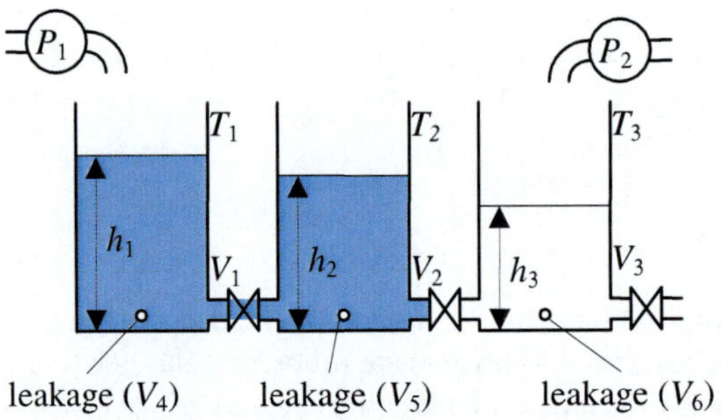

Fig. 1 Scheme of three tank system Amira DTS200

A simplified scheme of the system is shown in Fig. 1. The pump P_1 controls the inflow to tank T_1 while the pump P_2 controls the liquid inflow to tank T_2. There is no pump connected to the middle tank T_s. The characteristic of the flow between tank T_1 and tank T_s can be affected by valve V_1, flow between tanks T_s and T_2 can be affected by the valve V_2 and the outflow of the tank T_2 can be affected by valve V_3. The system also provides the capability of simulating leakage from individual tanks by opening the valves V_4, V_5 and V_6.

Pumps are controlled by analogue signals in range from -10V to 10V. Heights of water level are measured by pressure sensors. Each valve is operated by two digital signals which control motor of particular valve. First signal orders to start closing of the valve while the second signal is used for opening of the valve. If none of the signals is activated the valve remains in its current position. Each valve also provides three output signals: analogue voltage signal correspond to the current position of the valve and two informative logical signals which states that the valve is fully opened or fully closed respectively.

The overall number of inputs to the modeled plant DTS200 is 14:

- 2 analogues signals controlling the pumps,
- 12 digital signals (2 for each of the 6 valves) for opening / closing of the valves.

The plant provides 21 measurable outputs which can be used as a control feedback or for measurements of plant characteristics:

- 3 analogue signals representing level heights in the three tanks,
- 6 analogues signals representing position of the valves,
- 12 logical signals (2 for each of the 6 valves) stating that corresponding valve is fully opened / closed.

3 Modeling of the Plant

This section is focused to derivation of mathematical model of the three-tank system end its adaptation to the DTS200 plant using real-time experiments..

3.1 *Ideal Model*

This derivation of ideal plant model is based on ideal properties of individual components.The ideal flow of a liquid through a pipe can be derived from Bernoulli and continuity equations for ideal liquid.

Since the flow through a valve depends only on the level difference, the valve position and constants representing pipes and cylindrical tanks, the whole mathematical model can be written as follows:

$$
\begin{aligned}
\frac{dh_1}{dt} &= q_1 - k_1\sqrt{|h_1 - h_s|}\cdot \mathrm{sign}(h_1 - h_s) - k_4\sqrt{h_1} \\
\frac{dh_2}{dt} &= k_1\sqrt{|h_1 - h_s|}\cdot \mathrm{sign}(h_1 - h_s) + k_2\sqrt{|h_2 - h_s|}\cdot \mathrm{sign}(h_2 - h_s) - k_5\sqrt{h_s} \\
\frac{dh_3}{dt} &= q_2 - k_2\sqrt{|h_2 - h_s|}\cdot \mathrm{sign}(h_2 - h_s) - k_3\sqrt{h_2} - k_6\sqrt{h_2}
\end{aligned}
\tag{1}
$$

where symbols h_1, h_2 and h_3 represent water level heights in tanks T_1, T_2 and T_3 respectively, k is a parameter representing valve position and q represents inflow as change of water level in time:

$$
k_i = v_i \frac{S_{V\max}\sqrt{2g}}{S_T} \qquad i = 1,2,\dots,6 \qquad\qquad q_i = \frac{q_i'}{S_T} \qquad i = 1,2 \tag{2}
$$

The cross-sectional areas of all three tanks are the same and are symbolized by S_T This ideal model is successfully used in many control system studies as a demonstration example [6], [7]. Although the ideal model is based on simple equations, analytical solution of the outputs for a given course of inputs is complicated. The problem lies in nonlinearity of equations (1). Also computation of steady state for a given constant inputs is a very complicated task and leads to solution of higher order polynomial equation.

3.2 Enhanced Model

This section describes enhancement of the initial model and measurement of characteristics of individual parts of DTS200 system.

Characteristics of the Pumps

The characteristics of pumps were measured to refine on (1). The amount of water pumped within certain time was measured for different setting of driving signals u_1 and u_2. Characteristics of both pumps are similar and close to linear but do not start at a lower bound (u = -1MU) but at least the value of approx. u = -0.85 must be applied to the pump to obtain a nonzero output. The maximal pumping of approx. 6 mm/s of tank level rise represents 5.37 l/min. Dynamics of the pumps are very fast comparing to other time constants present in the system and therefore were neglected.

Characteristics of the Valves

As stated in Section 2, each of plant's 6 valves is driven by two dedicated logical signals. These signals are used for starting valve's motor in closing or opening direction respectively. If none signal is activated the valve remains in its current position. Each valve provides three output signals. The current valve position is determined by analogue signal. Higher values of signal represent closed valve and lower values represent opened valve. The other two signals are logical and state that valve is opened or closed respectively. Valve characteristics are studied in more detail in [9].

Valve flow parameters for valves

Valve flow parameters k_i as appeared in (1) were computed from measurements of draining through individual valves which are connected to outflow pipes (V_3, V_4, V_5 and V_6). The draining of a tank to the reservoir situated below the tanks is described by differential equation

$$\frac{dh(t)}{dt}=-k\sqrt{h(t)+h_0} \qquad h(t)=\frac{k^2}{4}\cdot t^2-k\sqrt{h(0)+h_0}\cdot t+h(0) \tag{3}$$

where $h(0)$ is initial water level and h_0 is the vertical length of outflow pipe. Due to mechanical configuration of the plant, the value of h_0 for outflow valves V_3, V_4, V_5, and V_6 cannot be measured directly. But it can be identified from draining course. The valve was partially closed to different positions at the beginning of draining experiment and relation between valve position and value of k was achieved.

Similar approach to obtaining values of k can be used also for valves V_1 and V_2 which interconnect tanks T_1 and T_2, and T_2 and T_3, respectively.

Valve hysteresis
Experiments unveiled a hysteresis present in all valves. The value of valve position itself does not give sufficient information about current value of parameter k_2. As can be seen from Fig. 2, if the position is 0 MU the value of k_2 can be anywhere in range 0.03 to 0.13. Especially in case of using the valve as an actuator the hysteresis should be taken into account. Otherwise control process can easily become unstable.

Modeling of valve characteristics
The course of relation between valve position in MU and k is similar to step responses of dynamical system and therefore it was modeled in similar way. A model based on transfer of 4th order aperiodic system produced satisfactory results. Thus relation between position and k border curve was as follows:

$$
\begin{aligned}
pos < pos_0: &\quad k = k_{max}\left[1 - \frac{1}{6} e^{\frac{-b}{a}} \frac{b^3 + 3b^2 a + 6ba^2 + 6a^3}{a^3}\right] \\
&\quad b = pos_0 - pos \\
pos \geq pos_0: &\quad k = 0
\end{aligned}
\tag{4}
$$

where pos is valve position in MU and parameters a and pos_0 were obtained by nonlinear regression. The regression for valve V_2 is presented in Fig. 2.

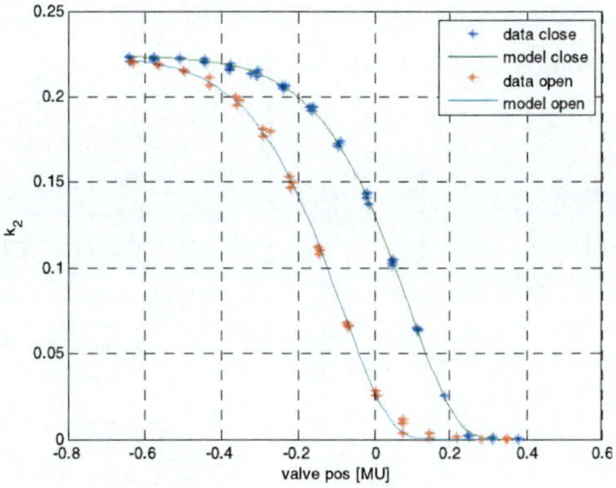

Fig. 2 Model of parameter k for valve V_2

4 Simulink Model

All the models of individual parts of the DTS200 plant were incorporated into single block in MATLAB / Simulink environment. The block has the same inputs and outputs as the real plant. Thus it contains all 14 inputs and 21 outputs described in Section 2. The Simulink block of resulting model of DTS200 plant is depicted in Fig. 3.

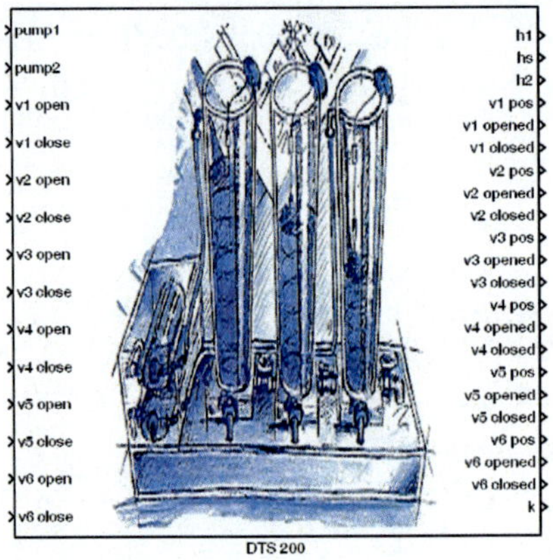

Fig. 3 Block of Simulink model of DTS200

The model is designed as masked subsystem where only necessary initial states can be entered by user. The initial states are: initial water level in individual tanks and initial valve positions and corresponding values of valve parameters k.

Masked subsystems and subsystems are used also to model individual parts of the plant. For example, internal structure of valve state subsystem is presented in Fig. 4. This hierarchical structure is useful to maintain lucidity.

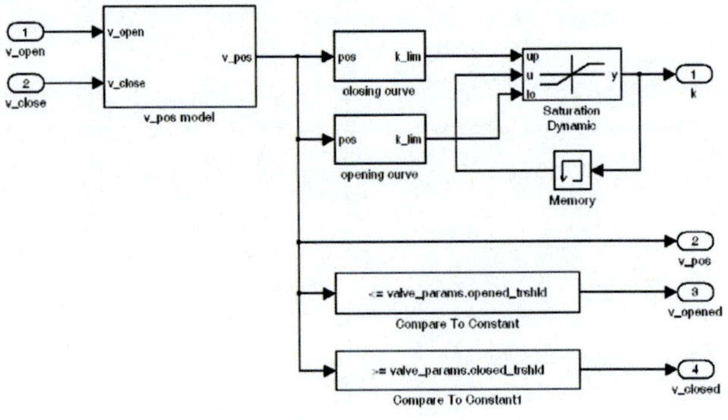

Fig. 4 Internal structure of valve state subsystem

Detailed description of the Simulink model can be found in [10].

5 PID and Model Predictive Control

The control experiment was carried out to verify the Simulink model and to find out how the controllers cope with nonlinearities in the plant.

The experiment configuration was following: there was constant inflow to tank T_1 by pump P_1. The valve V_1 between tanks T_1 and T_2 was opened to a constant position. Valves V_4 (leakage from tank T_1) and V_2 (interconnection of tanks T_2 and T_3) were closed. The goal of the control was to drive water level in tank T_2 using valve V_5 as an actuator. It should be noted that characteristics of valve V_5 is nonlinear and contains big hysteresis.

The plant was in steady state at the beginning of all experiments. The valve V5 was partly opened and initial water levels were as follows:

$$\begin{aligned} h_1(0) &= 235mm \\ y(0) = h_2(0) &= 151mm \\ u(0) &= 30\% \end{aligned} \tag{5}$$

The valve position was controlled by a simple internal controller connected to valve drive. This internal control loop is driven by required valve position signal. This control signal is in percentage where 0% corresponds to fully closed valve (i.e. step change of *v5 closed* signal) and 100% represents fully opened valve (i.e. step change of *v5 opened* signal). Sample time of T_s=0.1s was used for measurements but all the controller used sample period of T_c=5s.

Quadratic criterion was used to compare individual courses from the control error point of view:

$$J_e = \sum_{i=1}^{N} e^2(i) = \sum_{i=1}^{N} \left[w(i) - y(i)\right]^2 \tag{6}$$

Usually control signals differences are penalized of evaluated. But in case of DTS200 crucial problem is in starting and stopping the valve motor because often starting of the valve decreases its durability. The displacement of the valve is not as important as the fact that the valve motor had to be started. Valve starts were measured using following criterion:

$$J_u = \sum_{i=1}^{N-1} \left[v5\_open(i+1) - v5\_open(i) > 0\right] + \sum_{i=1}^{N-1} \left[v5\_close(i+1) - v5\_close(i) > 0\right] \tag{7}$$

PID Control

First of all the plant was controlled by classical PID controller. Controller parameters were tuned to minimize criterion (6). A MATLAB function *fminsearch* was used for this task. This function uses Nelder-Mead Simplex Method to find criterion minimum [11]. Resulting courses are depicted in Fig. 5. It can be seen that even the valve position (control signal) changes smoothly, the output is not so smooth. This corresponds to overcoming of valve's hysteresis.

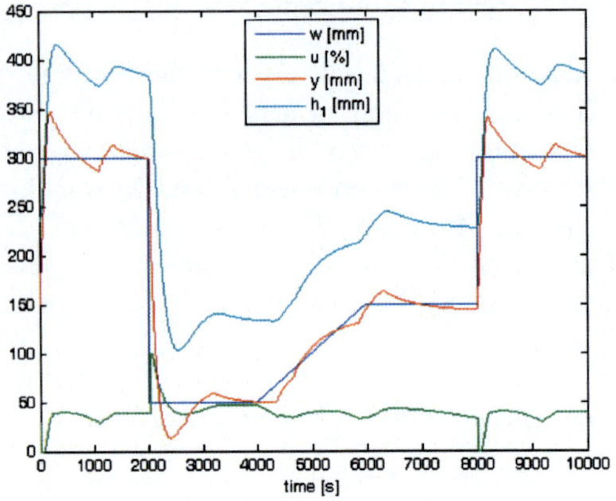

Fig. 5 Optimal PID control DTS200

Model predictive control
A linear model of the system was identified by applying random signal to its input and parameters of the linearized model were used in MPC controller. The Simulink control scheme is presented in Fig. 6.

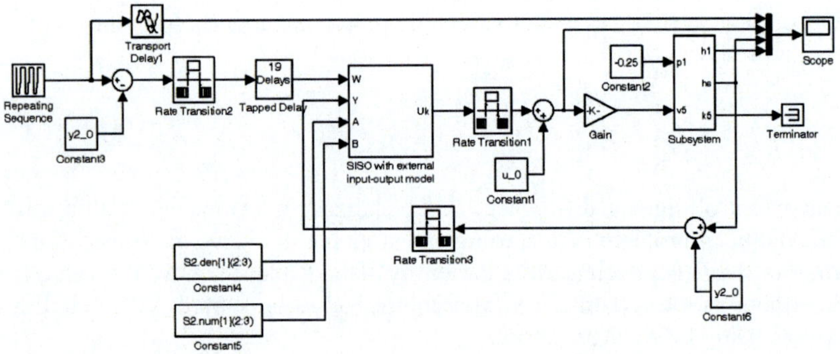

Fig. 6 MPC scheme

A controller from the STuMPCoL library [12] was used to perform the control task. A quadratic criterion with both control and prediction horizon equal to 20 samples was used to compute control signals and receding horizon strategy was used [13]. The resulting control courses are presented in Fig. 7.

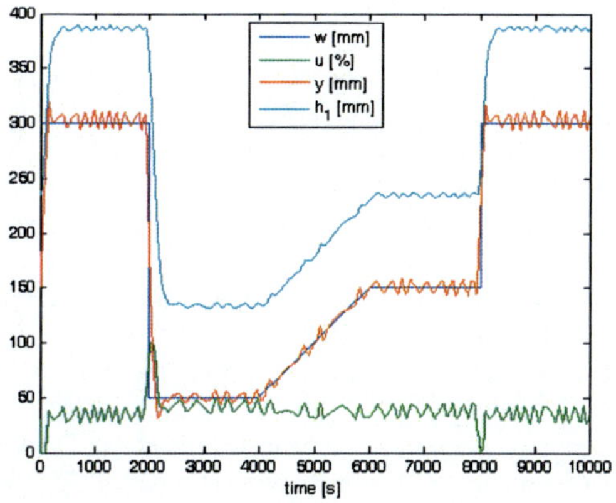

Fig. 7 Control courses using MPC

It was observed that the MPC copes better with crossing of the hysteresis but on the other hand control signal was oscillating around steady states .

Values of criteria defined by equations (6) and (7) are summed up in the Table 1. The MPC was more accurate but it also has slightly actuator demands.

Table 1 Control criterions

controller	J_e	J_u
PID	778	3118
MPC	290	5396

6 Conclusion

The paper presents Simulink model of hydraulic system. The Amira DTS200 three tank system was considered but used techniques can be easily generalized to a wide set of hydraulic systems. Despite simplicity of the ideal model, real-time system contains several nonlinearities which incorporate complexity to the system. Especially hysteresis of the valves plays a big role in the plant behavior.

A PID and MPC controllers were designed to control the system and verified usability of the model for the controller design. Further improvement of control courses can be reached by usage of nonlinear controller.

Acknowledgments. This work was supported by the Ministry of Education of the Czech Republic under grant 1M0567 and by the European Regional Development Fund under the project CEBIA-Tech No. CZ.1.05/2.1.00/03.0089.

References

1. Bobál, V., Böhm, J., Fessl, J., Macháček, J.: Digital Self-tuning Controllers: Algorithms, Implementation and Applications. Springer-Verlag London Ltd. (2005)
2. Liu, G.P.: Nonlinear identification and control – A neural network Approach. Springer-Verlag London Ltd., London (2001)
3. Ljung, L.: System identification: theory for the user. Prentice Hall PTR, Upper Saddle River (1999)
4. Himmelblau, D.M., Riggs, J.B.: Basic principles and calculations in chemical engineering. Prentice Hall, Upper Saddle River (2004)
5. Amira, DTS200 Laboratory Setup Three - Tank - System. Amira GmbH, Duisburg (2002)
6. Li, L., Zhou, D.: Fast and robust fault diagnosis for a class of nonlinear systems: detectability analysis. Computers & Chemical Engineering 28, 2635–2646 (2004)
7. Henry, D., Zolghadri, A.: Norm-based design of robust FDI schemes for uncertain system under feedback control: Comparison of two approaches. Control Engineering Practice 14, 1081–1097 (2006)
8. Humusoft, Real Time Toolbox. Humusoft, Praha, (2011), `http://www.humusoft.cz/produkty/rtt/`
9. Chalupa, P., Novák, J., Bobál, V.: Modeling of Hydraulic Control Valves. In: Proceedings of the 13th WSEAS International Conference (ACMOS 2011), Lanzarote, Canary Islands, Spain, May 27-29, pp. 195–200 (2011)
10. Chalupa, P., Novák, J., Bobál, V.: Comprehensive Model of DTS200 Three Tank System in Simulink. International Journal of Mathematical Models and Methods in Applied Sciences 6(2), 358–365 (2012)
11. Lagarias, J.C., Reeds, J.A., Wright, M.H., Wright, P.E.: Convergence Properties of the Nelder-Mead Simplex Method in Low Dimensions. SIAM Journal of Optimization 9(1), 112–147 (1998)
12. Chalupa, P.: STuMPCoL Self-Tuning Model Predictive Controllers Library (2012), `http://www.fai.utb.cz/people/chalupa/STuMPCoL`
13. Kwon, W.H., Han, S.: Receding Horizon Control, 380 pages. Springer, London (2005) 978-1-84628-024-5

Usage of the Evolutionary Designed Neural Network for Heat Demand Forecast

B. Chramcov and P. Vařacha

Tomas Bata University in Zlin, Faculty of Applied Informatics, nam. T.G. Masaryka 5555, 760 01 Zlin, Czech Republic
{chramcov,varacha}@fai.utb.cz

Abstract. This paper highlights the problem of forecast model design for time series of heat demand. We propose the forecast model of heat demand based on the assumption that the course of heat demand can be described sufficiently well as a function of the outdoor temperature and the weather independent component (social components). Time of the day affects the social components. Forecast of social component is realized by means of Box-Jenkins methodology. The weather dependent component is modeled as a heating characteristic (function that describes the temperature-dependent part of heat consumption). The principal aim is to derive an explicit expression for the heating characteristics. The Neural Network Synthesis is successfully applied here to find this expression. An experiment described in the paper was realized on real life data. We have studied half-hourly heat demand data, covering four month period in concrete district heating system (DHS) from Most agglomeration and heating plant situated in Komořany, Czech Republic.

1 Introduction

Due to the large operational costs involved, efficient operation control of the production sources and production units in a district heating system is desirable. Knowledge of heat demand is the base for input data for operation preparation of Centralized Heat Supply System (CHSS). Term "heat demand" is instantaneous required heat output or instantaneous consumed heat output by consumers. Term "heat demand" relates to term "heat consumption". It express heat energy, which is the customer supplied in a specific time interval (generally day or year). The course of heat demand and heat consumption can be demonstrated by means of heat demand diagrams. The most important one is the Daily Diagram of Heat Demand (DDHD) which demonstrates the course of requisite heat output during the day (See Fig. 2).

These diagrams are most important for technical and economic consideration. Therefore forecast of these diagrams course is significant for short-term and long-term planning of heat production. It is possible to judge the question of peak sources and namely the question of optimal distribution loading between cooperative

I. Zelinka et al. (Eds.): Nostradamus: Mod. Meth. of Prediction, Modeling, AISC 192, pp. 103–112.
springerlink.com

production sources and production units inside these sources according to time course of heat demand. The forecast of DDHD is used in this case (Balátě 1982). This forecast is necessary to control and optimize the operating schedule of a cogeneration plant in combination with the district heating system (DHS). The heat demand forecast implemented in an energy management system helps to increase the energy efficiency and supports the sustainable energy development. An analysis of the consumption data and of the main influence factors on the heat demand is necessary in order to obtain suitable forecast models.

2 State of the Art in Solved Problem

Many works solve the question of economical heat production and distribution in DHS. In the work (Park et al 2009), a model for operational optimization of the CHSS in the metropolitan area is presented by incorporating forecast for demand from customers. Some methods able to predict dynamic heat demand for space heating and domestic warm water preparation in DHS, using time-series analysis was presented (Popescu 2008). Other work present one step ahead prediction of water temperature returned from agglomeration based on input water temperature, flow and atmospheric temperature in past 24 hours (Vařacha 2009). The paper (Schellong and Hentges 2007) describes the data management as well as the process of the mathematical modeling. Most forecasting models and methods for prediction of heat load or heat demand have already been suggested and implemented with varying degrees of success. Most applications in the subject consider the prediction of electrical-power loads. Nevertheless there was created several works, which solve the prediction of DDHD and its use for control of District Heating System (DHS) or for a planning the most economical, technical and environmental optimal energy distribution system. A number of these works are based on mass data processing (Lehtoranta et al. 200), (Hippert et al. 2001). However these methods have a big disadvantage. It consists in out of date of real data. This problem solves for example (Bakker et al.) Out of date of real data problem can be eliminated using the forecast methods according to statistical method. The basic idea of this approach is to decompose the load into two components, whether dependent and whether independent. The weather dependent component is typically modeled as a polynomial function of temperature and other weather factors. The weather independent component is often described by a Fourier series, ARMA model, Box-Jenkins methodology or explicit time function. Previous works on heat load forecasting (Dotzauer 2002), (Arvastson 2001), show that the outdoor temperature, together with the social behaviour of the consumers, has the greatest influence on DDHD (with respect to meteorological influences). Other weather conditions like wind, sunshine etc. have less effect and they are parts of stochastic component. For example the papers (Pedersen et al. 2008) and (Chramcov 2011) use the statistical methods for heat demand prediction. The problem of heat demand prediction using heating characteristic function was solved in (Chramcov 2010). This work uses only expert heating characteristics estimation in the form cubic equation or a piecewice linear function.

In this paper we propose the forecast model of DDHD based on the statistical method approach with using the ANN for modeling the function of the temperature-dependent part of heat demand. To create ANN optimally performing heating characteristics Neural Network Synthesis can be used. The work (Chramcov and Vařacha 2012) outlined this possibility. The development of evolutionary methods aiming to design the ANN structure and weight values experienced boom at the end of the millennium with the introduction of sufficiently fast computers into common scientific practice. A comprehensive survey considering history of evolutionary computation methods of designing the ANN structure can be found in (Vonk et al. 1995). According to (Vonk et al. 1997) these methods can be used in the field of the ANN in several ways (train the weights of the ANN, analyze the ANN, generate the architecture of the ANN, generate both the ANN's architecture and weights). The Neural Network Synthesis was developed on the basis of Analytic Programming (AP) and Self Organizing Migration Algorithm (SOMA) algorithms and it was successfully tested on the real life problems (Koza et al. 1999), (Turner et al. 2010) as well as on widely recognized benchmark functions (Vonk et al. 1997) with respect to the function approximation, prediction and problems. The method was successfully used also in the field of head demand prediction (Vařacha and Jašek 2011) or identification of district heating network before (Král et al. 2010). Nevertheless the method provides very good results applied on broad variety of different task, for example on cancer classification (Vařacha 2011).

3 The Problem Formulation

As mentioned above, a number of works are based on mass data processing. But these methods have a big disadvantage. It consists in out of date of real data. From this point of view is available to use the forecast methods according to statistical method in our case the methodology of Box –Jenkins (Box and Jenkins 1976). This method works with fixed number of values, which are update for each sampling period. The course of time series of DDHD contains mostly two periodic components (daily and weekly period). But general model according to Box-Jenkins enables to describe only one periodic component. Therefore we can use two eventual approaches to calculation of forecast to describe both periodic components: The method that uses the model with double filtration or The method – superposition of models (Dostál 1986).

The method according to Box-Jenkins do not enable to model meteorological influences. But weather forecasts are an important input to many heat demand forecasting models. Therefore we have to include these influences in calculation of prediction. According to above mentioned reviews we assume only outdoor temperature influence. For inclusion of outdoor temperature influence in calculation of prediction of DDHD was proposed this general plan:

1. The influence of outdoor temperature filter off from time series of DDHD by means of heating characteristics (function that describes the temperature-dependent part of heat consumption)

2. Prediction of DDHD by means of Box-Jenkins method for this filtered time series
3. Filtration of predicted values for the reason of inclusion of outdoor temperature influence (on the base of weather forecast)

From the previous plan is evident that the principal aim is to derive an explicit expression for the temperature-dependent part of the heat load. Regarding to previous consideration we have to find the explicit function of heating characteristics in the form (1), where z_t^{kor} is correction value of heat consumption in time t including outdoor temperature influence, T_t is real value of outdoor temperature in time t. This paper solves the problem of finding this function using Neural Network Synthesis. The results for specific locality are presented in the next chapter.

$$z_t^{kor} = f(T_t) \tag{1}$$

Development of the ANN synthesis as a successful and effective method for the ANN performing heating characteristic function is the main aim of this paper. This part of paper explains what can be understood under the term ANN synthesis and how the method works.

Clause: Let there be a set of all neural networks with a forward running propagation $ANN_{all} = \{ANN_1, ANN_2, ..., ANN_i, ...\}$ and a set of all functions $F_{all} = \{f_1, f_2, ..., f_k, ...\}$. Then for each $ANN_i \in ANN_{all}$ there exists a function $f_k \in F_{all}$, alternatively a set of functions $F_k \subset F_{all}$ such, that holds $ANN_i \Leftrightarrow f_k$, alternatively $ANN_i \Leftrightarrow F_k$.

The Kolmonogorov theorem further shows the validity of the inverse clause: For every continuous function $f_k \in F_{all}$ there exists $ANN_i \in ANN_{all}$ such, that holds $f_k \Leftrightarrow ANN_i$.

Task: Design an algorithm, which will by the means of the symbolic regression methods, evolutionarily scan a set F_{all} in order to find: $f_k \Leftrightarrow ANN_i$; f_k, whose at least some subfunctions $\{f_1, f_2, ...\} \Leftrightarrow \{ANN_n, ANN_m, ...\}$ which solves the particular problem P with a global error $E_T < \xi$, where ξ is the user defined biased tolerance threshold.

AP can perform such evolutionary scanning above F_{all} set and provide the possibility to synthesize the ANN with an almost infinitely variable structure, complexity and scope. There is a very easy way of using AP for ANN synthesis. The most important part is to define items from which ANN will be composed. In this case General Function Set (GFS) contains only three items. Most important item of (2) is Artificial Neuron (AN) with hyperbolic tangent as transfer function (3).

$$GFSall = \{+, AN, K*Tt\} \tag{2}$$

$$AN(S) = w \cdot \frac{e^{2\lambda(S+\phi)} - 1}{e^{2\lambda(S+\phi)} + 1} \tag{3}$$

To allow more inputs into one ANN simple plus operator is used and finally, we can weight input data. Under such circumstances, translation of an individual to ANN can be easily grasped from Fig. 1. Whole process is cyclical. Individuals provided by EA) are translated to ANNs. ANNs are evaluated in accordance with training data set and their global errors are used to set fitness to these individuals. Consequently, a new generation is chosen and the whole process is repeated in next migration loop.

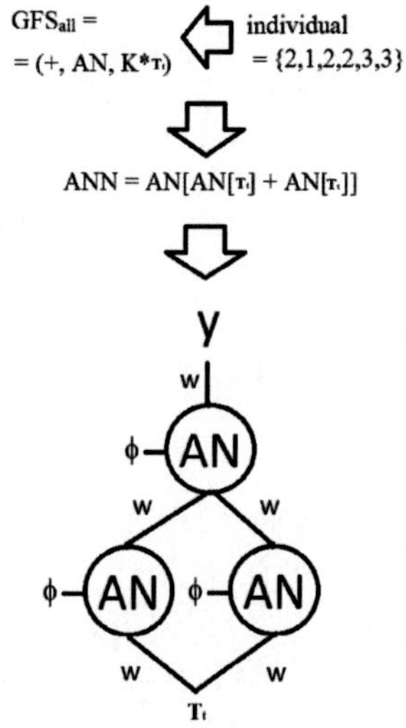

Fig. 1 Translation of an individual to ANN

3 Experiment for Data from Specific Agglomeration

An experiment was realized on real life data pursuant to the mentioned theory and literature. Firstly, the artificial neural network synthesis was applied for finding the explicit function of heating characteristics. As a technological background for applied method .NET framework and C# language were used. Parallel computing was employed to successfully synthetize a suitable ANN performing the heating characteristic function within a reasonable time. Subsequently some results of this searching were tested for heat demand forecast with usage of general plan described above. A program created in Matlab was used for calculation of prediction of DDHD course. The program is drawn in user's menu and by help of that it is possible to choose many parameters of forecast calculation. We realize the calculation of prediction by means of the method that uses model with double filtration. The results can be represented in concrete value form or in the form of resulting graphic window. Subsequent chapter present concrete results for specific locality and for selected heating characteristics.

4.1 Data for Experiments

It is necessary to stress that the real data are used for all experiments and tests of proposed forecast model. The real data were obtained due to close cooperation of our research workplace with energy plant operations. In our case it is close cooperation with company United Energy a.s. - Power and Heating plant Most-Komořany. Measured data from district heating system in the region Most, Czech Republic are used in our experiment. The system is situated to locality Most-Komořany. This system has a typical day load (winter day) of about 100-140 MW. This time series contains besides time and type of the day, the value of heat demand and outdoor temperature for every 30 minutes. Measured data of period November, 2008 – February, 2009 were available. Measured values of heat demand and outdoor temperature in the locality Most-Komořany for January, 2009 are presented in Fig. 2. The course of time series of DDHD displays only one periodic component (daily period). Therefore general model according to Box-Jenkins is used for forecast of social component.

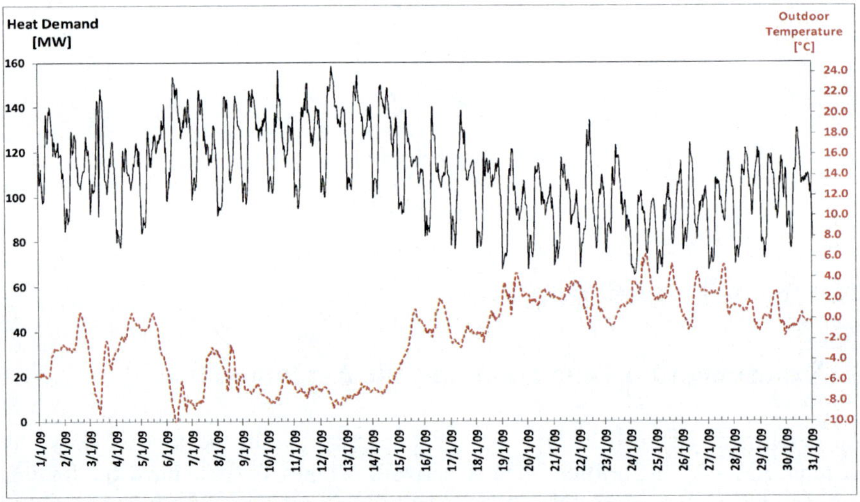

Fig. 2 DDHD for the concrete locality with outdoor temperature course

4.2 Determination of Heating Characteristic Function by Means of Neural Network Synthesis

As an EA used in interaction with AP to synthesize suitable ANN, asynchronously distributed Self-Organizing Migration Algorithm (SOMA) was chosen. SOMA scored great number of successful implementations since its first introduction by Springer (Zelinka 2004) and can be considered as well know algorithm these days. Table 1. shows how the algorithm was set within considered experiment and thus allows its repetition. Extensive explanation of these parameters and their application within SOMA can be found in several papers, e.g. (Vařacha 2011). The whole experiment was conducted in accordance with rules proposed in (Prechelt 1994) and GFS was constructed as described in (3).

Table 1 Setting of SOMA used as EA for AP

Parameter	SOMA used as EA for AP	SOMA used for ANN learning
Number of Individuals	48	number of K_n * 0.5 (at least 10)
Individual Parameters	100	100
Low	0	-10
High	3	10
PathLength	3	3
Step	0,11	0,11
PRT	1/ depth	1 / number of K_n
Divergence	0.01	0.01
Period	1	6

From 100 synthesized ANN the best solution was named as Neural function 1. Neural function 2 represents one of averagely obtained ANN. Their performance can be seen in Table 2.

4.3 Results of Heat Demand Forecast with Using a Specific Heating Characteristic

The model was tested on data from the locality Most-Komořany from two following weeks (13.1.2009 – 26.1.2009). 24 hours-ahead forecast was made twice a day at 6.00 AM and 6.00 PM. The models with inclusion of outdoor temperature using different heating characteristic function were used. Accuracy of the forecast is analyzed and summarized by means of Mean Absolute Percent Error (MAPE) and Root Mean Squared Error (RMSE). Table 2 presents results of heat demand prediction with inclusion of outdoor temperature for the locality Most-Komořany for different type of heating characteristic function. We use the cubic function according to (Chramcov 2010) and two expressions from previous chapter.

From the results, we conclude that the forecast model with inclusion of outdoor temperature for all types of heating characteristics achieves very good results for the locality Most-Komořany. MAPE for the test period is in most time less than 10 percent and RMSE seldom exceed the value of 10MW. Average value of MAPE in the test period is approximately 5-6% and average value of RMSE is approximately 6-7 MW. The average value of MAPE for forecast model without inclusion of outdoor temperature is approximately 8.5% and the average value of RMSE is approximately 10 MW (Chramcov 2010). Realized experiment for specific district heating system demonstrates possibility of using the ANN synthesis for determination of heating characteristic function. The results of forecast model using this approach of determination of heating characteristics are comparable with expert heating characteristics estimation.

Table 2 Accuracy of the forecast model for 24 ahead forecasts with inclusion of outdoor temperature for the locality Most-Komořany for different type of heating characteristics

Date, time	Cubic function		Neural function 1		Neural function 2	
	MAPE [%]	RMSE [MW]	MAPE [%]	RMSE [MW]	MAPE [%]	RMSE [MW]
13.1.2009, 6:00 AM	1.75	2.97	1.71	2.79	2.29	3.55
13.1.2009, 6:00 PM	2.95	5.24	3.07	5.40	2.96	5.26
14.1.2009, 6:00 AM	3.99	6.07	3.86	5.99	3.97	6.05
14.1.2009, 6:00 PM	4.35	6.26	3.46	5.21	6.63	8.59
15.1.2009, 6:00 AM	3.33	4.89	3.31	4.50	3.34	4.62
15.1.2009, 6:00 PM	3.62	4.90	3.85	5.08	4.21	5.35
16.1.2009, 6:00 AM	4.72	5.53	4.91	5.76	5.86	6.81
16.1.2009, 6:00 PM	6.90	9.38	7.37	9.73	6.24	9.44
17.1.2009, 6:00 AM	5.09	7.60	5.38	7.64	5.80	8.63
17.1.2009, 6:00 PM	4.35	6.36	4.15	5.96	4.37	6.24
18.1.2009, 6:00 AM	5.65	7.08	6.49	7.44	4.93	6.58
18.1.2009, 6:00 PM	5.79	7.01	6.87	8.23	4.89	5.96
19.1.2009, 6:00 AM	4.95	6.03	4.54	5.77	5.52	6.65
19.1.2009, 6:00 PM	3.33	4.18	3.33	4.26	3.69	4.66
20.1.2009, 6:00 AM	2.99	3.64	3.00	3.73	3.05	3.77
20.1.2009, 6:00 PM	2.70	3.28	2.69	3.26	2.77	3.33
21.1.2009, 6:00 AM	3.48	4.13	3.42	4.03	3.61	4.25
21.1.2009, 6:00 PM	6.55	7.37	6.24	7.07	7.12	8.07
22.1.2009, 6:00 AM	5.58	6.82	5.77	6.93	5.85	7.36
22.1.2009, 6:00 PM	7.51	9.78	7.62	9.62	7.17	9.69
23.1.2009, 6:00 AM	9.47	10.75	8.58	10.05	10.99	11.54
23.1.2009, 6:00 PM	8.26	8.29	7.00	7.33	9.36	8.79
24.1.2009, 6:00 AM	6.59	7.38	5.70	6.48	7.02	7.85
24.1.2009, 6:00 PM	4.15	5.18	4.13	5.42	3.99	5.14
25.1.2009, 6:00 AM	5.40	7.07	4.74	6.51	6.42	7.95
25.1.2009, 6:00 PM	9.57	11.16	8.58	10.22	11.17	12.66
26.1.2009, 6:00 AM	7.34	8.63	6.71	8.11	8.60	9.56
26.1.2009, 6:00 PM	5.39	6.29	4.96	6.00	7.01	7.29
Average value	5.21	6.55	5.05	6.38	5.67	6.99

5 Conclusions

This paper deals with design of a model for heat demand forecasting with usage of artificial neural network synthesis. We propose the forecast model of heat demand based on the assumption that the course of heat demand can be described sufficiently well as a function of the outdoor temperature and the weather independent component (social components). Forecast of social component is realized by means of Box-Jenkins methodology. For inclusion of outdoor temperature influence in calculation of prediction of heat demand is used the heating characteristic (function that describes the temperature-dependent part of heat consumption). Finding the explicit expression for the heating characteristics is the main aim of this paper. The function of heating characteristics is proposed by means of the Artificial Neural Network Synthesis. The paper present also comparison of accuracy of forecast model using expert heating characteristics estimation with model using heating characteristics obtained by means of ANN synthesis. The results of forecast model using this approach of determination of heating characteristics are comparable with expert heating characteristics estimation.

Heat demand forecast plays an important role in power system operation and planning. Accurate heat demand prediction saves costs by improving economic load dispatching, unit commitment, etc.

Acknowledgments. This work was supported in part by European Regional Development Fund within the project CEBIA-Tech No. CZ.1.05/2.1.00/03.0089.

References

Arvastson, L.: Stochastic modelling and operational optimization in district-heating systems. Doctoral thesis. Lund University, Centre for Mathematical Sciences (2001) ISBN 91-628-4855-0

Bakker, V., Bosman, M.G.C., Molderink, A., Hurink, J.L., Smit, G.J.M.: Improved heat demand prediction of individual households. In: Proceedings of IFAC Conference on Control Methodologies and Technology for Energy Efficiency, CMTEE 2010, Vilamoura, pp. 110–115 (2010) ISBN: 978-390266168-5

Balátě, J.: Design of Automated Control System of Centralized Heat Supply. Thesis of DrSc (Doctor of Science) Work. TU Brno, Faculty of Mechanical Engineering (1982)

Box, G.E., Jenkins, G.M.: Time series analysis: forecasting and control, Rev. edn. Holden-Day, San Francisco (1976) ISBN 0-8162-1104-3.

Chramcov, B.: Utilization of Mathematica environment for designing the forecast model of heat demand. WSEAS Transaction on Heat and Mass Transfer 6(1), 21–30 (2011)

Chramcov, B., Vařacha, P.: Design of a Model for Heat Demand Prediction Using the Neural Network Synthesis. In: Proceedings of the 6th International Conference on Applied Mathematics, Simulation, Modelling (ASM 2012), Athens, Greece (2012)

Chramcov, B.: Heat deman forecasting for concrete district heating system. International Journal of Mathematical Models and Methods in Applied Sciences 4(4), 231–239 (2010)

Dostál, P.: Machine Processing of Daily Diagram Course Prediction of Loading the Centralized Heat Supply System. Doctoral thesis. TU Brno, Faculty of Mechanical Engineering (1986)

Dotzauer, E.: Simple model for prediction of loads in district-heating systems. Applied Energy 73(3-4), 277–284 (2002) ISSN 0306-2619

Koza, J.R., Bennett, F.H., Andre, D., Keane, M.A.: Genetic Programming III; Darwinian Invention and problem Solving. Morgan Kaufmann Publisher (1999) ISBN 1-55860-543-6

Král, E., Dolinay, V., Vašek, L., Vařacha, P.: Usage of PSO Algorithm for Parameters Identification of District Heating Network Simulation Model. In: 14th WSEAS International Conference on Systems, Rhodes, Greece. Latest Trands on Systems, vol. II, pp. 657–659 (2010) ISBN/ISSN: 978-960-474-214-1

Lehtoranta, O., Seppälä, J., Koivisto, H., Koivo, H.: Adaptive district heat load forecasting using neural networks. In: Proceedings of Third International Symposium on Soft Computing for Industry, Maui, USA (2000)

Park, T.C., Kim, U.S., Kim, L.H., Kim, W.H., Yeo, Y.K.: Optimization of district heating systems based on the demand forecast in the capital region. Korean Journal of Chemical Engineering 26(6), 1484–1496 (2009) ISSN 0256-1115

Pedersen, L., Stang, J., Ulseth, R.: Load prediction method for heat and electricity demand in buildings for the purpose of planning for mixed energy distribution systems. Energy and Buildings 40(7), 1124–1134 (2008)

Popescu, D., Ungureanu, F., Serban, E.: Simulation of Consumption in District Heating Systems. In: Proceedings of the 1st WSEAS International Conference on Urban Rehabilitation and Sustainability (URES 2008), Bucharest, p. 195 (2008) ISBN 978-960-474-023-9, ISSN 1790-5095

Schellong, W., Hentges, F.: Forecast of the Heat Demand of a District Heating System. In: Proceedings of European Power and Energy Systems, Palma de Malorca, Spain (2007)

Torkar, J., Goricanec, D., Krope, J.: Economical heat production and distribution. In: Proceedings of 3rd IASME/WSEAS Int. Conf. on Heat Transfer, Thermal Engineering & Environment, Corfu, p. 445 (2005) ISBN 960-8457-33-5

Turner, S.D., Dudek, S.M., Ritchie, M.D.: Grammatical Evolution of Neural Networks for Discovering Epistasis among Quantitative Trait Loci. In: Pizzuti, C., Ritchie, M.D., Giacobini, M. (eds.) EvoBIO 2010. LNCS, vol. 6023, pp. 86–97. Springer, Heidelberg (2010)

Vařacha, P.: Impact of Weather Inputs on Heating Plant - Aglomeration Modeling. In: Recent Advances in Neural Networks: Proceedings of the 10th WSEAS International Conference on Neural Networks, Prague, p. 193 (2009) ISBN 978-960-474-065-9, ISSN 1790-5109

Vařacha, P.: Neural network synthesis dealing with classification problem. In: Recent ReSearches in Automatic Control: Proceedings of the 13th WSEAS International Conference on Automatic Control, Modelling & Simulation (ACMOS), Lanzarote, pp. 377–382 (2011) ISBN 978-1-61804-004-6

Vařacha, P., Jašek, R.: ANN synthesis for an agglomeration heating power consumption approximation. In: Recent Researches in Automatic Control: Proceedings of the 13th WSEAS International Conference on Automatic Control, Modelling & Simulation (ACMOS), Lanzarote, pp. 239–244 (2011) ISBN 978-1-61804-004-6

Zelinka, I.: SOMA - Self Organizing Migrating Algoritm. In: Batu, B.V., Onwubolu, G. (eds.) New Optimization Techniques in Engineering, ch. 7, p. 33. Springer (2004)

Prechelt, L.: Proben1 - A Set of Neural Network Benchmark Problems and Benchmarking Rules. Universität Karlsruhe, Germany (1994)

Vonk, E., Jain, L.C., Johnson, R.P.: Automatic Generation of Neural Network Architecture Using Evolutionary Computation. In: Advances in Fuzzy Systems – Applications and Theory, vol. 14, pp. 981–982. World Science (1997) ISBN: 981-02-3106-7

Vonk, E., Jain, L.C., Veelenturf, L.P.J., Hibbs, R.: Integrating Evolutionary Computation with Neural Networks. In: Electronic Technology Directions to the Year 2000, Adelaide, Australia, pp. 137–143 (1995)

Fuzzy Logic Decision Support for Long-Term Investing in the Financial Markets

Zdeněk Brož and Petr Dostál

Brno University of Technology, Faculty of Business and Management,
Institute of Informatics, Kolejní 2906/4, 612 00 Brno, Czech Republic

Abstract. This paper discusses the use of fuzzy logic and modeling as a decision support for long-term investment decisions in the financial markets. A simple model is proposed to calculate recommendations for investors. This research required thorough analysis of historical data that lead to the discovery of interesting dependencies between the Dow Jones index, currency pairs, oil price and the VIX volatility index. The fuzzy model uses several input variables that are used to simplify the complex conditions in the financial markets. The purpose of the model is to evaluate the current market situation, compare the current situation to similar situations in the past and to provide investment recommendations for long-term future investing.

1 Introduction

This paper researches the use of soft computing as a decision making support for long-term investment in the financial markets. It is very difficult to predict the development of financial markets [4], [11], [13] and [14]. Markets are dynamic and there are many complex factors and complicated relationships that influence indexes, currencies and commodities these makes investing complicated and risky [7]. The processes in economy are nonlinear. If the system is nonlinear and dynamic, it can generate randomly looking behaviour but it can include the permanent trends and cycles [6] and [16]. Investing in the financial markets is difficult because of globalized economies - there are different crises, bubbles, rising debts and prices of commodities, energy etc. These problems randomly escalate and create extreme imbalances in the market. These imbalances are both great opportunities and threats for the investors. Psychology also plays an important role in the financial markets - investors often do not recognize these opportunities because they are afraid of the future development [9] and [10]. This research is facing very actual and yet at the same time classic problem of investing – when to buy and sell stocks while minimizing the risk [8]. Understanding the markets and being able to predict what will happen in the near future are the key skills that every successful investor has to have. This research uses a simple model with a few variables that simplifies the complex market environment to make reliable recommendations for the investors and so provides a valuable decision making support tool.

This research has several objectives. The first objective is to analyse the past development of the Dow Jones index and to find extreme imbalances that occurred

I. Zelinka et al. (Eds.): Nostradamus: Mod. Meth. of Prediction, Modeling, AISC 192, pp. 113–121.
springerlink.com

in the past. These situations are opportunities for the investors. The second objective is to define a set of variables that reliably describe the situation in the market. The third objective is to research the dependencies and relationships between these variables. The final objective is to design a very simple and reliable fuzzy model that uses these variables to calculate recommendations for the investor.

This research focuses on the use of soft computing and fuzzy logic in finance. Investors and decision makers have to decide when, where and how to invest. This problem is very complex and decision makers always try to use methods, tools and algorithms that allow them to limit risk [7]. Fuzzy model designed in this research is intended as a decision making support tool for investors in the financial markets [12]. This research deals with extreme situations that occur in the financial markets and that are very difficult to predict [24], [25] and [26]. Many researchers in the past used soft computing in business and finance [2], [3], [21], [22] or [27]. This research helps to identify current imbalances in the market based on similarity to past known events. The fuzzy model then processes several input variables to calculate recommendations for investors. Instead of promoting short term speculation this research aims to provide a decision making support model that helps to identify long-term critical imbalances and helps the investor to find possibilities for making long-term profits with low risk.

2 Methods

The proposed model is based on fuzzy logic and fuzzy sets. A fuzzy set A is defined as (U, μ_A), where U is the relevant universal set and μ_A: U $\rightarrow \langle 0,1 \rangle$ is a membership function, which assigns each elements from U to fuzzy set A. The membership of the element $x \in$ U of a fuzzy set A is indicated μ_A(x). We call $\boldsymbol{F}$(U) the set of all fuzzy set. Then the "classical" set A is the fuzzy set where: μ_A: U $\rightarrow \{0, 1\}$. Thus $x \in A \Leftrightarrow \mu_A(x) = 0$ and $x \notin A \Leftrightarrow \mu_A(x) = 1$. Let U_i, $i = 1, 2, ..., n$, be universals. Then the fuzzy relation R on U = U1×U2 ×...× Un is a fuzzy set R on the universal U.

The creation of the model was preceded by a thorough statistical analysis of the historical data. It was necessary to identify ideal moments in the past when the investor could buy or sell stocks to generate profit – these moments are often characterized as maximum or minimum values of the Dow Jones index. To increase the reliability of the designed model it was necessary to use not only the data from the long-term time series of the Dow Jones index but also variables that are not directly related to the index itself. Therefore it was necessary to find more variables that have a relationship with the Dow Jones index and describe the situation in the financial market. After the maximum and minimum values were found it was necessary to collect information about all the variables that are used in the model. The fuzzy model uses five input variables: Dow Jones industrial index, past trend of this index, EUR/USD currency pair, oil price and VIX index. Due to the constraints of this paper the input variables are not discussed in greater detail.

The fuzzy model requires a set of rules that capture the important relationships between the input variables [30]. These relationships have to be researched from the past data. A key step in this research even before the work on the fuzzy model began was to find extreme imbalances of the Dow Jones index in the past two decades. To find these imbalances it was necessary to analyze carefully the historical time series [14], [19] and [29]. The long-term time series was analyzed differently several times in order to find key moments in the past that were opportunities for the investors [18]. A very simple and reliable method to find these past moments is to compare the historical prices with the long term averages. When the historical price is very high in a certain time period and above both calculated long-term averages then this moment in time is a good opportunity for the investor to sell. When the price is the lowest in a certain time period and well below both long-term averages it is an ideal opportunity to buy. These two rules were used to determine the moments in time that were used in the model. When these dates were found the values for all the input variables were calculated. After all the values of the input variables had been determined the work on the fuzzy model started.

Fig. 1 Long-term statistical analysis of the Dow Jones index

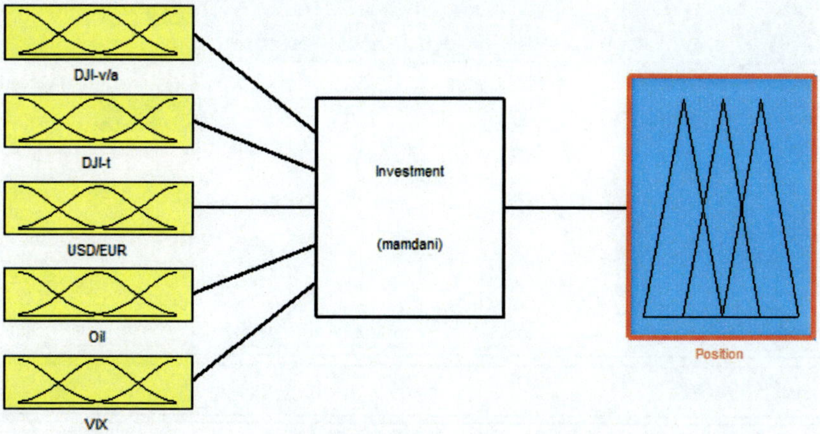

Fig. 2 FIS editor – input and output variables

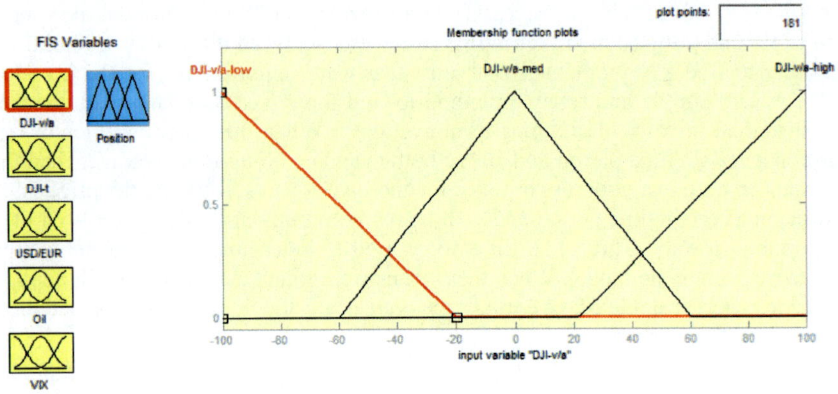

Fig. 3 Membership function editor

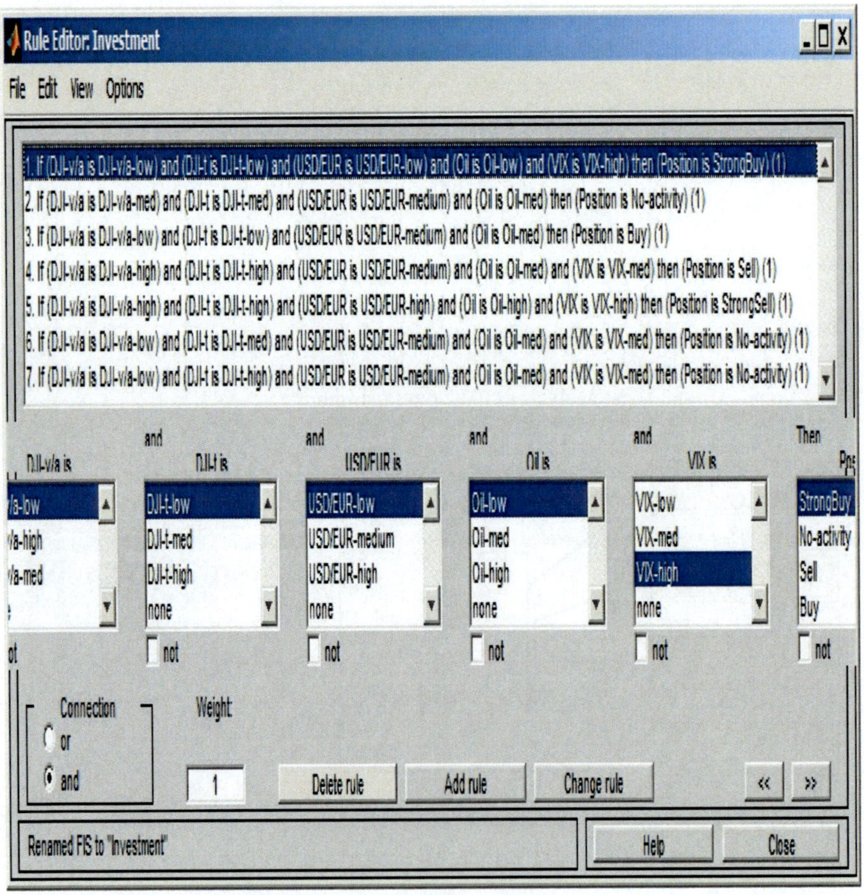

Fig. 4 Rules of the rule editor

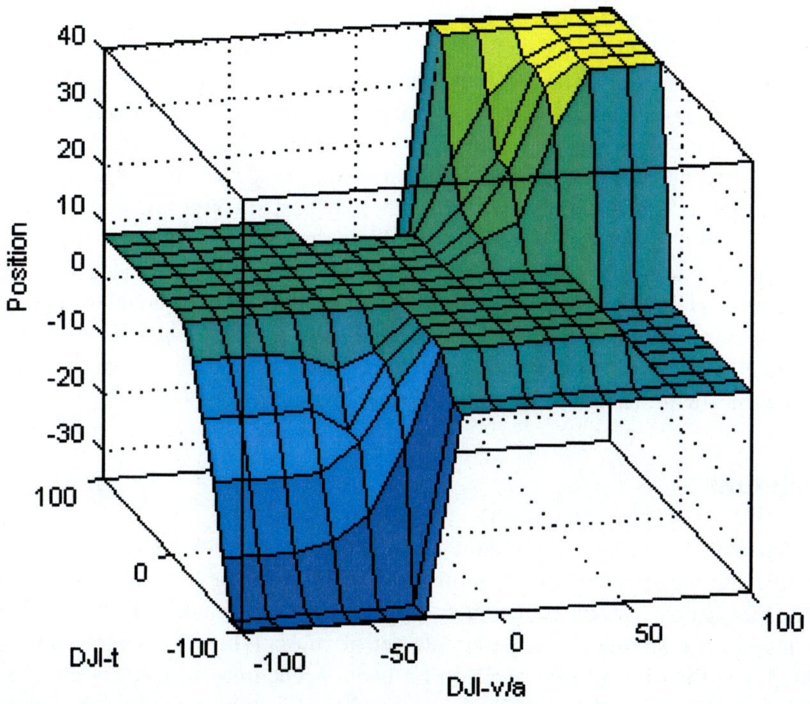

Fig. 5 Visualization of dependence Position = f(DJI-t, DJI-v/a)

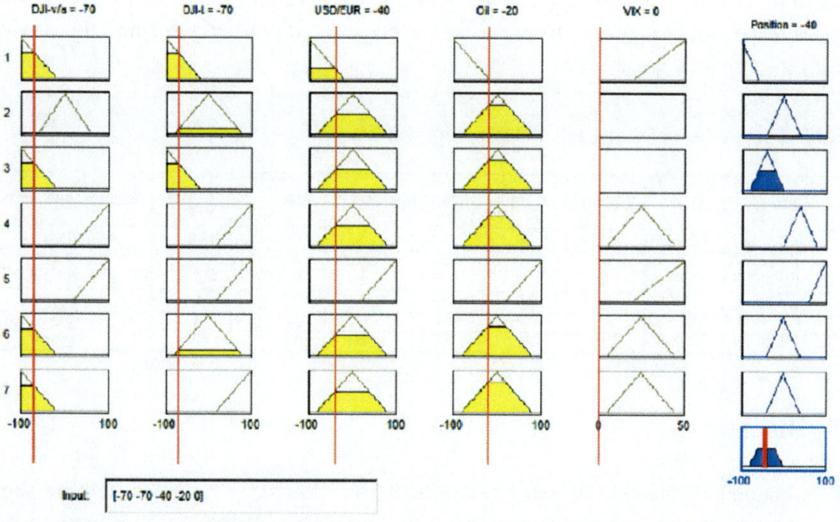

Fig. 6 Rule viewer and calculated output variable

The fuzzy model is implemented in the Fuzzy logic toolbox in MATLAB. First the input variables are defined. The model then calculates the value from these input variables based on the defined rules and returns an output variable called Position. This variable is the recommendation for the investor. In order to keep the model as simple as possible only three attributes were used for each variable (low, medium and high and a small number of rules were defined. These rules record the basic relationships that have been determined from the analysis of the past data and from the values of input variables in the key past situations. Before the work on this research the authors were looking for an algorithm or method that could be used by investors who are not skilled enough to use complicated financial software. That led the authors to this research with a simple model with a few input variables. After the rules have been defined the surface viewer can be used to visualize the dependency between input variables and the output variable. Further information about modeling in MATLAB can be found in [28].

3 Results

This chapter contains a simple table showing the input values for the five input variables for the selected key situations determined from the long-term Dow Jones index time series. The fuzzy model calculated the output value in each case from the input variables based on the simple set of rules. When the model outputs a value it can then be clearly translated to recommendations for the investor. Even without the fuzzy model very interesting and useful information can be learned from the values of the input variables. There are some relatively strong dependencies and relationships between the input variables. It would of course be possible to add more input variables and make the model more sophisticated but the objective of this research was to keep the set of input variables and the model itself as simple as possible.

Table 1 Input variables and calculated recommendations

Date	DJI	DJI-v/a	DJI-t	USD/EUR	Oil	VIX	Calculated value	Recommendation
30.6.1998	7640	-10	-40	-40	-30	45	-87,3	strong buy
1.1.2000	11722	90	90	50	50	25	40	strong sell
11.2.2002	9739,81	-70	-70	-40	-20	22	-40	strong buy
1.7.2007	13800	95	90	50	90	30	84	strong sell
9.3.2009	6625,74	-90	-70	-50	-100	80	-84	strong buy
1.5.2011	12600	80	80	80	-30	18	40	strong sell

4 Discussion

This chapter discusses the results obtained from the fuzzy model. It can be seen that the input variables have very different values for all the key moments in the past. These selected imbalances of the market were chosen to demonstrate the model. The financial market is dynamic and a very complex system so there is no simple way to predict the future development [1], [5] and [20]. The objective of the model is not to predict the future development but merely to identify opportunities and calculate recommendations from the input variables. Because this model focuses on the extreme imbalances of the market it can identify them safely. When the calculated recommendations are combined with other information and investing skills of the individual investor this decision support model is very valuable. This model promotes long-term investing strategy with low risk. That is a major difference when compared to most other methods that promote short-term speculation with high risk [15] or [17].

The designed model is intended as a decision making support for long-term investment. The financial market is a complex, dynamic and chaotic environment. A large number of factors influence the developments in the financial market each day. The reliability of the model would decrease significantly if it were used for short-term investing. Another limitation is that the model is designed for investing in large mutual funds that are highly correlated with the Dow Jones index. The model is designed to be as simple as possible and easy to use. The objective of this research is to develop an easy to use model that has a few simple input variables and yet is able to provide reliable recommendations to the investor. The research has shown that at some point the model becomes too complex and the reliability of the model decreases [31]. This research shows that it is possible to reliably detect the long-term major imbalances of the financial markets which can then be used by investors to generate profits while maintaining low risk.

5 Conclusions

Investment decision making support based on the fuzzy model can prove to be very useful for investors who are looking for a path to manage risk when dealing with their long-term investment portfolio. The proposed model uses several input variables to evaluate the current situation in the market and calculate recommendations for the investor. The objective of this research is of course to limit risk and safely identify opportunities. This research does not promote risky short-term speculations. The designed model has been tested extensively on the historical data and it has proved to provide correct investment recommendations with high statistical probability. This research will be continued in the near future.

References

1. Abel, H.: Digital Day Trading, Dearborn, USA, 269p. (1999) ISBN 0-7931-3113-8
2. Aliev, A., Aliev, R.: Soft Computing and Its Applications, 444 p. World Scientific Pub. Ltd., UK (2002) ISBN 981-02-4700-1

3. Altrock, C.: Fuzzy Logic & Neurofuzzy – Applications in Business & Finance, 375 p. Prentice Hall, USA (1996) ISBN 0-13-591512-0
4. Armstrong, S.J.: Principles of Forecasting, 849s p. Kluwer Academic Publishers, USA (2001) ISBN 0-7923-7930-6
5. Baird, B., Burney, C.: Electronic Day Trading to Win, 220 p. John Wiley & Sons Inc., USA (1999) ISBN 0-471-35072-9
6. Bernice, C.: The Edge of Chaos - Financial Booms, Bubbles, Crashes and Chaos, 379 p. John Wiley & Sons Ltd., UK (1997) ISBN 0-471-96907-9
7. Dostál, P., Sojka, Z.: Financial Risk Management. In: Zlín 2005, Skripta, 60 p. (2005) ISBN 80-7318-343-9
8. DostÁL, P.: Prediction of Competitive Environment in Business. In: 28th International Symposium on Forecasting, Nice, France, p. 171, 6p., ISSN 1997-4116, ISSN 1997-4124
9. DostÁL, P., Rais, K.: Stock Market Decision Machine. In: 29th International Symposium on Forecasting, Hong-Kong, China, p. 171, 5 p. (2009) ISSN 1997-4116
10. Dostál, P., Budík, J.: System for Intelligent Investment Portfolio Making. In: 30th International Symposium on Forecasting, San Diego, USA, 6 p. (2010) ISSN 1997-4116
11. Dostál, P.: Advanced Decision Making in Business and Public Services, 168 p. CERM, Brno (2011) ISBN 978-80-7204-747-5
12. Dostál, P., Brož, Z.: Fuzzy Logic Investment Support on the Financial Market. In: 31th International Symposium on Forecasting, Praha, CZ, 6 p. (2011) ISSN 1997-4116
13. Ehlers, F.J.: Cybernetic Analysis for Stock and Futures, 256 p. John Wiley & Sons Inc, USA (2004) ISBN 0-471-46307-8
14. Franses, P.H.: Time Series Models for Business and Economic Forecasting, Cambridge, 213 p. University Press, UK (2001) ISBN 0-521-58641-0
15. Frost, A.J., Pretcher, R.: Elliott Wave Principle, 245 p. New Classic Library, USA (2000) ISBN 0-932750-43-5
16. Gleick, J.: Chaos, 330 p. Ando Publishing (1996) ISBN 80-86047-04-0
17. Gozzhelf, P.: Techno Fundamental Trading, 267 p. Probus Publishing Company, USA (1995) ISBN 1-55738-541-6
18. Hand, D., Mannila, H.: Principles of Data Mining, 546 p. The MIT Press, USA (2001) ISBN 0-262-08290-X
19. Herbst, F.: Analyzing and Forecasting Futures Prices, 238 p. John Wiley & Sons Inc., USA (1992) ISBN 0-471-53312-2
20. Howard, A.: Digital Day Trading, Dearborn, USA, 269 p. (1999) ISBN 0-7931-3113-8
21. Chen, S., Wang, P., Wen, T.: Computational Intelligence in Economic and Finance. Springer (2004) ISBN 978-3-540-72820-7
22. Chen, S., Wang, P., Wen, T.: Computational Intelligence in Economic and Finance, vol. II. Springer (2007) ISBN 978-3-540-72820-7
23. Klir, G.J., Yuan, B.: Fuzzy Sets and Fuzzy Logic, Theory and Applications, 279 p. Prentice Hall, New Jersy (1995) ISBN 0-13-101171-5
24. Peters, E.E.: Fractal Market Analysis – Applying Chaos Theory to Investment & Economic, 315 p. John Wiley & Sons Inc, USA (1994) ISBN 0-471-58524-6
25. Peters, E.E.: Chaos and Order in the Capital Markets: A New View of Cycles, Prices, Wiley Finance Edition, USA, 274 p. (1996) ISBN 0-471-13838-6
26. Trippi, R.R.: Chaos & Nonlinear Dynamic in the Financial Markets, 505p. Irwin Professional Publishing, USA (1995) ISBN 1-55738-857-1

27. Ribeiro, R., Yager, R.: Soft Computing in Financial Engineering, 590 p. A Springer Verlag Company (1999) ISBN 3-7908-1173-4
28. THE MATHWORKS. MATLAB – User's Guide, The MathWorks, Inc. (2010)
29. Weigend, A.: Time Series Prediction: Forecasting the Future and Understanding the Past, 643 p. Addison-Wesley, Massachusetts (1993) ISBN 0-201-62602-0
30. Williams, L.V.: Information Efficiency in Financial and Betting Markets, 353 p. Cambridge University Press (2005) ISBN 0–521-81603-3
31. Wisniewski, M.: Quantitative Methods for Decision Makers, 4th edn., 592 p. Pearson Education Limited, England (2006) ISBN 0 27368 789 1

Possibilities of Industrial Data Description Using Fractal Geometry

Vlastimil Hotař

Technical University of Liberec, Department of Glass Producing Machines and Robotics, Studentska 2, 460 17 Liberec 6

Abstract. The continuous growth of competitive pressure to increase the quality of products necessitates the requirement for objective measurement and control methods for materials, processes and productions. However, many structures (e.g. defects, surfaces, cracks, time series from dynamic processes) can hardly be described by conventional methods because they are complex and irregular. A new approach is the application of fractal geometry that is successfully used in science. Even though applications in industry are only sporadic and experimental, fractal geometry in connection with statistics can be used as a useful and powerful tool for an explicit, objective and automatic description of production process data (laboratory, off-line and potentially on-line). Despite the fact that this research focuses on data from the glass industry, the methodology and principles for data evaluation from industry can be applied to industry generally.

1 Introduction

Off-line and automatic on-line quality monitoring and control are a standard part of production lines. The choice of how to monitor is dependent on the analyses used and should correspond with the character of obtained data. In all industrial areas the character of data set obtained from production processes or from products can be highly structured. For this kind of data set a powerful tool for analysis of complexity – fractal geometry (especially a fractal dimension) should be used [1, 2, 3]. The fractal dimension (FD) with a combination of statistic tools is experimentally used and is an interesting and powerful tool for complex data quantification, for poor quality source searching, production optimalization and non-stability of production process subsystems searching in industrial applications. The application of FD to the industry is generally experimental [4], but application to production is possible and brings benefits. Currently, there are tools to monitor three basic data format types: digitalized photos (evaluation of 2D pictures of surface defects), time series (analysis for control systems) and topological one dimensional dividing lines (application for the corrugation test of windows glass shields and surface roughness - iron aluminides in comparison with the carbide-nickel steel in contact with glass melt) [5].

I. Zelinka et al. (Eds.): Nostradamus: Mod. Meth. of Prediction, Modeling, AISC 192, pp. 123–132.
springerlink.com

2 Fractal Dimension Used for Industrial Data Description

The fractal dimension (also named the Hausdorff-Besicovitch dimension) is closely connected to fractals that were defined by Benoit Mandelbrot [1], though scientists found some geometric problems with specific objects before him (e.g. the measurement of coast lines per different length of rulers by Richardson). A potentially powerful property of the FD is describing complexity by using a single number that defines and quantifies structures. The number is mostly a no integer value and the FD is higher than the topological dimension. For example, the Koch curve (one of the most famous mathematical deterministic fractal) has the topological dimension $D_T = 1$, but the FD $D_F = 1.2619$. A smooth curve as a line has the topological dimension $D_T = 1$ and the FD $D_F = 1$. The FD can be computed for set of points, curves, surfaces, topological 3D objects, etc. and if the FD is higher than the topological dimension, we name the objects fractals.

A comparison of statistical tools and the FD is possible, but should be done with care. The FD gives added information about the character of describing data sets and to say that the FD is better than statistics and vice versa is impossible. Furthermore, the FD should not be used separately because the dimension does not give all the information about data set captures. Using added parameters (statistics, topology, spectral analysis, etc.) together with the dimension brings benefits and is recommended. A decisive number (a testing number) for production control or quality monitoring (for example) can be computed from obtained parameters (including the FD) by weight coefficients. This text shows methods that use the "cooperation" between the FD and statistics and gives examples of their application."

3 Description of Time Series

A lot of production processes have dynamic subsystems, which have an influence on the production. The production process can be influenced by the unpredictable of the subsystems and measured data (time series) from production sensors containing this dynamic influence. The time series are structured and typical statistic data evaluations are not often sufficient. Good quality products are the reason of accurate settings of a production process and sensors, which often produce time series in time monitors. Our research target is to develop software tools which quickly and accurately describe time series from production processes sensors using fractal geometry and statistic analyses. Fractal geometry uses the FD which describes the character (complexity) of time series with one number. To analyses time series we use statistical methods, power spectral analysis and an estimate of the FD. The estimation of the FD is calculating using the rescaled range method, the aggregated variance method and the box counting method from an "iso-set" [6].

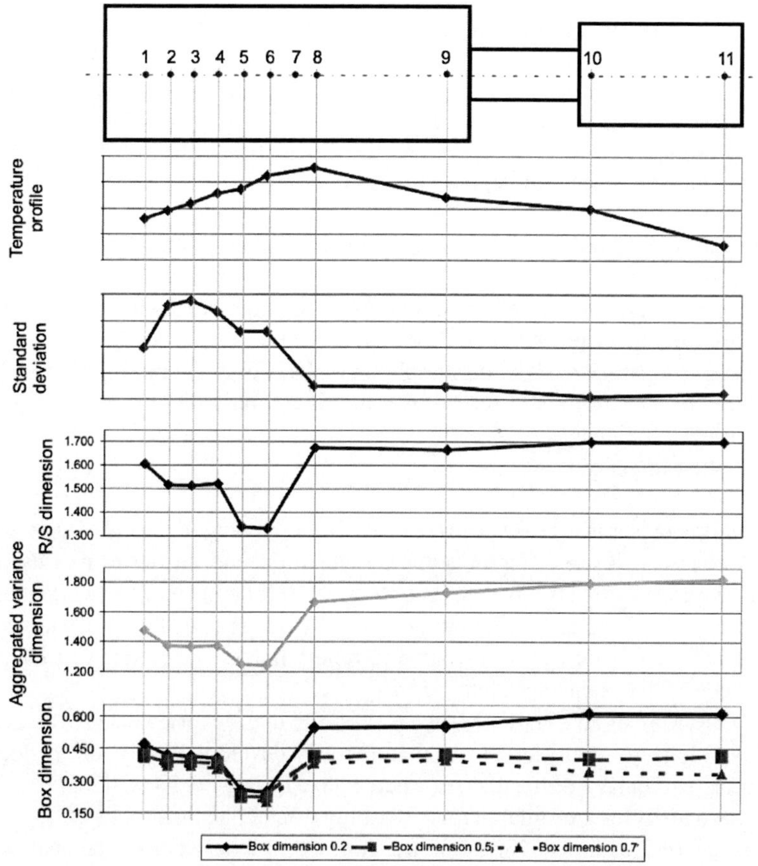

Fig. 1 Results of a tank furnace siege time series analysis

The application possibilities of a fast and accurate description of time series from production process sensors using fractal geometry and statistic analyses are numerous. Fig. 1 shows the results of a tank furnace siege time series analysis, where the standard deviation, the R/S dimension, the aggregated variance dimension and the box dimension were used. The average temperature indicates a temperature profile of the siege and an implicit temperature profile of glass melt during the siege. The FD result is a relatively large decrease from positions 5 and 6 and a subsequent ascension between positions 6 and 8. The large decrease indicates simple time series and the higher dimension represents complex time series. The decrease is in position, where a change of longitudinal currents is expected. The complex time series represent changes in temperature and the movement of melt causes the changes. It appears that fractal analysis can be used for the detection of glass melt currents [5, 7].

4 Evaluation of 2D Pictures and Defects

The explicit, objective and automatic description of image complexity can be made by different methods, both statistic and the FD. Only some of the possibilities are presented below.

In practice, the process of description has five steps:

- Preparation of samples – the structure must be visible, costume jewellery is cut, fig. 2, A.
- Taking photographs. For example photos of the hole cracks in costume jewellery are from an electron microscope, fig. 2.
- Software preparation of the digital photographs, fig. 2, C (cutting of the photographs, because only some parts of the photos are important for analysis).
- Analyses of the images.
- Evaluation of analyses results.

A digital image is a matrix (or matrixes) of pixels (rectangular array of points, fig. 2, D). Pixels can reach different numbers which depend on the format used for digital images. The pixels have numbers between 0 (black) and 255 (white) for the grey 8-bit palette bitmap and the bitmap has only one matrix. Fig. 2, C shows two typical poor quality surfaces of costume jewellery holes. The cutting C-1 has deep cracks and C-2 has a thin structure.

Software Matlab and HarFa [8] were used for these experimental evaluations. A methodology for analyses of the pictures was developed based on: histogram evaluation, percentage of black pixels, percentage of large defects, the FD, ... Only the last two analyses are suitable for describing these kinds of structures.

The grey images in fig. 2 must be transformed to binary before the analyses of the percentage of large defects and the FD. The technique for this transformation is based on what is called "thresholding"; the binary image can be determined from a grey 8-bit palette bitmap, where all the black pixels fulfil conditions e.g. $0 \leq \text{black} \leq 100$ and all the other pixels become white ($100 < \text{white} \leq 255$). We used the threshold 35.

The percentage of large defects is suitable for the detection of single, relatively large cracks and defects. This method computes the percentage of pixels with neighbouring pixels of the same value. The analysis searches for black pixels in binary images, which have five or more neighbouring black pixels. Large defects contain black pixels with five or more neighbouring black pixels that represent defects, structures, cracks, etc (in fig. 3).

The FD must be estimated for natural fractals and, of course, for data from production processes or products. For this reason many analyses can be used but the most suitable for these kinds of data and structures is the box counting method (in fig. 4). The estimated FD is named the box dimension [3] and for research the dimension was multiplied by 1000 [9].

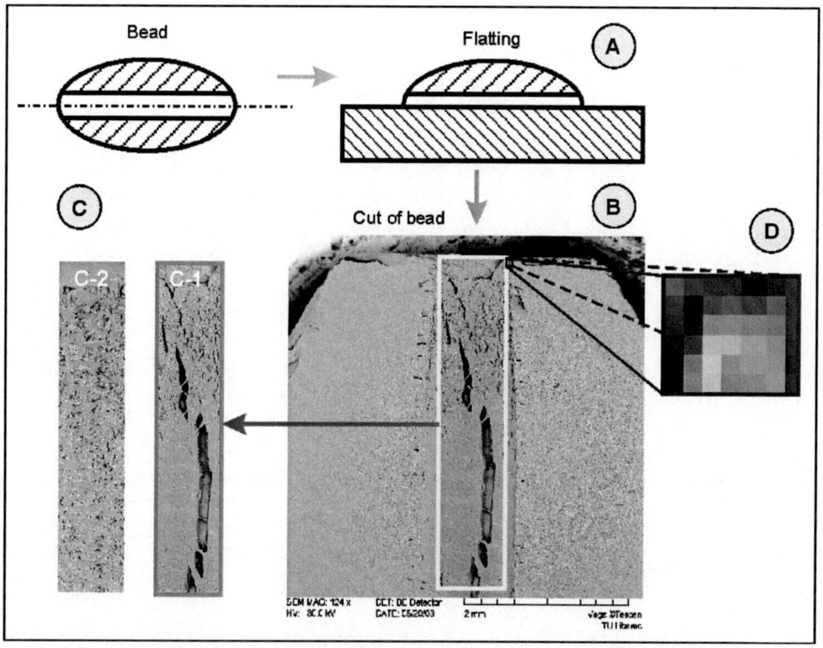

Fig. 2 Preparation of samples, taking photographs, software preparation

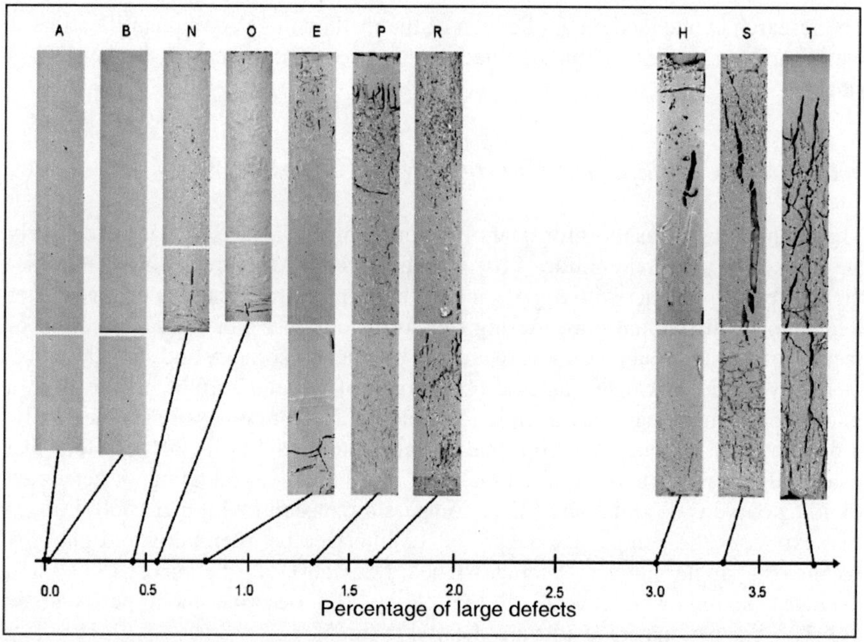

Fig. 3 Computation of the percentage of pixels with neighboring pixels of the same value - percentage of large defects

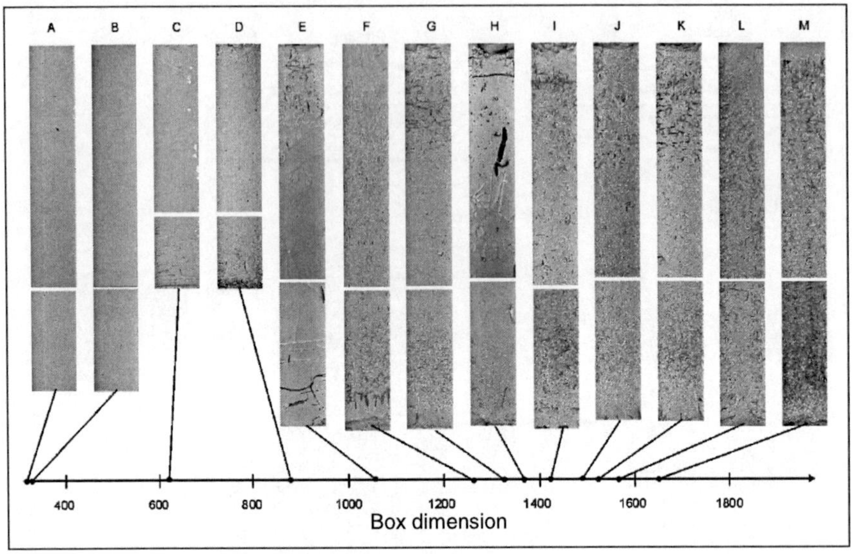

Fig. 4 Computation of box dimension

5 Classification of Topological One-Dimensional Dividing Lines

The research is intended on application of the methodology for a quantification of metal surface changes and on an objectification corrugation test of windows glass shields.

5.1 Surface Roughness Changes after Corrosion Tests

The methodology was used for quantification of metal surface changes. Relatively new materials: iron aluminides [10] are compared with currently used chrome-nickel steels in contact with a glass melt. The samples roughness is changed after interaction with molten glass during the corrosion tests. For a quantification of metal surfaces the roughness was evaluated by fractal geometry and statistic tools.

Firstly, a digital camera takes a photograph of a metal surface profile from a microscoped metallographic sample (5 photographs from one sample). Secondly, a dividing line is generated from the digital photography by a software tool that exactly defines the curve between material alloys and a surrounding. A generated dividing curve is described by FD (a compass dimension multiplied 1000, $D_{C\ 1000}$) that expresses the complexity degree of the interface between alloy and glass by means of a single number. Average standard deviation all the curves (STD) and average maximum roughness all the curves (R) describe them per statistic [11, 12]. The parameters results are showed in fig. 5.

	Chrome-nickel steel EN 10095 (AISI 310) Dividing line - roughness profile (size in pixels)	R [mm] s [mm] $D_{o\,1000}$ [-]		FA - iron aluminide on base Fe3Al Dividing line - roughness profile (size in pixels)
Ground state	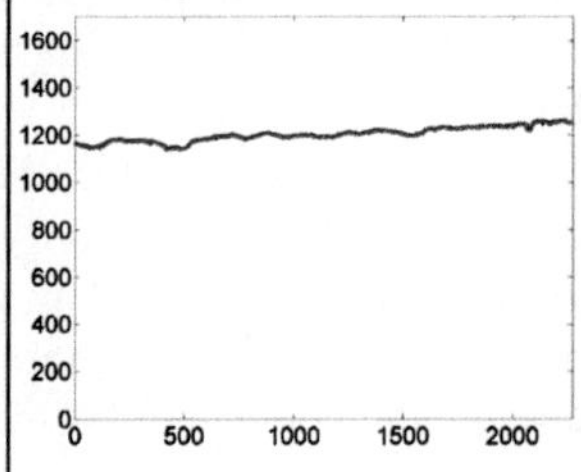	3.07 0.79 1029.2	1.56 0.29 1055.8	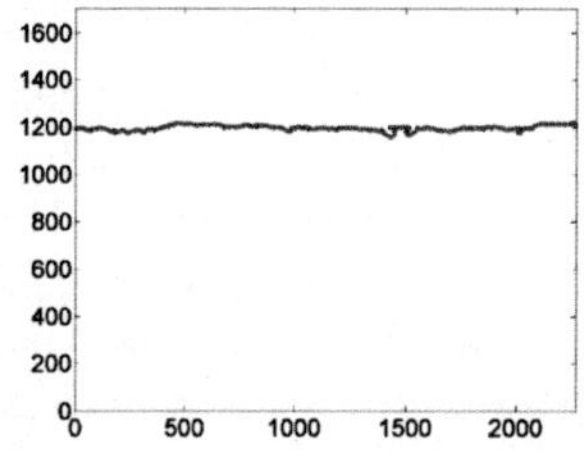
1100OC, 24 hour	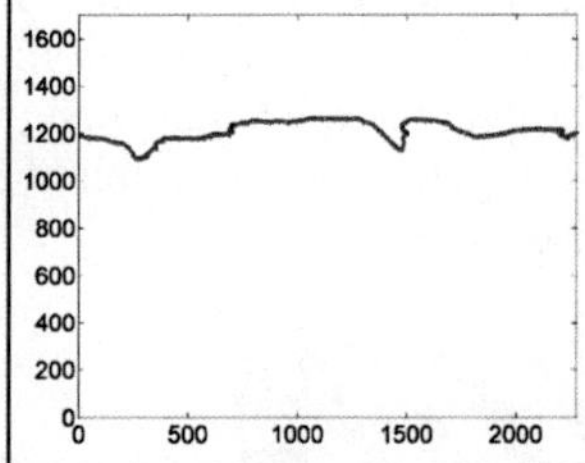	4.28 1.01 1056.4	1.54 0.47 1012.0	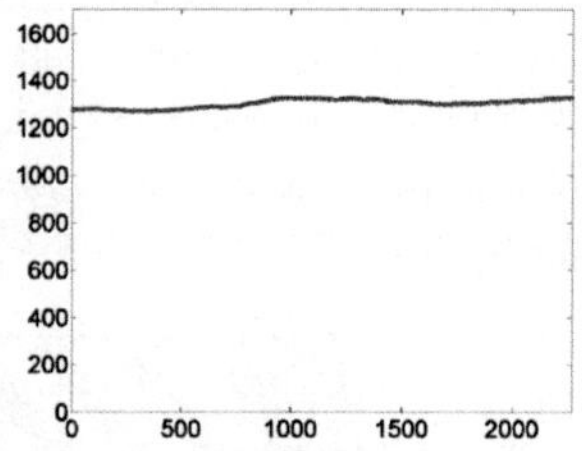
1250OC, 96 hour	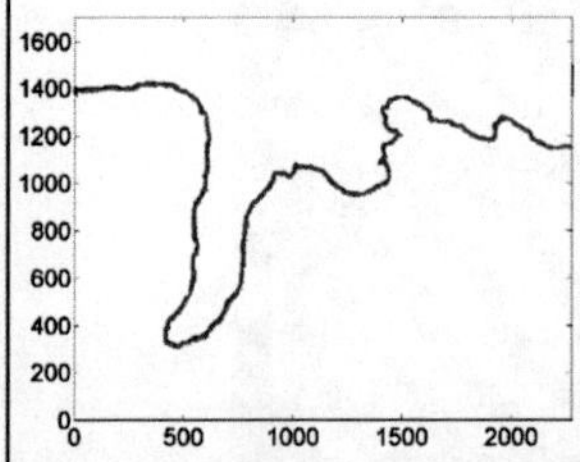	28.11 8.59 1121.6	5.52 1.39 1100.3	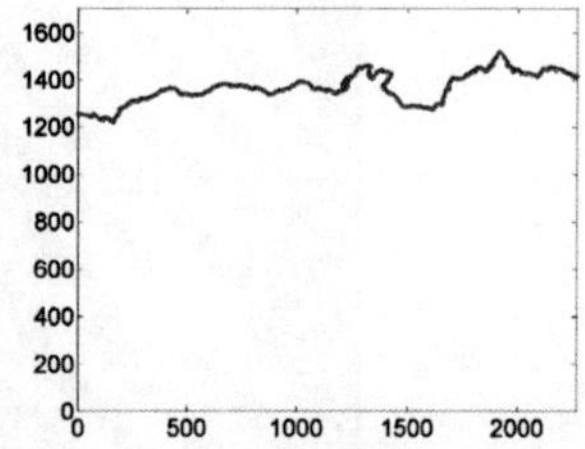
1350OC, 96 hour	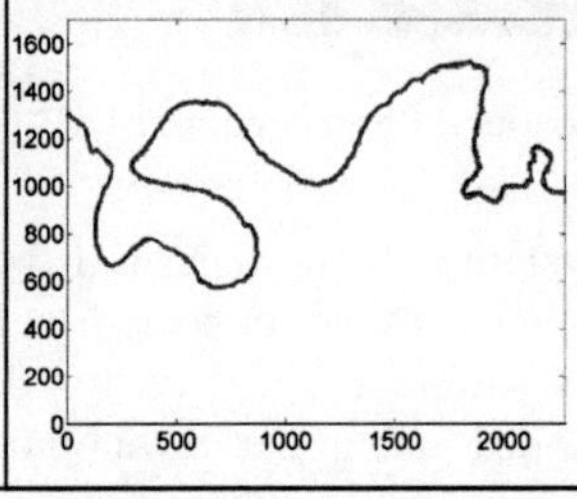	24.08 6.09 1185.4	22.64 8.09 1124.0	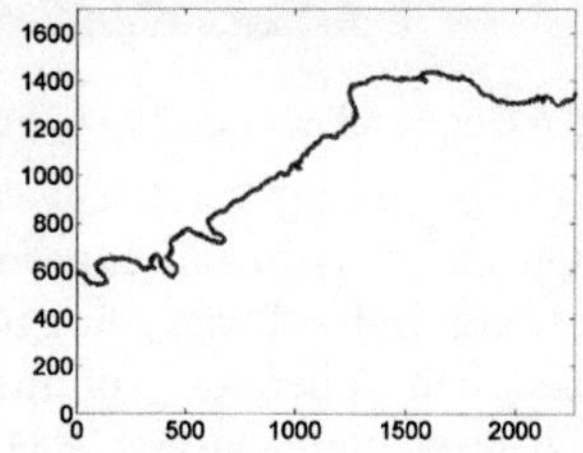

Fig. 5 Examples of dividing lines chrome-nickel steel material and iron aluminide, after static glass melt effects in different temperatures and results of analyses

5.2 Corrugation Test

The optical measurement using a zebra plate is one of many important and widely used measurements for mass production and it is used in a wide range of situations: by manufacturers of float glass as a production control; by glass processors as a quality control on the glass they buy and as a production control of products (laminated glass, thermal treated glass, etc.); and as a production control, by the final customer as a quality control on the glass they buy.

The Corrugation Test is based on the reflection of light off a glass sample sheet from a skew striped plate. The test is focused on the reflection while another type of tests are specialized on passage of light though the glass. The zebra plate is 1 x 2 m with 25 mm wide black strips at an angle of 45 degrees and it is 4 m from the sheet. An observer is 4 m from the table with the sheet and the quality of the sheet is subjectively evaluated on the basis of comparison with etalons. The quality of the sheet is classified using a rate from 1.5 to 3.5. The evaluation is conducted in a dark-room and it is performed off-line. Samples of flat glass are obtained from an on-line production process and they are cut from the whole width. Fig. 6 shows good and poor quality of a glass sheet during the corrugation test. The relatively extreme "distortion" is caused by using a small angle of observation.

Fig. 6 Good and poor quality of glass sheets with a specification of measured parts

On the basis of our experiments a measuring system, which consists of both hardware and software, has been developed. A digital camera connected to PC replaces the operator. For the corrugation analysis a unique software for data acquisition and analysis was designed and developed using the development software Matlab, based on image analysis [13, 14]. The code is compiled as an independent program named Corrugation. An observer (an operator) lays a sheet to be tested on the table in the defined place, fills in a form in the Corrugation software and starts the evaluation by pressing a button. This initiates the communication with the camera and the image is taken on request from the software.

After downloading the image from the camera to the computer, the image analysis starts. The analysis uses the developed method and its principle is based on the detection of boundary curves between the light and the dark areas of the sheet in the image (using standard image analysis). The obtained curves are evaluated by a statistical approach, a measurement of curve length and an estimation of its FD (the compass dimension mentioned above) [1, 3]. After 30 to 60 seconds the operator can read an evaluated quality on the screen of the computer.

The tests have proved, that one parameter only cannot describe all types of the corrugation. For the assessments of a flat glass quality the following three parameters of curves generated from the reflected image are important: first complexity (smoothness) of separated curves from contours (using the FD), second the range of waviness from an ideal line (using range) and third the rate of deformation (using a length of curves). All parameters are measured in pixels. The above mentioned measurements were chosen after an extensive analysis and they reached the best conformity between a subjective evaluation done by an experienced operator and the evaluated quality using computed parameters.

The system has been successfully tested for two years on the production line with the accuracy of 0.1 – 0.3 in last version and the system has been enhanced with new features. During this test, the same image format and methodology has been used. The results of over a thousand of measurements from the real production process show good potential for the application of this methodology on-line [15].

6 Conclusion

The FD is widely used in science, but industrial applications have been rather rare. Image analysis using the FD has a great potential in combination with statistical and other measurements in industry. This methodology has been used in our research for monitoring three basic data format types: digitalized photos (e.g. digital, classical photographs, images from electron microscopes), time series (e.g. analyses for monitoring and control systems) and topological one dimensional dividing lines (e.g. profiles, roughness and the dividing line of light and shadow). Modified methods using common findings of the fractal geometry are applied to each data type.

Despite of the text focusing on data from glass industry an accent was put on possibilities of using of obtained knowledge, the methodology and principles for data evaluation from industry and products in industry generally.

References

[1] Mandelbrot, B.B.: The fractal geometry of nature. W. H. Freeman and Co., New York (1982)
[2] Peitgen, H.O., Juergens, H., Saupe, D.: Chaos and Fractals: New Frontiers of Science. Springer, Heidelberg (1992)
[3] Evertsz, C.J.G., Peitgen, H.O., Voss, R.F.: Fractal Geometry and Analysis. World Scientific Publishing Co.Pte. Ltd., Singapore (1996)

[4] Levy, V.J., Lutton, E., Tricot, C.: Fractals in Engineering. Springer, Heidelberg (1997)
[5] Hotař, V.: Methodology of Industrial Data Description Using Fractal Geometry (In Czech language.) Technical university of Liberec, Liberec (2008)
[6] Philpott, D., Barnard, H.D., Bullen, P.: Some Fractal Measures Used with the Noviscam Project. In: Proceedings: ISQVPFD 2000, Faculty of Mechanical engineering, Ljubljana (2000)
[7] Hotař, V., Novotný, F.: Some Advanced Analyse for Quality Monitoring. In: Proceedings: 21st International Congress on Glass. International Commission on Glass, Strasbourg (2007)
[8] Zmeska, O., Nezadal, M., Buchnicek, M., Sedlak, O.: HarFA and HarFA: e-journal (2002), `http://www.fch.vutbr.cz/lectures/imagesci` (accessed May 30, 2012)
[9] Hotar, V., Novotny, F.: Evaluation of Surface Defects by Fractal Geometry and Statistical Analysis. Glastech Ber. Glass Sci. Technol. 77 C, 230–237 (2004)
[10] Hotař, A., Kratochvíl, P.: The corrosion resistance of iron aluminide Fe28Al3Cr0.02Ce (at%) in molten glass. Intermetallics 15, 439–441 (2007)
[11] Hotař, V., Novotný, F.: Surface Profile Evaluation by Fractal Dimension and Statistic Tools. In: Proceedings: 11th International Conference on Fracture, p. 588. CCI Centro Congressi Internazionale s.r.l., Turin (2005)
[12] Hotař, A., Kratochvíl, P., Hotař, V.: The Corrosion Resistance of Fe3Al Based Iron Aluminides in Molten Glasses. Kov Materiály - Met Mater, 247–252 (2009)
[13] Sonka, M., Hlavac, V., Boyle, R.: Image Processing, Analysis, and Machine Vision. Books/Cole Publishing Company, Pacific Grove (1998)
[14] Seul, M., O'Gorman, L., Sammon, M.J.: Practical algorithms for image analysis: description, examples, and code. Cambridge University Press, New York (2005)
[15] Hotař, V., Novotný, F., Reinischová, H.: Objective evaluation of corrugation test. Glass Technology: Eur. J. Glass Sci. Technol. Part A 52(6), 197–202 (2011)

Influence of Number of Neurons in Time Delay Recurrent Networks with Stochastic Weight Update on Backpropagation Through Time

Juraj Koščák, Rudolf Jakša , and Peter Sinčák

Department of Cybernetics and Artificial Intelligence, Technical University Košice, Slovakia
jurajkoscak@gmail.com, jaksa@neuron.tuke.sk, peter.sincak@tuke.sk

Abstract. We will examine the various modifications of backpropagation through time algorithm (BPTT) done by stochastic update in the recurrent neural networks (RCNN) including the influence of the different numbers of recurrent neurons. The general introduction involving the stochasticity into neural network was provided by Salvetti and Wilamowski in 1994 in order to improve probability of convergence and speed of convergence. The implementation is simple for arbitrary network topology. In stochastic update scenario, constant number of weights and neurons (neurons selected before starting learning phase) are randomly selected and updated. This is in contrast to classical ordered update, where always all weights or neurons are updated. Stochastic update is suitable to replace classical ordered update without any penalty on implementation complexity and with good chance without penalty on quality of convergence. We have provided first experiments with stochastic modification on backpropagation algorithm (BP) used for artificial feed-forward neural network (FFNN) in detail described in our paper [1]. We will present experiment results on simple toy-task data of time shifted and skewed signal as a verification of our implementation of different algorithm modifications.

Keywords: artificial neural networks, stochastic weight update, stochastic learning, recurrent neurons, backpropagation through time.

1 Introduction

The stochastic weight update was introduced by Salvetti and Wilamowski in 1994 [2] in order to improve probability of convergence and speed of convergence of the backpropagation algorithm. Besides the stochastic weight update, they examined another two stochastic methods: random pattern selection and randomized learning rate. On the XOR problem they demonstrated significant

I. Zelinka et al. (Eds.): Nostradamus: Mod. Meth. of Prediction, Modeling, AISC 192, pp. 133–141.
springerlink.com

improvement in the learning speed and probability of convergence for every one from these methods, especially for randomized learning rate.

The backpropagation algorithm (Werbos, 1974; Rumelhart, McClealland, 1986) is one of most used learning algorithms today. Plain vanilla implementation and momentum implementation (Silva, Almeida, 1990) [3] are the predominant implementations on feed-forward topologies. Multilayer perceptron (MLP) is most used feed-forward topology, it is a layered architecture with fully interconnected layers. The BPTT [4] is most used backpropagation variant for recurrent topologies. It is well suited algorithm for recurrent topologies with full connectivity. Although the BPTT implementation is not much more complex compare to feed-forward vanilla backpropagation, the feed-forward time-delay networks are preferably used for time-related problems. Time-delay networks are of the MLP design, so layered topology is most used topology with the backpropagation algorithm. Sparse topologies became recently popular with the recurrent echo state networks (ESN) (Jaeger, 2001) [5]. However these are usually not based on backpropagation algorithm. ESN networks are based on so called reservoir.

2 Stochastic Weight Update

Salvetti and Wilamowski did incorporate stochastic processes into backpropagation algorithm in order to improve convergence — to avoid to get stuck in a local minima on error surface [2]. Stochastic Weight Update was one from several ways how to do it. During learning, maybe it is not necessary to update all weights. Not all of them have the same influence on the network error. Some of them are more important, maybe have higher values, some of them are close to zero, maybe their influence is lower, or none. If we don't update all weights, we can also save some computation time and thus maybe speed up the learning process. We identified and evaluated several types of weight update stochasticity:

1. Ordered Update: Classic BPTT algorithm, where all weights are update in one iteration. Error signal is computed for all neurons without any preferences.
2. Stochastic Update: Modified BPTT algorithm, where not all weights are updated in the same iteration. The chance of weight update is controlled by r, where $0 \leq r \leq 100$, which represents the probability of weight update in percentage. The weight is selected for update if the generated number is $rand \leq r$. Otherwise the weight is not going to be updated. The RCNN keeps this way information from the past.
3. Stochastic Δ Weight Update: This approach combines the Stochastic Update with the stochastic selection of δweights, that are going to update. After the back propagated error signal is computed, the δ weights are going to be update. The stochastic approach is the same as with Stochastic Update. δ weight is updated if the $rand \leq r$.

4. Stochastic δ Update: Combines Stochastic Update with stochastic selection of δ Update. Stochastic factor is the same, as in the rest of stochastic modifications. The δ is selected to be update with probability. This selective updating of δ cases, that also here some old δ from $t-1, ..., t-x$ (x - not known the past, that haven't been update yet) acts in the learning process and have influence on the whole process.
5. Stochastic Complex Update: The stochastic factor of probability is the same as in previous stochastic methods. It combines Stochastic Update, Stochastic δWeight Update and Stochastic δ Update.
6. Shuffle δ Update: This approach simplifies the implementation BPTT algorithm. It is more simple because you do not have to care about layers and the order of the neurons. Simply one neuron is randomly selected from the "pool of neurons", error signal is computed and all incoming weights are updated. The "pool of neurons" represents all neurons in all layers, that contribute on δ calculation.

If we want to increase weight selection chance over 100% (all weights will be updated) we can add another loop with parameter, which control how many times this random selection should be done. If set to 1 it is equal to the loop missing.

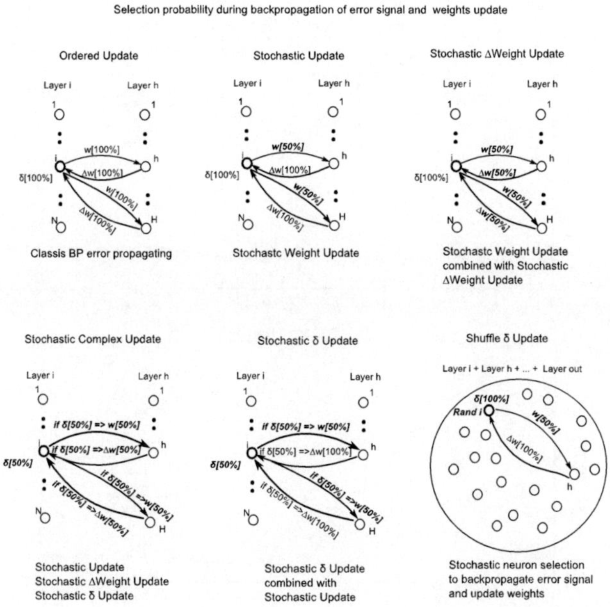

Fig. 1 Algorithms visualization. Each Stochastic approach uses 50% ratio that the selected weightδ is going to be updated. The stochasticity is implemented on the different levels of learning algorithm as they are describe on the Figure.

3 Experiments

The Time Delay RCNN (TDRCNN) is an RCNN with one of six types of implemented algorithms provided (see Figure 1) with a window consisting of time series data as the input. The experiment was set up based on the results of the Temperature Daily Profile (TDP) prediction experiment [6] using the RCNN to discover the influence of a different number of recurrent neurons in the RCNN. In the TDP prediction experiment the RCNN was getting stacked in local minima and due to a more complicated dynamic system of RCNN than the FFNN, it was difficult to find out a suitable topology and network

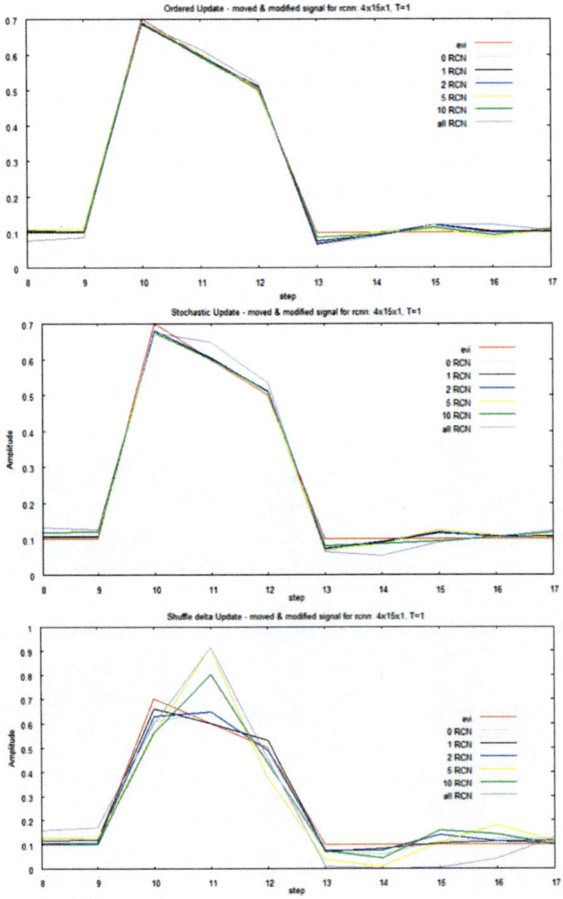

Fig. 2 The influence of recurrent neurons on the learning process using Ordered Update, Stochastic Update and Shuffle δ Update. The *evi* is expected output of RCNN.

settings. By the recurrence in the network the network becomes more simple, which provides more room for manipulation of network parameters and topology. Without recurrent neurons the RCNN becomes regular TDNN and adding recurrent neurons, the RCNN becomes more complicated. A regular RCNN is more likely get stacked in local minima with a full connection than with simplified topology. The experiment was realized with the same topology as previous experiments. The RCNN consists of 15 hidden neurons, 1 output neuron and 4 input neurons.

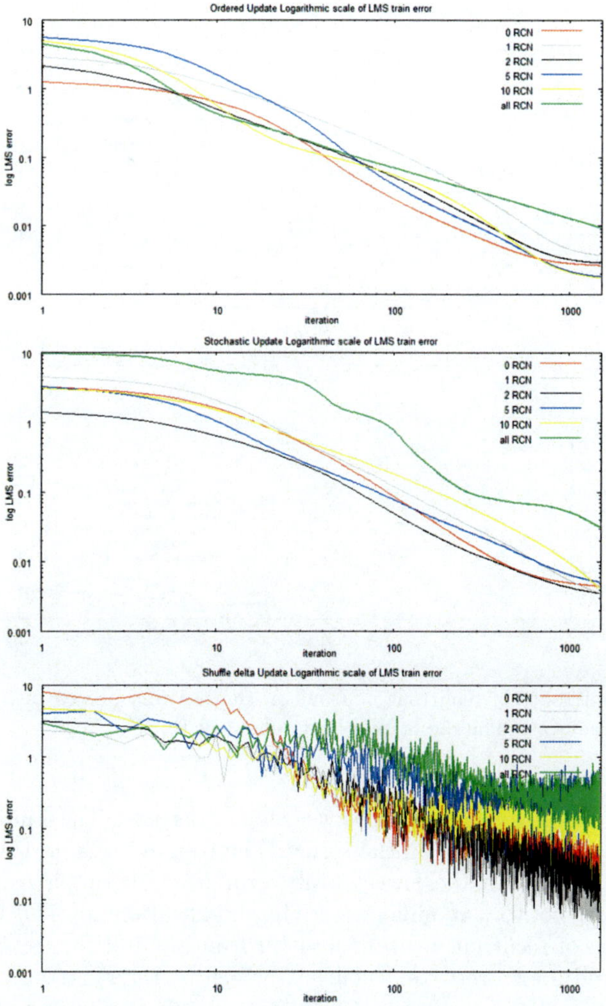

Fig. 3 LMS error of the influence of recurrent neurons on the learning process using Ordered Update. , Stochastic Update and Shuffle δ Update.

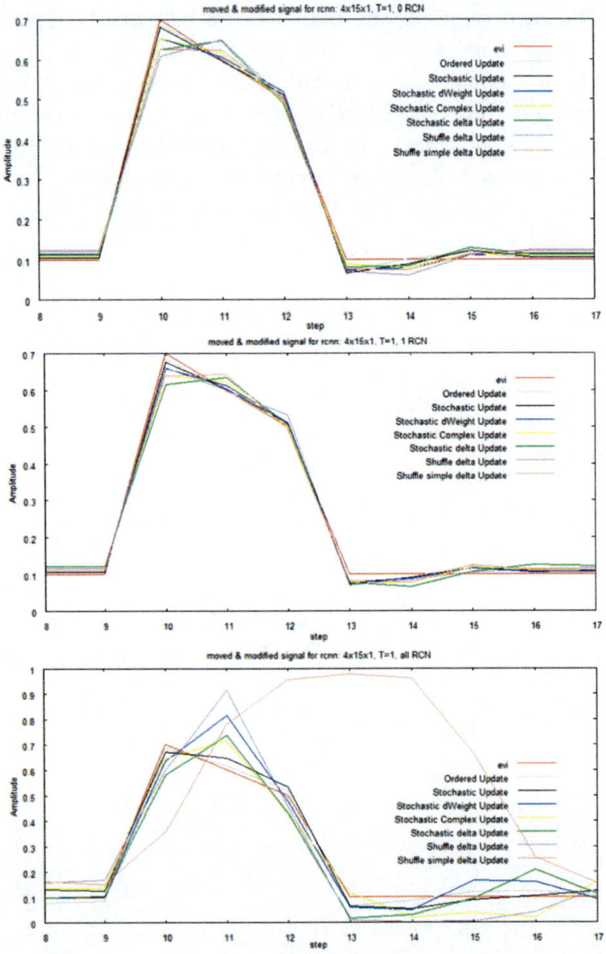

Fig. 4 The influence of recurrent neurons on the learning process using 0 , 1 and all recurrent neurons. The *evi* is expected output of RCNN.

Toy-data the time-shifted and skewed signal was used. The 4 input neurons represent the time window on the signal. The results in Table 1 were chosen as an experiment with the lowest LMS error from 10 running experiments of one type of network modification. The modifications of TDRCNN lie in the reduction of recurrent neurons number from the full connection through 10, 5, 2, 1 and 0 - no recurrent neurons, which represents a regular TDNN, and *evi* is expected output of RCNN. The recurrent neurons were chosen randomly from a pool of all neurons. Once the neuron was chosen, it was not taken into consideration for the next random neuron choice. The selected

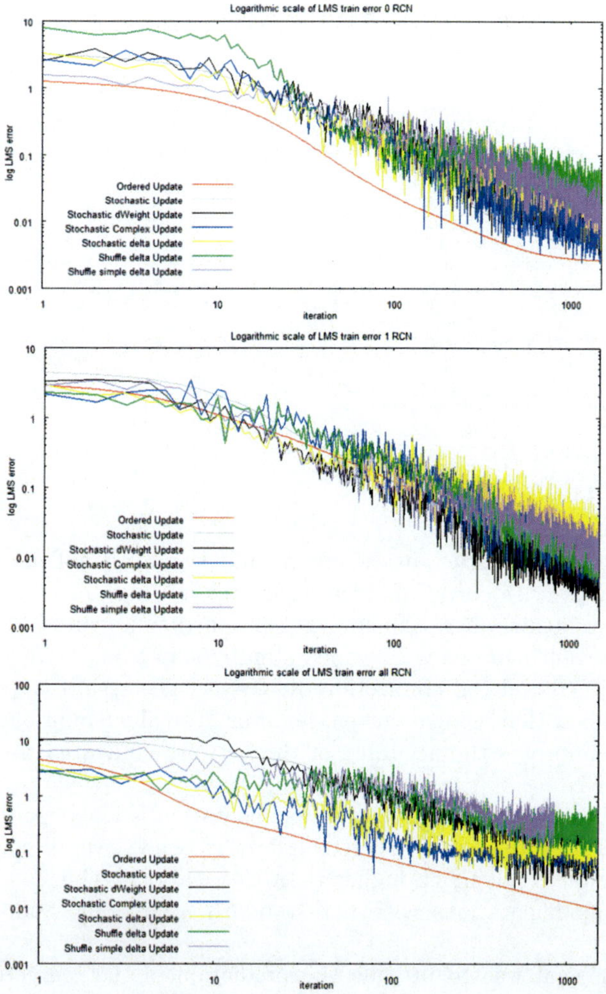

Fig. 5 LMS error of the influence of recurrent neurons on the learning process using 0 , 1 and all recurrent neurons

neurons were taken into consideration in the learning phase - the adaptation of weights.

The results are in Table 1. Provided figures illustrate two kinds of comparisons: the comparison of selected neurons in one type of algorithm (Figure 2) with the logarithmic LMS error in Figure 3, compared to the selected number of neurons through all types of algorithm modifications (Figure 4) with the logarithmic LMS error in Figure 5. More details about this experiment and related experiments can be found in doctoral dissertation proposal [7].

Table 1 Influence of different numbers of recurrent neurons in Time delay RCNN (TDRCNN)

Method	Ordered Update	Stoch. Update	Stoch. ΔWeight Update	Stoch. Complex Update	Stoch. δUpdate	Shuffle δUpdate
Selection Ratio	100%	50%	50%	50%	50%	50%
γ	0.03	0.03	0.03	0.03	0.03	0.03
λ	0.65	0.65	0.65	0.65	0.65	0.65
α momentum	0.0	0.0	0.0	0.0	0.0	0.0
Unfolding history T	1	1	1	1	1	1
Topology	4-15-1	4-15-1	4-15-1	4-15-1	41-15-1	4-15-1
Adaptation in one cycle	1	1	1	1	1	1
Iterations	1500	1500	1500	1500	1500	1500
LMS 0 RCN*	$19{,}51\times10^{-4}$	$33{,}0\times10^{-4}$	$41{,}29\times10^{-4}$	$34{,}22\times10^{-4}$	$59{,}52\times10^{-4}$	$293{,}9\times10^{-4}$
LMS 1 RCN*	$27{,}45\times10^{-4}$	$25{,}43\times10^{-4}$	$26{,}37\times10^{-4}$	$65{,}21\times10^{-4}$	$41{,}47\times10^{-4}$	$57{,}12\times10^{-4}$
LMS 2 RCN*	$21{,}6\times10^{-4}$	$26{,}38\times10^{-4}$	64.4×10^{-4}	$35{,}85\times10^{-4}$	$48{,}53\times10^{-4}$	$115{,}9\times10^{-4}$
LMS 5 RCN*	$13{,}4\times10^{-4}$	$37{,}68\times10^{-4}$	$66{,}95\times10^{-4}$	$14{,}72\times10^{-4}$	$25{,}69\times10^{-4}$	1321×10^{-4}
LMS 10 RCN*	$13{,}39\times10^{-4}$	$28{,}77\times10^{-4}$	$116{,}5\times10^{-4}$	$575{,}1\times10^{-4}$	$48{,}49\times10^{-4}$	$619{,}4\times10^{-4}$
LMS all RCN*	$73{,}29\times10^{-4}$	$262{,}9\times10^{-4}$	$539{,}7\times10^{-4}$	$337{,}1\times10^{-4}$	$443{,}7\times10^{-4}$	2527×10^{-4}

RCN* - number of chosen recurrent neurons during the learning process.

4 Conclusion

The RCNN can easily stuck in the local minima and it is difficult to find out proper topology with corresponding parameter settings. That was reason to provide experiments where the RCNN contained time window as an input and the recurrent neurons were added step by step from 0 - regular TDNN through 1,2,5,10 and full connection of RCNN. The results, represented in the Table 1 show that using recurrent neurons in small amount of all neurons, 13% - 33%, improves the accuracy of the network compared to the TDNN and full RCNN.

Complexity of stochastic update implementation is less dependent on the network topology than the Ordered update is, while overall complexity of implementation is not much higher than classical one. This can be utilized in BPTT implementations with non-standard, sparse, or random network topologies.

Our main goal was to provide the prediction of the temperature daily profile using the implemented BPTT stochastic update, mainly the Shuffle Update, due to the ability of RCNN to use the history of previous steps stored inside. We already prepared the testing dataset for temperature prediction, that consist of the measured temperature from the year 2000 till the end of February 2011. The data are measured each half hour, which provides us wide range of options to create suitable training and testing data for RCNN. This wide range also allows the RCNN to create internal model of daily temperature profile.

Acknowledgements. This research was supported by the National Research and Development Project Grant 1/0667/12 Incremental Learning Methods for Intelligent Systems 2012-2015 and Center of Competence of knowledge technologies for product system innovation in industry and service, with ITMS project number: 26220220155 for years 20012-2015.

References

1. Koščák, J., Jakša, R., Sinčák, P.: Stochastic weight update in the backpropagation algorithm on feed-forward neural networks. In: The 2010 International Joint Conference on Neural Networks, IJCNN (2010)
2. Salvetti, A., Wilamowski, B.M.: Introducing stochastic processes within the backpropagation algorithm for improved convergence. In: Dagli, C., Fernández, B., Ghosh, J., Kumara, R. (eds.) ANNIE 1994 - Artificial Neural Networks in Engineering (Intelligent Engineering Systems Through Artificial Neural Networks), vol. 4, pp. 205–209. ASME PRESS, New York (1994)
3. Silva, F., Almeida, L.: Acceleration Technique for the Backpropagation Algorithm. In: Almeida, L.B., Wellekens, C. (eds.) EURASIP 1990. LNCS, vol. 412, pp. 110–119. Springer, Heidelberg (1990)
4. Werbos, P.: The Roots of Backpropagation: From Ordered Derivation to Neural Networks and Political Forecasting. John Wiley and Sons, Inc. (February 1994), ISBN 0-471-59897-6
5. Jaeger, H.: Gmd report 148, the "echo state" approach to analysing and training recurrent neural networks. GMD - German National Research Institute for Computer Science, Tech. Rep. (2001)
6. Koščak, J., Jakša, R., Sincák, P.: Prediction of temperature daily profile by stochastic update of backpropagation through time algorithm. In: ISF 2011, pp. 1–8 (2011)
7. Koščak, J.: Stochastic weight selection for backpropagation through time learning (theoretical study of stochastic neural networks learning). Subscribed Doctoral Dissertation Proposal (2012)

Multiple Model Predictive Control of a Styrene Polymerization Process

Jakub Novák and Petr Chalupa

Tomas Bata University in Zlin, Faculty of Applied Informatics,
T.G. Masaryka 5555, 760 01 Zlin, Czech Republic
{jnovak,chalupa}@fai.utb.cz

Abstract. In this paper, multiple piecewise linearizations of a nonlinear process in different operating points are used within a Kalman filter bank which computes the conditional probabilities of various hypotheses that are modeled by the filters. State estimates provided by the Kalman Filters and local model parameters are weighted using conditional probabilities and then used within the predictive control framework. The proposed strategy is tested on the complex model of styrene polymerization process.

1 Introduction

Processes in the chemical industry usually exhibit nonlinear behavior. The nonlinearities arise from the dynamics in chemical reactions, thermodynamic relationships, etc. Such processes are relatively complex and difficult to control. Several methods have been mainly used for predictive control of these processes. The first one is based on the direct use of nonlinear models [1] and involves the online solution of a higher order nonlinear optimization problem with constraints, which is usually computationally expensive and may even be unable to guarantee a feasible solution for real-time control. Another method is to use local linearization approach for representing a nonlinear plant by applying an off-line estimated globally nonlinear and locally linear model to solve a Quadratic Programming (QP) problem online in order to obtain optimal control. A global process model is formed by blending of a number of local models which have been identified over the operating range of the process. Multiple model networks have been more or less independently developed in different disciplines like neural networks, fuzzy logic, statistics and artificial intelligence with different names such as local model networks, Takagi -Sugeno fuzzy models or neuro-fuzzy models [2],[3].

In Continuous Stirred Tank Reactor (CSTR) the product properties are rarely measured accurately on-line, thus much effort has been put to the development of model-based estimation and control techniques. Multivariable nonlinear quadratic dynamic matrix control strategy was applied in [4] to control the polymerization

I. Zelinka et al. (Eds.): Nostradamus: Mod. Meth. of Prediction, Modeling, AISC 192, pp. 143–152.
springerlink.com

process. Both parameter and states of the process are estimated using the Extended Kalman Filters (EKF) and multivariable nonlinear quadratic dynamic matrix control is applied for control of number molecular weight (NAMW). In [5] linear parameter varying (LPV) model which interpolates second order output error models is used to represent the polymerization process. Maner et al. in [6] presented an MPC algorithm based on second-order Volterra models where the model parameters are obtained by discretizing the bilinear Taylor series approximation of the fundamental model. In the paper predictive control is based on a mixture distribution of state-space models [7] with probabilities estimated with a Kalman filter bank.

2 Multiple Model Estimation

The primary feature of Multiple Model Adaptive Estimation (MMAE) is a bank of Kalman filters operating in parallel, using vectors of measurements y and control commands u as their input. Each Kalman filter has the same structure based on the linearized description of the process (Fig. 1). The model is assumed to be linear and of the form:

$$\begin{aligned} \boldsymbol{x}(k+1) &= \boldsymbol{A}\boldsymbol{x}(k)+\boldsymbol{B}\boldsymbol{u}(k) \\ \boldsymbol{y}(k) &= \boldsymbol{C}\boldsymbol{x}(k)+\boldsymbol{D}\boldsymbol{u}(k) \end{aligned} \tag{1}$$

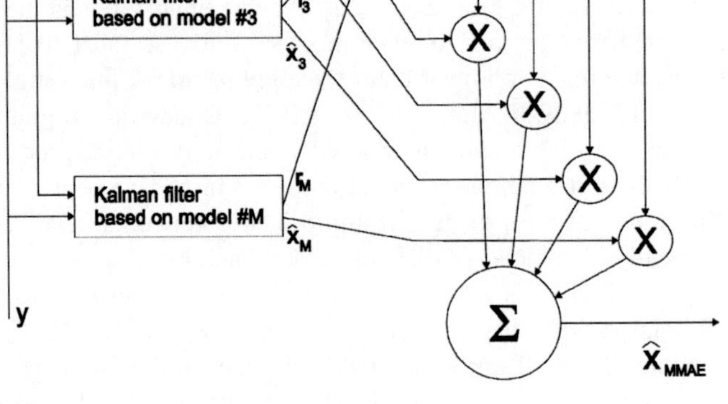

Fig. 1 Multiple Model Adaptive Estimation scheme

Output of each KF is weighted by its corresponding conditional probability based on the measurement history. At every sampling period, each of these Kalman filters is producing its estimate of the state and residual. The idea is that the model with well-behaved residuals contains the parameters that best matches true parameters of the system. Testing the hypothesis which model is the correct one is evaluated in the hypothesis testing block. The initial probability of each hypothesis being correct is distributed evenly:

$$\alpha_i(0) = 1/M \tag{2}$$

The output prediction is given by the mixture of conditional probability density functions:

$$p(y(k)|u(k)) = \sum_{i=1}^{M} \alpha_i p_i(y(k)|u(k)) \tag{3}$$

where α_i are the probabilities of each model being the correct one and normalized to 1.

$$\alpha_i = p(m(t) = i), \sum_{i=1}^{M} \alpha_i = 1 \tag{4}$$

The predictive conditional probability density functions p_i are given by the state-space model as:

$$p_i(y(k)|u(k)) = p_i(y(k)|x(k))p_i(x(k)|u(k)) \tag{5}$$

where $p_i(x(k)|u(k))$ is the state-estimate provided by the i-th Kalman filter. One step of the Kalman filter can be written as:

$$\begin{aligned} K_i(k) &= A_i P_i(k) C^T \left(I + C_i P_i(k) C_i\right)^{-1} \\ P(k+1) &= A_i P_i(k) A_i + Q_i - K_i \left(C_i P_i(k) C_i + I\right) K_i^T \\ \hat{x}(k+1) &= A_i \hat{x}(k) + B_i u(k) + E_i + \\ &\quad + K_i(y - C_i \hat{x}(k) - D_i u(k) - F_i) \end{aligned} \tag{6}$$

The conditional probability density function for known measurement noise has normal distribution [7] and can be computed using:

$$\begin{aligned} & p(y(k)|x(k), \sigma_{ei}^2) = N(y, (1 + C_i P_i C_i^T)\sigma_{ei}^2) \\ & \Sigma = (1 + C_i P_i C_i^T)\sigma_{ei}^2 \\ & p(y(k)|x(k), \sigma_{ei}^2) = \frac{1}{(2\pi)^{m/2}\sqrt{|\Sigma|}} \exp\left(-\frac{1}{2}\left(\frac{r_i(k)^T r_i(k)}{\Sigma}\right)\right) \\ & r_i = y(k) - \hat{y}(k) \end{aligned} \tag{7}$$

The estimate of variance can be updated with exponential forgetting with factor φ as:

$$\sigma_{ei}^2(k+1) = \frac{S_i^2(k+1)}{\upsilon(k+1)} \tag{8}$$

where variables S_i^2 and $\upsilon(k+1)$ are updated at each step:

$$\begin{aligned} S_i^2(k+1) &= \varphi\left(S_i^2(k) + \frac{r^T r}{1 + C_i P_i(k+1) C_i}\right) \\ \upsilon(k+1) &= \varphi(\upsilon(k)+1) \end{aligned} \tag{9}$$

3 Description of the Styrene Polymerization Process

The proposed control strategy is tested on the model of the free radical solution polymerization of styrene in a jacketed CSTR. The reactor is controlled around the low conversion stable steady-state point. Two controlled variables are considered: number average molecular weight (y_1) and reactor temperature (y_2). The initiator flow (u_1) and the cooling flow rate (u_2) were selected as manipulated variables. This work uses a mathematical model of the styrene polymerization process shown in Fig. 2. The nonlinear process is modeled by the following differential equations:

$$\begin{aligned} \frac{d[I]}{dt} &= \frac{\left(Q_i[I_f] - Q_t[I]\right)}{V} - k_d[I] \\ \frac{d[M]}{dt} &= \frac{\left(Q_m[M_f] - Q_t[M]\right)}{V} - k_d[M][P] \\ \frac{dT}{dt} &= \frac{Q_t\left(T_f - T\right)}{V} + \frac{\left(-H_r\right)}{\rho C_p} k_p[M][P] - \frac{hA}{\rho C_p V}\left(T - T_c\right) \\ \frac{dT_c}{dt} &= \frac{Q_c\left(T_{cf} - T_c\right)}{V_c} + \frac{hA}{\rho C_{pc} V_c}\left(T - T_c\right) \\ \frac{dD_0}{dt} &= 0.5 k_t [P]^2 - \frac{Q_t D_0}{V} \\ \frac{dD_1}{dt} &= M_m k_p [M][P] - \frac{Q_t D_1}{V} \\ u_1 &= Q_i, u_2 = Q_c, y_1 = \frac{D_1}{D_0}, y_2 = T \end{aligned} \tag{10}$$

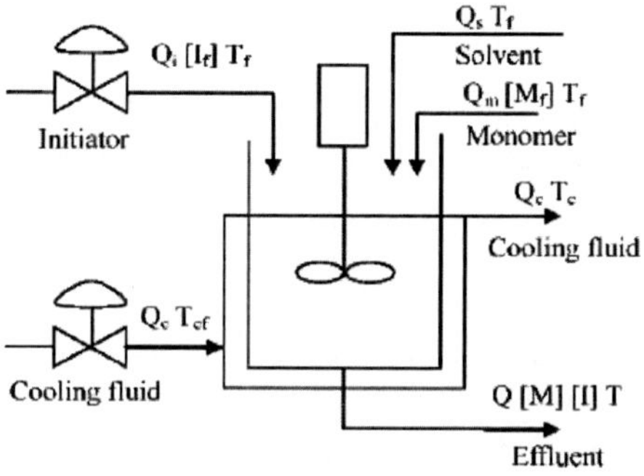

Fig. 2 CSTR reactor

where the rate constants are given as:

$$k_i = A_i \exp\left(-E_i / T\right), i = d, p, t$$
$$\left[P\right] = \sqrt{\frac{2fk_d\left[I\right]}{k_t}}, Q_t = Q_i + Q_s + Q_m \tag{11}$$

The goal of the control system is to drive the polymerization system to a new state to produce polymers with different number average molecular weights while keeping the temperature at its setpoint. The kinetic parameters and thermodynamic parameters can be found in [5].

4 State-Space Predictive Control

The state-space model based predictive control is based on the time-invariant model:

$$x(k+1) = \boldsymbol{A}x(k) + \boldsymbol{B}u(k)$$
$$y(k) = \boldsymbol{C}x(k) + \boldsymbol{D}u(k) \tag{12}$$

The model of the process is obtained at every sampling interval and its parameters are used for the entire prediction horizon *Hp*. The discrete model contains also the affine part that results from linearization around non-zero steady-state:

$$x(k+1) = \boldsymbol{A}x(k) + \boldsymbol{B}u(k) + \boldsymbol{E}$$
$$y(k) = \boldsymbol{C}x(k) + \boldsymbol{D}u(k) + \boldsymbol{F}$$
$$\boldsymbol{A} = \sum_{i=1}^{M} p_i \boldsymbol{A}_i, \boldsymbol{B} = \sum_{i=1}^{M} p_i \boldsymbol{B}_i, \boldsymbol{E} = \sum_{i=1}^{M} p_i \boldsymbol{E}_i \tag{13}$$
$$\boldsymbol{C} = \sum_{i=1}^{M} p_i \boldsymbol{C}_i, \boldsymbol{D} = \sum_{i=1}^{M} p_i \boldsymbol{D}_i, \boldsymbol{F} = \sum_{i=1}^{M} p_i \boldsymbol{F}_i$$

The H_p -step ahead output prediction can be deduced:

$$\begin{aligned}
&\hat{Y} = \Phi_{yx}x(k) + \Phi_{yu}U + \Phi_{y0} \\
&\hat{Y} = \begin{bmatrix} \hat{y}(k+1) \\ \hat{y}(k+2) \\ \vdots \\ \hat{y}(k+H_p) \end{bmatrix}, U = \begin{bmatrix} u(k) \\ u(k+1) \\ \vdots \\ u(k+H_p) \end{bmatrix}, \Phi_{yx} = \begin{bmatrix} CA \\ CA^2 \\ \vdots \\ CA^{H_p} \end{bmatrix} \\
&\Phi_{yu} = \begin{bmatrix} CB & D & 0 & 0 & 0 \\ CAB & CB & D & 0 & 0 \\ \vdots & \vdots & \ddots & \ddots & \vdots \\ CA^{H_p-1}B & CA^{H_p-2}B & \cdots & CB & D \end{bmatrix} \\
&\Phi_{y0} = \begin{bmatrix} CE+F \\ CAE+CE+F \\ \vdots \\ CA^{H_p-1}E+\ldots+CE+F \end{bmatrix}
\end{aligned} \tag{14}$$

The computation of a control law of MPC is based on minimization of the following criterion:

$$J_{MPC} = (\hat{y} - w)^T Q (\hat{y} - w) + \Delta u R \Delta u \tag{15}$$

where $\hat{y}(k+j|k)$ is a j steps ahead prediction of the system, $w(k+j)$ is a future reference trajectory and Q, R are positive definite weighting matrices. The minimization of the criterion can be transformed into a quadratic programming problem:

$$J_{MPC} = u^T H u + f u \tag{16}$$

where matrix H and vector f are derived from model parameters given by (14)equation reference goes here. The quadratic problem is usually solved numerically. As formulated, the nonlinear model predictive controller will exhibit steady – state offset in the presence of plant/model mismatch due to a lack of integral action. To introduce an integral action to remove steady-state error an integrator state must be added to the system:

$$\begin{aligned}
&v(k) = v(k-1) + (w(k) - y(k)) \\
&\xi(k) = \begin{bmatrix} x(k) \\ v(k) \end{bmatrix}
\end{aligned} \tag{17}$$

System matrices (12) are updated as follows:

$$\begin{aligned}
&\xi(k+1) = \tilde{A}\xi(k+1) + \tilde{B}u + \tilde{E} \\
&y(k) = \tilde{C}\xi(k+1) + \tilde{D}u + \tilde{F} \\
&\tilde{A} = \begin{bmatrix} A & 0 \\ -C & I \end{bmatrix}, \tilde{B} = \begin{bmatrix} B \\ -D \end{bmatrix}, \tilde{E} = \begin{bmatrix} E \\ W(k) - F \end{bmatrix} \\
&\tilde{C} = [C \quad 0], \tilde{D} = D, \tilde{F} = F
\end{aligned} \tag{18}$$

In order to minimize the augmented state the cost criterion for MPC (22) is transferred to:

$$J_{MPC} = \left(\widehat{Y} - W\right)^T Q\left(\widehat{Y} - W\right) + \Delta U^T R \Delta U + \widehat{X}^T S \widehat{X} \tag{19}$$

where

$$\widehat{X} = \Phi_{xx} x(k) + \Phi_{xu} U + \Phi_{x0} \tag{20}$$

$$\Phi_{xx} = \begin{bmatrix} A \\ A^2 \\ \vdots \\ A^{H_p} \end{bmatrix}, \Phi_{x0} = \begin{bmatrix} E \\ AE + E \\ \vdots \\ A^{H_p-1}E + \ldots + E \end{bmatrix}$$
$$\Phi_{xu} = \begin{bmatrix} B & 0 & 0 & 0 & 0 \\ AB & B & 0 & 0 & 0 \\ \vdots & \vdots & \ddots & \ddots & \vdots \\ A^{H_p-1}B & A^{H_p-2}B & \cdots & B & 0 \end{bmatrix} \tag{21}$$

5 Implementation

Multiple-model predictive control described in Section 4 is applied for control of the NAMW and temperature of CSTR. Three linear models were obtained through linearization in steady-state operating points for input signals $u_1 = 72{,}108$ and 144 l/h. The sampling period was set to 1h due to the dynamics of the process. In practice the NAMW is measured with sampling period of 30 minutes and the time delay of the measurement is also 30 minutes. The Kalman filter bank with these local models was constructed with initial condition $x = \begin{bmatrix} 0.0784 \\ 3.3516 \\ 324.9078 \\ 305.6496 \\ 0.0004 \\ 19.4179 \end{bmatrix}$ which corresponds to the third operating point. The initial estimate covariance matrix was chosen to be:

$$P = 10^4 I \tag{22}$$

and saturation constraints in the manipulated variables are imposed to take into account the minimum/maximum aperture of the valve regulating the flow rates. The prediction horizon was set to 20 samples as a result of using different values and comparing control performances. The weighting matrices Q,R,S associated with the error from set-point, control output increment and integrator gain were set to:

$$Q = \begin{bmatrix} 0.0001 & 0 \\ 0 & 100 \end{bmatrix}, R = \begin{bmatrix} 10 & 0 \\ 0 & 10 \end{bmatrix}, S(7,7) = 0.01, S(8,8) = 100 \tag{23}$$

The results of a closed-loop simulation for a set-point changes from 51 494 to 64 463 g/mol which corresponds to operating point #3 and #2 are shown in Fig. 3, where they are compared with the results obtained from the linear MPC controller and the nonlinear MPC controllers without the integral behavior. At t= 150h an disturbance is introduced by increasing the value of monomer flow-rate by 10%. model. The linear model is only valid in the vicinity of the operating point #3 and as the system is driven out this point the steady-state error increases. If both models of the system at the operating points are available and the current model is identified by Kalman filters correctly (Fig. 4) then the zero steady state-error can be obtained at the operating point #2. However, the system is not able to cope with the disturbance that is not included in the model. Zero off-set is reached when multiple models also contain augmented state.

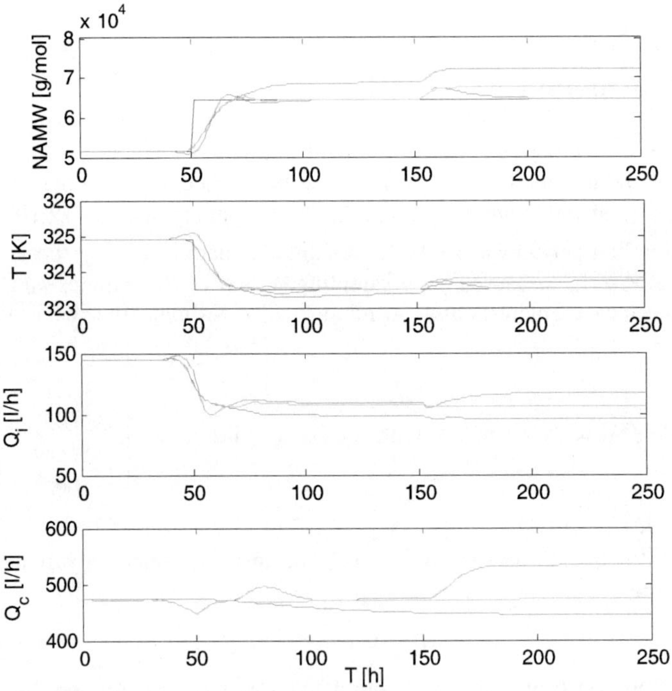

Fig. 3 Closed loop simulation for transition control (black – reference, blue – linear MPC with single model, green - Multiple model MPC, red - Multiple model MPC with integrator state)

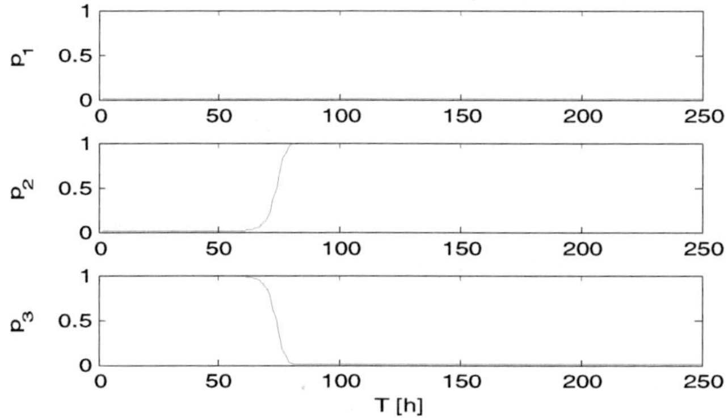

Fig. 4 Probability of each of the model during the closed loop simulation

6 Conclusion

In the paper, a multiple-model predictive control methodology is applied to a styrene polymerization system. The correct model at the current sampling point is estimated using the residuals provided by a bank of Kalman filters. The obtained state is in the form of a mixture of states provided by the filters. The parameter of the linearized model and the states are then used within the predictive control approach for prediction of the future plant behavior. To remove steady-state error for model/plant mismatch an augmented state is added to the state space description. The simulations show that this type of predictive control can be applied to the styrene polymerization system effectively.

Acknowledgments. This work was supported by European Regional Development Fund under the project CEBIA-Tech No. CZ.1.05/2.1.00/03.0089.

References

[1] Grune, L., Pannek, J.: Nonlinear Model Predictive Control. Springer, London (2011)

[2] Murray-Smith, R., Johansen, T.A.: Multiple Model Approaches to Modelling and Control. Taylor and Francis, London (1997)

[3] Gregorcic, G., Lightbody, G.: Nonlinear system identification: From multiple-model networks to Gaussian processes. Engineering Applications of Artificial Intelligence 21, 1035–1055 (2008)

[4] Schley, M., Prasad, V., Russo, L.P., Bequette, B.W.: Nonlinear Model Predictive Control of a Styrene Polymerization Reactor. Progress in Systems and Control Theory 26, 403–417 (2000)

[5] Xu, Z., Zhao, J., Qian, J.: Nonlinear MPC using identified LPV model. Industrial & Engineering Chemistry Research 48, 3043–3051 (2009)

[6] Maner, B.R., Doyle, F.J., Ogunnaike, B.A., Pearson, R.K.: Nonlinear Model Predictive Control of a Simulated Multivariable Polymerization Reactor using second-order Volterra Models. Automatica 32, 1285–1301 (1996)
[7] Pekar, J., Havlena, V.: Control of CSTR using model predictive controller based on mixture distribution. In: Proceedings of the 6th IFAC Symposium on Nonlinear Control Systems (2004)
[8] Rupp, D., Ducard, G., Shafai, E., Geering, H.P.: Extended Multiple Model Adaptive Estimation for the Detection of Sensor and Actuator Faults. In: Proceedings of the IEEE Conference on Decision and Control, pp. 3079–3084 (2005)

Impact of Various Chaotic Maps on the Performance of Chaos Enhanced PSO Algorithm with Inertia Weight – An Initial Study

Michal Pluhacek, Roman Senkerik, and Ivan Zelinka

Tomas Bata University in Zlin, Faculty of Applied Informatics, T.G. Masaryka 5555,
760 01 Zlin, Czech Republic.
{pluhacek,senkerik,zelinka}@fai.utb.cz

Abstract. In this paper, it is proposed the utilization of three different chaotic maps based number generators to enhance the performance of PSO algorithm with inertia weight. This initial study presents results of performance testing on several test functions. Results obtained for different chaotic generators are compared and briefly analyzed.

1 Introduction

PSO algorithm is one of evolutionary algorithms, which are a class of soft computing methods inspired by nature. In past years and decades, various evolutionary algorithms were designed, modified and used with great results in many areas of optimization [1-5]. More recently, some studies hinted, that using chaotic number generators may improve the performance of evolutionary optimization algorithms on such tasks as PID controller design [6]

This initial study investigates on the chaos enhanced PSO algorithm with inertia weight, specifically on the influence of chaotic map that is used as the number generator to the performance of such algorithm. Three different chaotic maps are used. Performance tests were done on four different test functions.

2 Particle Swarm Optimization Algorithm

PSO (Particle swarm optimization) algorithm is based on the natural behavior of birds and fishes and was firstly introduced by R. Eberhart and J. Kennedy in 1995 [1-3]. As an alternative to genetic algorithms [4] and differential evolution [5], PSO proved itself to be able to find better solutions for many optimization problems. Term "swarm intelligence" [2,3] refers to the capability of particle swarms

I. Zelinka et al. (Eds.): Nostradamus: Mod. Meth. of Prediction, Modeling, AISC 192, pp. 153–166.
springerlink.com

to exhibit surprising intelligent behavior assuming that some form of communication (even very primitive) can occur among the swarm particles (individuals).

Basic PSO algorithm disadvantage is the rapid acceleration of particles which causes abandoning the defined area of interest. In each generation, a new location of a particle is calculated based on its previous location and velocity (or "velocity vector"). For this reason, several modifications of PSO were introduced to handle with this problem. Main principles of PSO algorithm and its modifications are well described in [1-3].

Within this research, chaos driven PSO strategy with inertia weight was used. The selection of inertia weight modification of PSO was based on numerous previous experiments. Default values of all PSO parameters were chosen according to the recommendations given in [2,3]. Inertia weight is designed to influence the velocity of each particle differently over the time [8]. In the beginning of the optimization process, the influence of inertia weight factor w is minimal. As the optimization continues, the value of $w$is decreasing, thus the velocity of each particle is decreasing, since w is always the number < 1 and it multiplies previous velocity of particle in the process of new velocity value calculation. Inertia weight modification PSO strategy has two control parameters w_{start} and w_{end}. New w for each generation is then given by Eq. 1, where i stand for current generation number and n for total number of generations.

$$w = w_{start} - \frac{\left(\left(w_{start} - w_{end}\right) * i\right)}{n} \tag{1}$$

Chaos driven number generator is used in the main PSO definition (Eq. 2) which determines a new "velocity", thus the position of each particle in the next generation (or migration cycle).

$$\begin{aligned} v(t+1) = v(t) + c_1 \cdot Rand \cdot (pBest - x(t)) + \\ c_2 \cdot Rand \cdot (gBest - x(t)) \end{aligned} \tag{2}$$

Where:
$v(t+1)$ – New velocity of particle.
$v(t)$ – Current velocity of particle.
c_1,c_2 – Priority factors.
pBest – Best solution found by particle.
gBest – Best solution found in population.
$x(t)$ – Current position of particle.
Rand – Random number, from the interval <0,1>. Within Chaos PSO algorithm, the basic inbuilt computer (simulation software) random generator is replaced with chaotic generator (Dissipative standard map or Lozi map or Arnold´s cat map).

New position of a particle is then given by Eq. 3, where x(t+1) represents the new position:

$$x(t+1) = x(t) + v(t+1) \quad (3)$$

3 Dissipative Standard Map

The Dissipative Standard map is a two-dimensional chaotic map. The parameters used in this work are $b = 0.1$ and $k = 8.8$ as suggested in [9]. The Dissipative standard map is given in Fig. 1. The map equations are given in Eq. 4 and 5.

$$X_{n+1} = X_n + Y_{n+1} (\bmod 2\pi) \quad (4)$$

$$Y_{n+1} = bY_n + k \sin X_n (\bmod 2\pi) \quad (5)$$

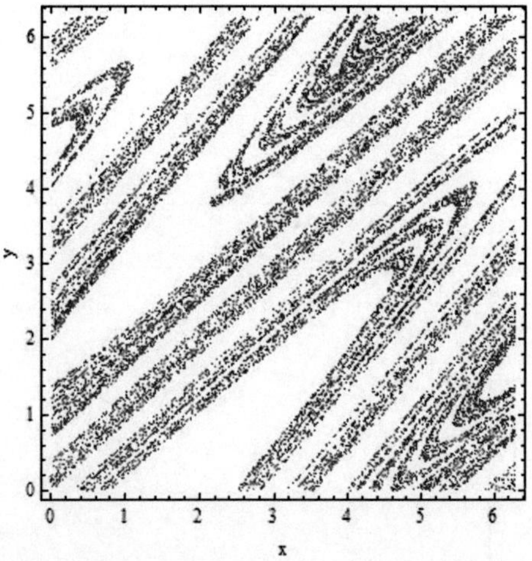

Fig. 1 Dissipative standard map

4 Lozi Map

The Lozi map is a simple discrete two-dimensional chaotic map. The Lozi map is depicted in Fig. 2. The map equations are given in Eq. 11 and 12. The parameters used in this work are: $a = 1.7$ and $b = 0.5$ as suggested in [9].

$$X_{n+1} = 1 - a|X_n| + bY_n \quad (6)$$

$$Y_{n+1} = X_n \quad (7)$$

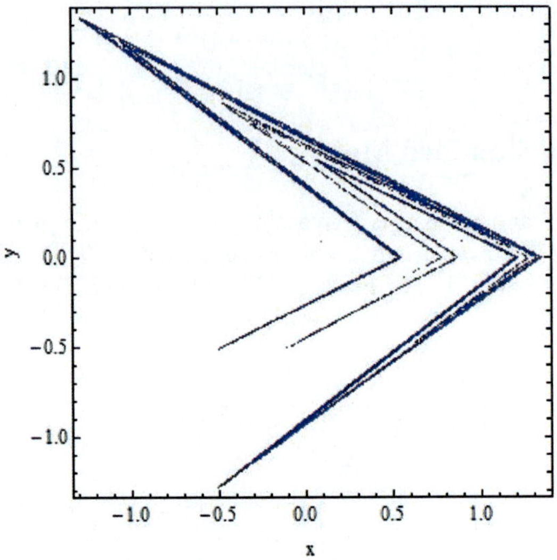

Fig. 2 Lozi map

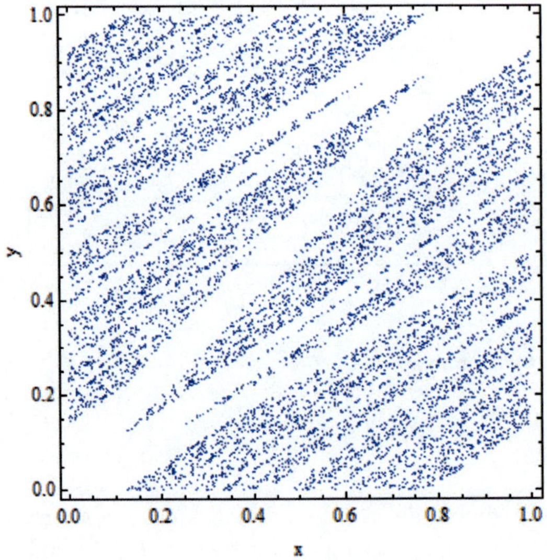

Fig. 3 Arnold´s cat map

5 Arnold's Cat Map

The Arnold's Cat map is a simple two dimensional discrete system that stretches and folds points (x, y) to (x+y , x+2y) mod 1 in phase space. The map equations are given in Eq 3 and 4. This map uses parameter k = 2.0 as suggested in [9]. The x-y plot ofthis map is depicted in Fig. 3.

$$X_{n+1} = X_n + Y_n (\mathrm{mod}\, 1) \tag{8}$$

$$Y_{n+1} = X_n + kY_n (\mathrm{mod}\, 1) \tag{9}$$

6 Test Functions

Four test functions were used in this study.

The First De Jong`s function is given by Eq. 10.

$$f(x) = \sum_{i=1}^{dim} x_i^2 \tag{10}$$

Function minimum:
Position for E_n: $(x_1,x_2 \ldots x_n)$ = E_n: $(x_1,x_2 \ldots x_n) = (0,0,\ldots, 0)$
Value for E_n: y = 0

The Second De Jong`s function is given by Eq. 11.

$$f(x) = \sum_{i=1}^{dim-1} 100\,(x_i^2 - x_{i+1})^2 + (1 - x_i)^2 \tag{11}$$

Function minimum:
Position for E_n: $(x_1,x_2 \ldots x_n) = (1,1,\ldots,1)$
Value for E_n: y = 0

Rastrigin`s function is given by Eq. 12.

$$f(x) = 10dim \sum_{i=1}^{dim} x_i^2 - 10\cos(2\pi x_i) \tag{12}$$

Function minimum:
Position for E_n: $(x_1,x_2 \ldots x_n) = (0,0,\ldots,0)$
Value for E_n: y = 0

Schwefel`s function is given by Eq. 13

$$f(x) = \sum_{i=1}^{dim} -x_i \sin\left(\sqrt{|x_i|}\right) \quad (13)$$

Function minimum:
Position for E_n: $(x_1, x_2 \ldots x_n) = (420.969, 420.969, \ldots, 420.969)$
Value for E_n: y = -418.983 * dimension

7 Experiment Setup

Within all performance testing four PSO versions were used. The first one was the canonical PSO with inertia weight without chaos number generator (noted PSO Weight in Tables), and the others were PSO with inertia weight and enhanced by chaos number generator that uses a) Lozi map (noted PSO Lozi) b) Disipative standard map (noted PSO Disi) and c) Arnold´s cat map (noted PSO Arnold).

In the first part of the performance testing, all four test functions, which are described in the section 6, were used for all versions of PSO algorithm. Results obtained within these tests are given in Tables 1 – 4. For each version, 30 separate runs were performed and analysed for the worst found solution (noted max), the best found solution (noted min) and median of all 30 final solutions.

Control parameters were set up based on the previous numerous experiments and literature [1-3,8] as follows:

Population size: 30
Iterations / generations 10 * dimension
w_{start}: 0.9
w_{end}: 0.4
Dimension: 2, 5, 10, 20, 40

In second part of the performance investigation, all four PSO versions were more extensively tested on Schwefel´s function. The history of the best individual solutions was tracked over 2000 generations in order to analyze the speed of convergence and overall behaviour of the algorithms. As in the previous case, 30 separate runs of each PSO version were run and tracked. The graphical results are displayed in Figures. 4 – 7. Mean values are then depicted in Figure. 8.

8 Results and Analysis

Results for the 1st De Jong´s function are shown in Table 1. From these results, it follows, that for almost all tested dimensions, the PSO with chaos generator that uses Lozi chaotic map obtained the best result (sometimes very significantly). Nevertheless, for the higher dimensions also PSO Arnold and PSO Disi have given the best results in some of the observed categories.

Table 1 Results for the 1st De Jong´s function

Dim = 2	max	min	median
PSO Weight	0.00238765	1.12672E-05	0.000207606
PSO Lozi	**0.000160378**	**2.04E-07**	**2.36845E-05**
PSO Disi	0.00366504	0.000012003	0.000393726
PSO Arnold	0.00355144	1.84201E-05	0.000133626
Dim = 5	**max**	**min**	**median**
PSO Weight	0.00392897	1.65228E-05	0.000174885
PSO Lozi	**2.56645E-05**	**2.34E-07**	**2.82E-06**
PSO Disi	0.00472208	8.05912E-05	0.000617638
PSO Arnold	0.000806617	1.90396E-05	0.000160753
Dim = 10	**max**	**min**	**median**
PSO Weight	0.0375076	0.000145797	0.00625555
PSO Chaos	**0.0208165**	**4.21E-07**	**0.000630301**
PSO Disi	0.0319727	0.000802657	0.00959562
PSO Arnold	0.0290758	9.65377E-05	0.00318635
Dim = 20	**max**	**min**	**median**
PSO Weight	1.62833	0.0842995	0.478742
PSO Lozi	1.95377	0.0460756	**0.405318**
PSO Disi	**1.2292**	0.155438	0.4703
PSO Arnold	1.92954	**0.0431876**	0.63985
Dim = 40	**max**	**min**	**median**
PSO Weight	9.06273	0.2401	4.49434
PSO Lozi	8.60618	**0.00243866**	4.25399
PSO Disi	**7.34072**	1.71549	**4.02592**
PSO Arnold	8.54468	0.980315	4.72424

Table 2 contains results for the 2nd De Jong´s function. Very similar trends as in the previous case (see Table 1 - 1st De Jong´s function) can be observed here. Overall the Lozi map seems to have the most significant positive impact on the performance of chaos enhanced PSO algorithm with this test function.

Table 2 Results for the 2nd De Jong´s function

Dim = 2	max	min	median
PSO Weight	0.0270018	6.30078E-05	0.0034439
PSO Lozi	**0.0159781**	2.91655E-05	**0.00084753**
PSO Disi	0.0450083	3.48167E-05	0.0047833
PSO Arnold	0.0584033	**1.08442E-05**	0.00559849
Dim = 5	**max**	**min**	**median**
PSO Weight	0.164005	0.00125174	0.0339318
PSO Lozi	**0.0115352**	**5.10981E-05**	**0.000797568**
PSO Disi	0.233613	0.0119405	0.105771
PSO Arnold	0.149441	0.000187331	0.0232775
Dim = 10	**max**	**min**	**median**
PSO Weight	0.192185	0.00335088	0.0387091
PSO Lozi	**0.0170628**	**8.64266E-05**	**0.00116313**
PSO Disi	0.333229	0.0105965	0.111803
PSO Arnold	0.230833	0.00306061	0.0294698
Dim = 20	**max**	**min**	**median**
PSO Weight	0.260444	0.00529196	0.0801234
PSO Lozi	**0.31728**	**0.000786091**	**0.0367927**
PSO Disi	0.648495	0.0484751	0.20016
PSO Arnold	0.449775	0.00226948	0.0658723
Dim = 40	**max**	**min**	**median**
PSO Weight	**1.03241**	0.0793424	0.374946
PSO Lozi	1.33416	0.022454	**0.218751**
PSO Disi	1.38807	0.0838238	0.823501
PSO Arnold	1.21392	**0.0125984**	0.2779

Following Table 3 contains the results of testing the PSO versions with Rastrigin´s function. As a contrast to the previous two test functions, where using of Lozi map helped to obtain the best results in almost all cases, the presented results show that all proposed versions were able to achieve the best result for some categories and dimension values, thus the performance of all PSO versions is comparable in this study case.

Table 3 Results for Rastrigin´s function

Dim = 2	**max**	**min**	**median**
PSO Weight	2.40482	0.0032492	0.539905
PSO Lozi	1.99328	**0.0014524**	**0.379036**
PSO Disi	**1.37104**	0.00391877	0.424698
PSO Arnold	1.49584	0.00513813	0.677154
Dim = 5	**max**	**min**	**median**
PSO Weight	13.5637	1.0756	6.50172
PSO Lozi	9.98535	**0.00222202**	**3.07613**
PSO Disi	**8.06491**	0.997318	4.98575
PSO Arnold	10.5125	1.08479	4.93331
Dim = 10	**max**	**min**	**median**
PSO Weight	25.1741	**0.259855**	**9.51575**
PSO Lozi	27.2665	3.00979	10.4711
PSO Disi	**22.7636**	4.8055	10.6585
PSO Arnold	29.6662	1.6683	11.8762
Dim = 20	**max**	**min**	**median**
PSO Weight	74.9295	27.4552	47.5319
PSO Lozi	**68.5083**	17.2733	46.651
PSO Disi	80.9735	**8.73105**	**38.9205**
PSO Arnold	69.6355	25.5659	41.2236
Dim = 40	**max**	**min**	**median**
PSO Weight	**186.578**	100.044	152.065
PSO Lozi	192.493	47.8408	159.009
PSO Disi	224.106	114.104	162.564
PSO Arnold	195.753	**10.2981**	**151.983**

As can be seen in Table 4, the results obtained for Schwefel´s function lend weight to the argument that from all the tested chaotic maps the Lozi map as a chaotic number generator has the best impact on the performance of PSO algorithm with inertia weight.

Table 4 Results for Schwefel´s function

Dim = 2	**max**	**min**	**median**
PSO Weight	-673.386	-837.868	**-820.347**
PSO Lozi	-697.479	**-837.937**	-819.427
PSO Disi	-660.777	-837.437	-819.055
PSO Arnold	**-699.735**	-837.595	-813.001
Dim = 5	**max**	**min**	**median**
PSO Weight	-1191.59	-1760.3	-1458.34
PSO Lozi	**-1262.01**	**-1942.73**	**-1525.28**
PSO Disi	-1224.87	-1916.65	-1433.96
PSO Arnold	-1239.27	-1687.44	-1506.39
Dim = 10	**max**	**min**	**median**
PSO Weight	-1915.02	-2740.49	-2234
PSO Lozi	**-1980.15**	**-3037.15**	**-2561.77**
PSO Disi	-1870.47	-2667.19	-2178.76
PSO Arnold	-1850.76	-2604.74	-2237.06
Dim = 20	**max**	**min**	**median**
PSO Weight	-2886.88	-5372.33	-3477.26
PSO Lozi	**-3133.19**	-5722.76	**-4217.83**
PSO Disi	-2647.19	**-5736.72**	-3263.8
PSO Arnold	-2810.32	-4437.79	-3450.35
Dim = 40	**max**	**min**	**median**
PSO Weight	-4839.42	-8400.45	-6558.88
PSO Lozi	**-6112.29**	**-9935.4**	**-7527.7**
PSO Disi	-4840.88	-8414.88	-6318.99
PSO Arnold	-5291.19	-9474.16	-6260.92

Within the second part of the performance testing, the history of the best solutions (CF values) was analyzed for the Schwefel`s function. This function was selected due to the results from previous testing (see table 4), where chaos enhanced PSO that uses Lozi map as chaotic number generator has achieved significantly better results for higher dimensions than other PSO versions. Figures 4 – 7 show the history, which is the time-evolution of CF values of 30 separate runs of each PSO version and for the Schwefel´s function over 2000 generations for 20 dimensions.

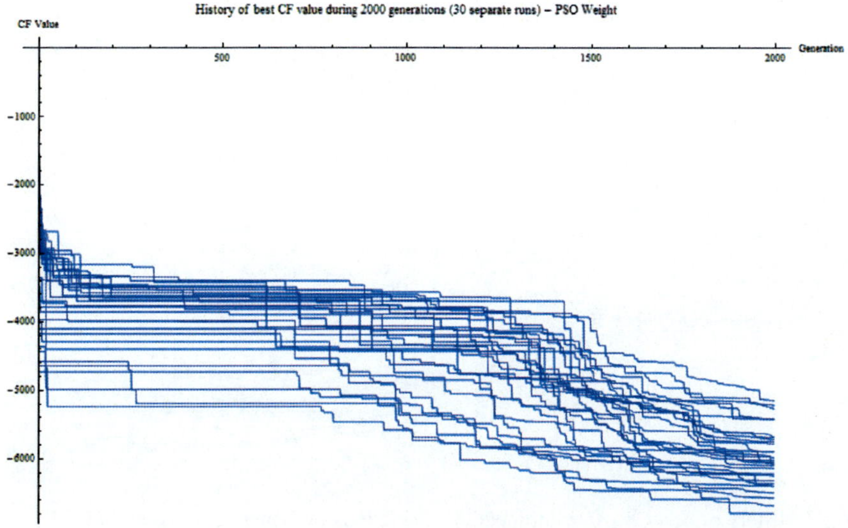

Fig. 4 History of best CF value during 2000 generations (30 separate runs) - PSO Weight

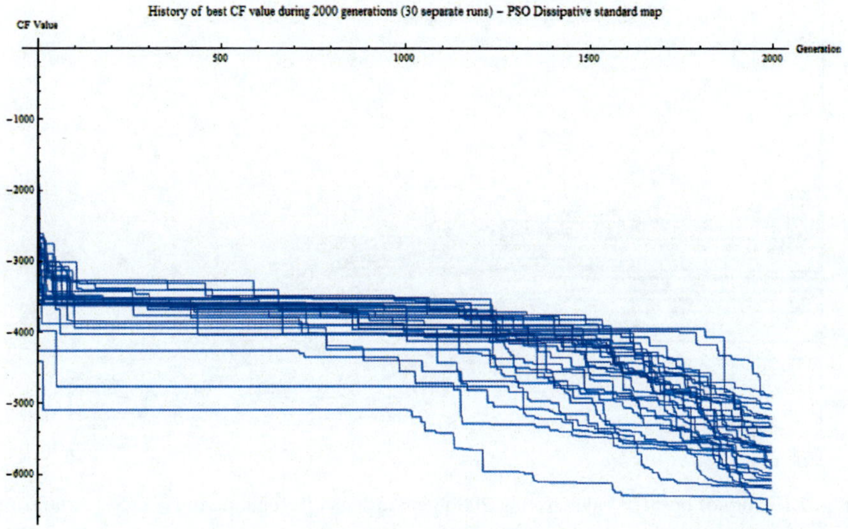

Fig. 5 History of best CF value during 2000 generations (30 separate runs) - PSO Disi

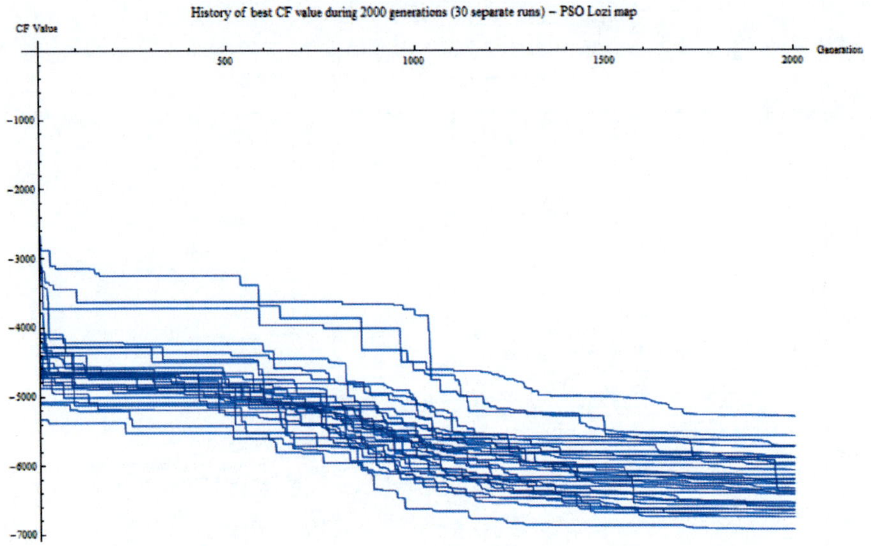

Fig. 6 History of best CF value during 2000 generations (30 separate runs) - PSO Lozi

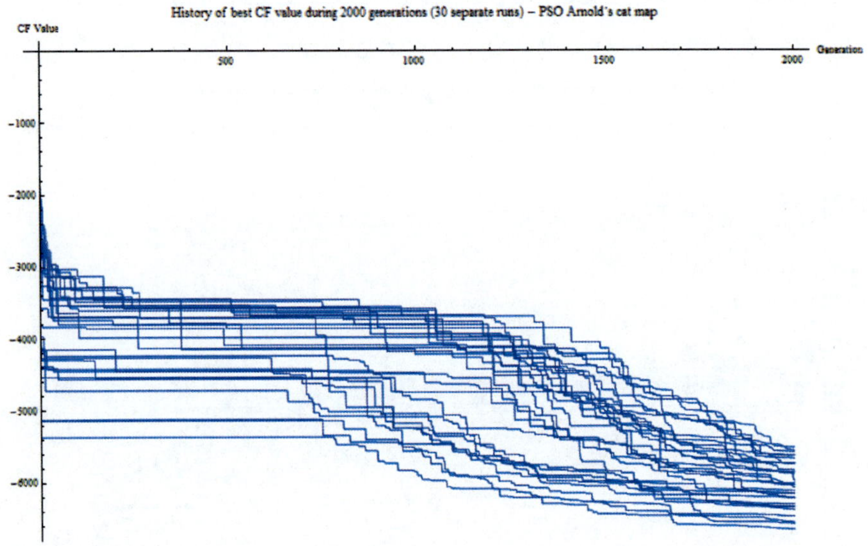

Fig. 7 History of best CF value during 2000 generations (30 separate runs) - PSO Arnold

For the better imagination and understanding of the results presented in Figures 4 – 7, the average of 30 runs of each version is depicted in Figure 8.This averaged history confirms that using Lozi map as chaotic number generator may improve the speed of convergence and overall performance of PSO algorithm with inertia weight on such tasks as Schwefel`s function in higher dimensions.

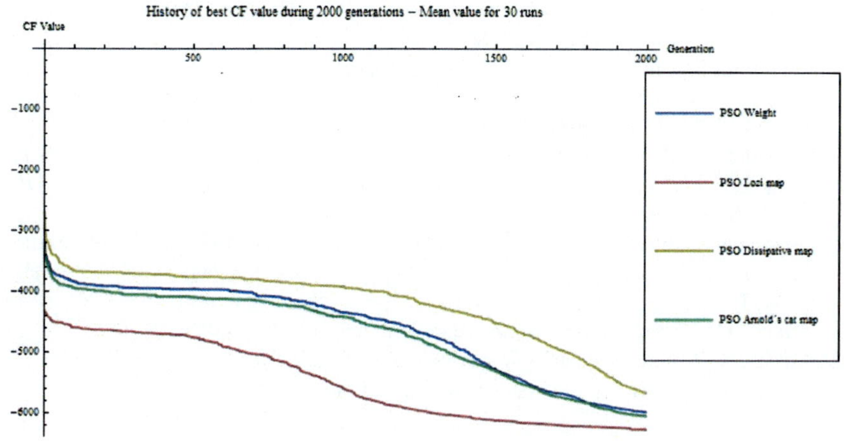

Fig. 8 History of mean best CF value for 30 runs - comparison

9 Conclusion

In this research three different chaotic maps were used as the chaotic number generator for PSO algorithm with inertia weight. Proposed versions of PSO algorithm were tested with satisfactory results that give weight to claims that using chaotic number generator may improve the performance of evolutionary algorithms. Within this initial study, the Lozi map seems to have the most significant impact on the performance of PSO algorithm with inertia weight. Future research should examine more extensively these findings and provide more data and analysis to confirm the claims about the influence of chaos number generator on the performance of PSO algorithm.

Acknowledgements. This work was supported by European Regional Development Fund under the project CEBIA-Tech No. CZ.1.05/2.1.00/03.0089, and by Internal Grant Agency of Tomas Bata University under the project No. IGA/FAI/2012/037.

References

1. Kennedy, J., Eberhart, R.: Particle Swarm Optimization. In: Proceedings of IEEE International Conference on Neural Networks, vol. IV, pp. 1942–1948 (1995)
2. Eberhart, R., Kennedy, J.: Swarm Intelligence, The Morgan Kaufmann Series in Artificial Intelligence. Morgan Kaufmann (2001)
3. Dorigo, M.: Ant Colony Optimization and Swarm Intelligence, Springer (2006)
4. Storn, R., Price, K.: Differential evolution—a simple and efficient heuristic for global optimization over continuous spaces. Journal of Global Optimization 11, 341–359 (1997)

5. Goldberg, D.E.: Genetic Algorithms in Search Optimization and Machine Learning. p. 41. Addison Wesley (1989) ISBN 0201157675
6. Araujo, E., Coelho, L.: Particle swarm approaches using Lozi map chaotic sequences to fuzzy modelling of an experimental thermal-vacuum system. Applied Soft Computing 8(4), 1354–1364 (2008) ISSN 1568-4946
7. Davendra, D., Zelinka, I., Senkerik, R.: Chaos driven evolutionary algorithms for the taskof PID control. Computers & Mathematics with Applications 60(4), 1088–1104 (2010) ISSN 0898-1221
8. Nickabadi, A., Ebadzadeh, M.M., Safabakhsh, R.: A novel particle swarm optimization algorithm with adaptive inertia weight. Applied Soft Computing 11(4), 3658–3670 (2011) ISSN 1568-4946
9. Sprott, J.C.: Chaos and Time-Series Analysis. Oxford University Press (2003)
10. Astrom, K.: Control System Design. University of California, Santa Barbra (2002)
11. Landau, Y.: Digital Control Systems. Springer, London (2006)

Extended Initial Study on the Performance of Enhanced PSO Algorithm with Lozi Chaotic Map

Michal Pluhacek, Vera Budikova, Roman Senkerik, Zuzana Oplatkova, and Ivan Zelinka

Tomas Bata University in Zlin, Faculty of Applied Informatics,
Nam T.G. Masaryka 5555, 760 01 Zlin, Czech Republic
{pluhacek,budikova,senkerik,oplatkova,zelinka}@fai.utb.cz

Abstract. In this paper, it is proposed the utilization of discrete Lozi map based chaos random number generator to enhance the performance of PSO algorithm with inertia weight. Performance tests and results are presented. Results are analyzed and compared with another evolutionary algorithm. Tuning experiment was performed.

1 Introduction

PSO algorithm is one of evolutionary algorithms (class of soft computing methods that are inspired by nature). In recent years, various evolutionary algorithms were designed and used with great results in many areas of optimization [1-5]. More recently some studies indicated that using of chaotic number generators may improve the performance of evolutionary optimization algorithms on such tasks as PID controller design [6] or fuzzy modelling of an experimental thermal-vacuum system [7].

This initial study is focused on the investigation on the performance of PSO algorithm with implemented chaotic Lozi map as a random number generator. Firstly, Particle swarm optimization algorithm is explained. The next sections are aimed on the description of used chaotic systems and benchmark test functions. Results and conclusion follow afterwards.

2 Particle Swarm Optimization Algorithm

Particle swarm optimization (PSO) algorithm is based on the natural behaviour of birds and fishes. It was firstly introduced by Eberhart and Kennedy in 1995 [1,2] as an alternative to genetic algorithms [4] and differential evolution [5],

In each generation, a new location of a particle is calculated based on its previous location and velocity (or "velocity vector"). One of PSO algorithm disadvantages is the rapid acceleration of particles which causes them abandoning the

I. Zelinka et al. (Eds.): Nostradamus: Mod. Meth. of Prediction, Modeling, AISC 192, pp. 167–177.
springerlink.com

defined area of interest. For this reason, several modifications of PSO were introduced to handle with this problem. Main principles of PSO algorithm and its modifications are well described in [1-3].

Within this research, chaos driven PSO strategy with inertia weight was used. The selection of inertia weight modification of PSO was based on numerous previous experiments. Default values of all PSO parameters were chosen according to the recommendations given in [2, 3].

Inertia weight is designed to influence the velocity of each particle differently over the time [8]. In the beginning of the optimization process, the influence of inertia weight factor w is minimal. As the optimization continues, the value of w is decreasing, thus the velocity of each particle is decreasing, since w is always the number < 1 and it multiplies previous velocity of particle in the process of new velocity value calculation. Inertia weight modification PSO strategy has two control parameters w_{start} and w_{ends}. New w for each generation is then given by Eq. 1, where i stand for current generation number and n for total number of generations.

$$w = w_{start} - \frac{((w_{start} - w_{end}) * i)}{n} \tag{1}$$

Chaos driven random number generator is used in the main PSO formula (Eq. 2) that determines new "velocity" and thus the position of each particle in the next generation (or migration cycle).

$$v(t+1) = v(t) + c_1 \cdot Rand \cdot (pBest - x(t)) + c_2 \cdot Rand \cdot (gBest - x(t)) \tag{2}$$

Where:
v(t+1) – New velocity of particle.
v(t) – Current velocity of particle.
c1,c2 – Priority factors.
pBest – Best solution found by particle.
gBest – Best solution found in population.
x(t) – Current position of particle.
Rand – Random number, interval <0,1>. Within Chaos PSO algorithm, the basic inbuilt computer (simulation software) random generator is replaced with chaotic generator (in this case, by using of Lozi map).

New position of particle is then given by Eq. 3, where x(t+1) is the new position:

$$x(t+1) = x(t) + v(t+1) \tag{3}$$

3 Chaotic Lozi Map

The Lozi map is a simple discrete two-dimensional chaotic map. The Lozi map is depicted in Fig. 1. The map equations are given in Eq. 4 and 5. The parameters used in this work are: a = 1.7 and b = 0.5 as suggested in [9].

$$X_{n+1} = 1 - a|X_n| + bY_n \tag{4}$$

$$Y_{n+1} = X_n \tag{5}$$

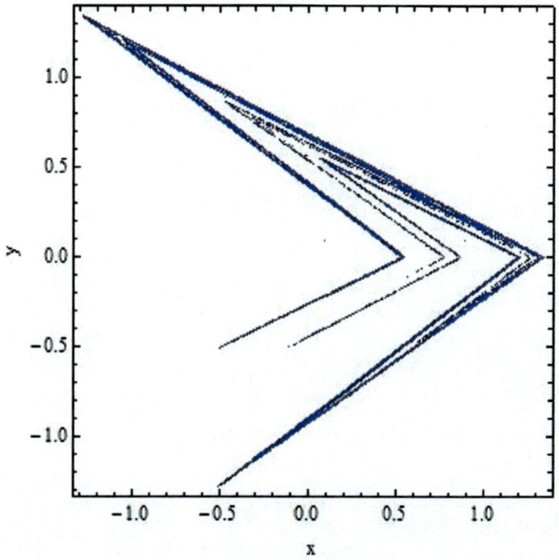

Fig. 1 Lozi map

4 Experiment Setup

In section 2 the PSO with inertia weight was described. To compare the impact of using Lozi map as a chaotic number generator, performance tests were performed for both PSO with chaotic and non-chaotic random number generator. Classic version of PSO with inertia weight modification is labelled PSO Weight. The proposed novel Lozi map enhanced PSO with inertia weight is labelled PSO Lozi. For further comparison with different heuristic, tests were also performed for differential evolution (DE) and its strategy DERand1Bin.

Basic PSO control parameters were set based on previous experiments and literature [1-4,7] as follows:

Population size: 30 (for sections 5.1 – 5.4), 50, 75, 100, 150, 200, 300, 400
Iterations / generations: 10 * dimension
w_{start}: 0.9
w_{end}: 0.4
Dimension: 2, 5, 10, 20, 40

The algorithms were tested on 4 different benchmark functions. From the statistical reasons, optimization for each dimension value was repeated 30 times.

4.1 Test Functions

The First De Jong`s function is given by Eq. 6.

$$f(x) = \sum_{i=1}^{dim} x_i^2 \tag{6}$$

Function minimum:
Position for E_n: $(x_1,x_2 \ldots x_n)$ = E_n: $(x_1,x_2 \ldots x_n) = (0,0,\ldots, 0)$
Value for E_n: $y = 0$

The Second De Jong`s function is given by Eq. 7.

$$f(x) = \sum_{i=1}^{dim-1} 100\,(x_i^2 - x_{i+1})^2 + (1 - x_i)^2 \tag{7}$$

Function minimum:
Position for E_n: $(x_1,x_2 \ldots x_n) = (1,1,\ldots,1)$
Value for E_n: $y = 0$

Rastrigin`s function is given by Eq. 8.

$$f(x) = 10dim \sum_{i=1}^{dim} x_i^2 - 10\cos\,(2\pi x_i) \tag{8}$$

Function minimum:
Position for E_n: $(x_1,x_2 \ldots x_n) = (0,0,\ldots,0)$
Value for E_n: $y = 0$
Schwefel`s function is given by Eq. 9.

$$f(x) = \sum_{i=1}^{dim} -x_i \sin\,(\sqrt{|x_i|}) \tag{9}$$

Function minimum:
Position for E_n: $(x_1,x_2 \ldots x_n) = (420.969, 420.969,\ldots, 420.969)$
Value for E_n: $y = -418.983 * \text{dimension}$

5 Results and Brief Analysis

The results of experiments and brief commentary on these results follow in this section. Following Tables 1 – 8 contain the best, the worst and the median of obtained final results for all 30 runs of evolutionary algorithm. For the comparison between algorithms, the best individual results are highlighted by bold number in all tables. Results of PSO algorithm are also compared with the performance of DE.

5.1 The First De Jong`s Function

Following tables 1 and 2 contain results for 1st De Jong´s function. Proposed implementation of chaotic Lozi map to PSO algorithm seems to have improved the performance of the algorithm. Values for PSO Lozi (chaos number generator) are better in all cases than values for PSO Weight (classic number generator). Furthermore those results are better or comparable with those of DE. However PSO seems to be less efficient in solving higher dimension with the setting used than DE.

Table 1 Results for the first De Jong`s function for dim = 2, 5 and 10

	dim = 2			dim = 5			dim = 10		
	PSO Weight	PSO Lozi	DE	PSO Weight	PSO Lozi	DE	PSO Weight	PSO Lozi	DE
The worst result	2.3877E-03	**1.6038E-04**	2.3719E-04	3.9290E-03	**2.5665E-05**	1.0020E-03	3.7508E-02	2.0817E-02	**2.8169E-03**
The best result	1.1267E-05	**2.0428E-07**	2.4734E-07	1.6523E-05	**2.3375E-07**	7.4393E-05	1.4580E-04	**4.2146E-07**	5.2609E-04
Median	2.0761E-04	2.3685E-05	**2.3118E-05**	1.7489E-04	**2.8169E-06**	1.9458E-04	6.2556E-03	**6.3030E-04**	1.1871E-03

Table 2 Results for the first De Jong`s function for dim = 20 and 40

	dim = 20			dim = 40		
	PSO Weight	PSO Lozi	DE	PSO Weight	PSO Lozi	DE
The worst result	1.6283E+00	1.9538E+00	**4.0143E-02**	9.0627E+00	8.6062E+00	**1.6867E+00**
The best result	8.4300E-02	4.6076E-02	**1.0478E-02**	2.4010E-01	**2.4387E-03**	6.5759E-01
Median	4.7874E-01	4.0532E-01	**1.8644E-02**	4.4943E+00	4.2540E+00	**9.9788E-01**

5.2 The Second De Jong`s function

Results in tables 3 and 4 were obtained by optimizing the 2nd De Jong´s function. Almost similar trends as in previous section (1st De Jong´s function) can be seen in those results with the exception being the worse performance of DE in comparison with PSO Lozi for the higher dimensions.

Table 3 Results for the second De Jong`s function for dim = 2, 5 and 10

	dim = 2			dim = 5			dim = 10		
	PSO Weight	PSO Lozi	DE	PSO Weight	PSO Lozi	DE	PSO Weight	PSO Lozi	DE
The worst result	2.7002E-02	**1.5978E-02**	5.5802E-02	1.6401E-01	**1.1535E-02**	3.7692E+00	1.9219E-01	**1.7063E-02**	1.9067E+01
The best result	6.3008E-05	**2.9166E-05**	2.0637E-04	1.2517E-03	**5.1098E-05**	6.2450E-01	3.3509E-03	**8.6427E-05**	7.7341E+00
Median	3.4439E-03	**8.4753E-04**	1.0980E-02	3.3932E-02	**7.9757E-04**	2.1158E+00	3.8709E-02	**1.1631E-03**	1.4072E+01

Table 4 Results for the second De Jong`s function for dim = 20 and 40

	dim = 20			dim = 40		
	PSO Weight	PSO Lozi	DE	PSO Weight	PSO Lozi	DE
The worst result	**2.6044E-01**	3.1728E-01	5.2821E+01	1.0324E+00	**1.3342E+00**	1.3381E+02
The best result	5.2920E-03	**7.8609E-04**	2.5292E+01	7.9342E-02	**2.2454E-02**	5.0176E+01
Median	8.0123E-02	**3.6793E-02**	3.7204E+01	3.7495E-01	**2.1875E-01**	7.7725E+01

5.3 *Rastrigin`s Function*

In the case of the Rastrigin`s benchmark function there is no significant improvement in PSO performance, however both PSO algorithms were significantly worse for dim = 40 than those of DE (see tables 5 and 6). This finding was one of those that encouraged the extended analysis in section 6.

Table 5 Results for second Rastrigin`s function for dim = 2, 5 and 10

	dim = 2			dim = 5			dim = 10		
	PSO Weight	PSO Lozi	DE	PSO Weight	PSO Lozi	DE	PSO Weight	PSO Lozi	DE
The worst result	2.4048	1.9933	**1.7853**	13.5637	9.9854	**9.6667**	**25.1741**	27.2665	26.7314
The best result	0.0032	**0.0015**	0.0056	1.0756	**0.0022**	0.1615	**0.2599**	3.0098	6.2952
Median	0.5399	**0.3790**	0.5707	6.5017	**3.0761**	4.0811	**9.5158**	10.4711	15.7813

Table 6 Results for Rastrigin`s function for dim = 20 and 40

	dim = 20			dim = 40		
	PSO Weight	PSO Lozi	DE	PSO Weight	PSO Lozi	DE
The worst result	74.9295	**68.5083**	75.9887	186.5780	192.4930	**155.9540**
The best result	27.4552	**17.2733**	18.7197	100.0440	47.8408	**38.8948**
Median	47.5319	**46.6510**	50.0244	152.0650	159.0090	**119.5870**

5.4 Schwefel`s Function

Presented results in Tables 7 and 8 show increasing difference between values of median for PSO Weight and PSO Lozi together with increasing dimension in favour of Lozi map enhanced PSO. As in the previous case, both algorithms were surpassed by DE (most significantly in higher diemensions).

Table 7 Results for Schwefel`s function for dim = 2, 5 and 10

	dim = 2			dim = 5			dim = 10		
	PSO Weight	PSO Lozi	DE	PSO Weight	PSO Lozi	DE	PSO Weight	PSO Lozi	DE
The worst result	-673.39	-697.48	**-702.79**	-1191.59	-1262.01	**-1635.06**	-1915.02	-1980.15	**-3174.74**
The best result	-837.87	**-837.94**	-837.91	-1760.30	-1942.73	**-1964.83**	-2740.49	-3037.15	**-3562.67**
Median	**-820.35**	-819.43	-816.07	-1458.34	-1525.28	**-1842.63**	-2234.00	-2561.77	**-3372.21**

Table 8 Results for Schwefel`s function for dim = 20 and 40

	dim = 20			dim = 40		
	PSO Weight	PSO Lozi	DE	PSO Weight	PSO Lozi	DE
The worst result	-2886.88	-3133.19	**-5830.19**	-4839.42	-6112.29	**-9104.93**
The best result	-5372.33	-5722.76	**-6482.53**	-8400.45	-9935.40	**-11411.20**
Median	-3477.26	-4217.83	**-6358.48**	-6558.88	-7527.70	**-10308.10**

From the results presented in tables 1 - 8 it may be stated, that the majority of results obtained by proposed PSO Lozi algorithm were better than results of classic PSO Weight. The observed median of final cost function values for all 30 runs was better in 17 of 20 experiments - 4 benchmark functions x 5 dimensions values (see Fig. 2).

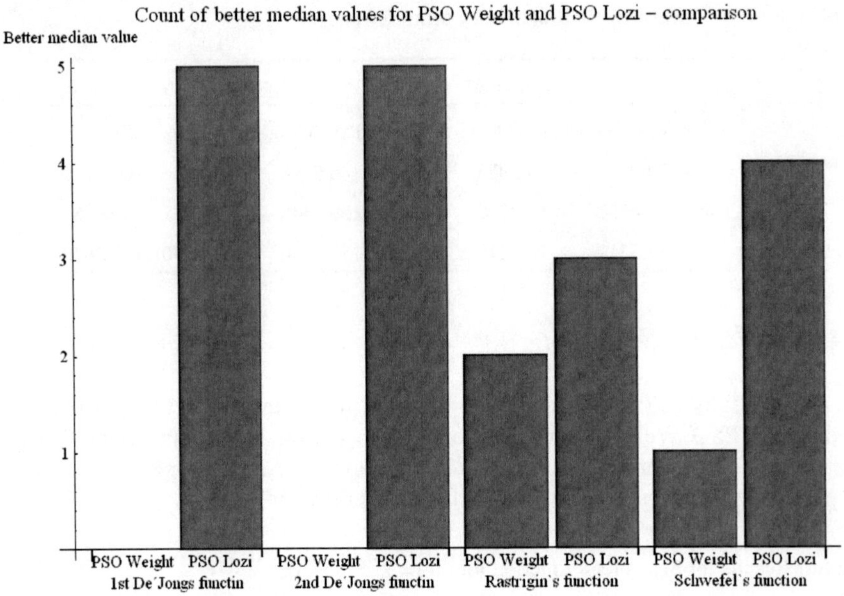

Fig. 2 PSO Weight and PSO Lozi results comparison

6 Additional Tests and Analysis

The analysis of the results presented in previous section shows an interesting phenomenon, that the performance of DE in comparison with both PSO algorithms is sometimes much better (especially in the case of Schwefel´s function). This section presents an investigation on this phenomenon. Based on the previous experiences, the PSO algorithm is able to achieve better results for higher population size (NP). To prove this theory an experiment was set-up as follows:

Population size: 30, 50, 75, 100 (Schwefel´s f. only), 150, 200, 300, 400
Iterations / generations: 200
w_{start}: 0.9
w_{end}: 0.4
Dimension: 20
Benchmark functions: Schwefel´s, 1st De Jong

Results are shown in following tables (9 and 10) and depicted on figures 3 and 4. Analysis follows afterwards.

From Tables 9 and 10, it follows, that increasing the size of population (NP) led to significant improvement in the performance of both PSO algorithms with inertia weight. Nevertheless the performance of DE had opposite trend. These trends can be seen on figures 3 and 4. Presented results in this section support the claims, that using Lozi map as a chaotic number generator could lead to improvement of the performance of PSO algorithm thus to achieve better or at least similar results when comparing PSO algorithms with different evolutionary algorithm, which was DE.

Table 9 Mean value for 30 runs; Schwefel`s function; dim = 20; generations = 200

NP:	30	50	75	100	150	200	300	400
PSO Weight	-3697.63	-3873.62	-4140.99	-4255.84	-4329.57	-4866.41	-5316.41	-5377.72
PSO Lozi	-4340.06	-4560.42	-5032.82	-5241.78	**-5801.99**	**-5998.05**	**-6174.55**	**-6225.63**
DE	**-6100.9**	**-5737.68**	**-5649.1**	**-5500.01**	-5635.55	-5651.5	-5673.33	-5651.23

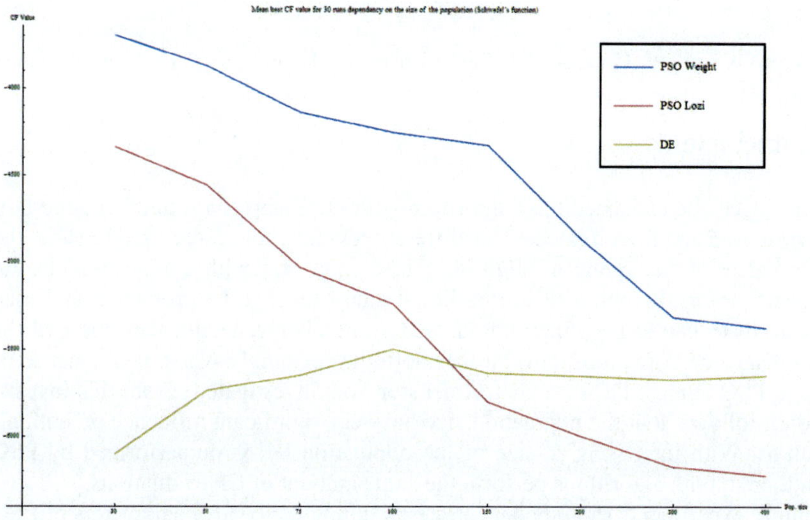

Fig. 3 PSO Weight, PSO Lozi and DE – influence of population size – Schwefel´s function

Table 10 Mean value for 30 runs; 1st De Jong`s function; dim = 20; generations = 200

NP:	150	200	300	400
PSO Weight	7.33004*10^-6	3.45288*10^-7	3.7368*10^-9	2.60345*10^-10
PSO Lozi	2.77017*10^-6	**4.12126*10^-8**	**3.66273*10^-11**	**1.03193*10^-14**
DE	**4.25565*10^-7**	4.94665*10^-7	6.11666*10^-7	7.07218*10^-7

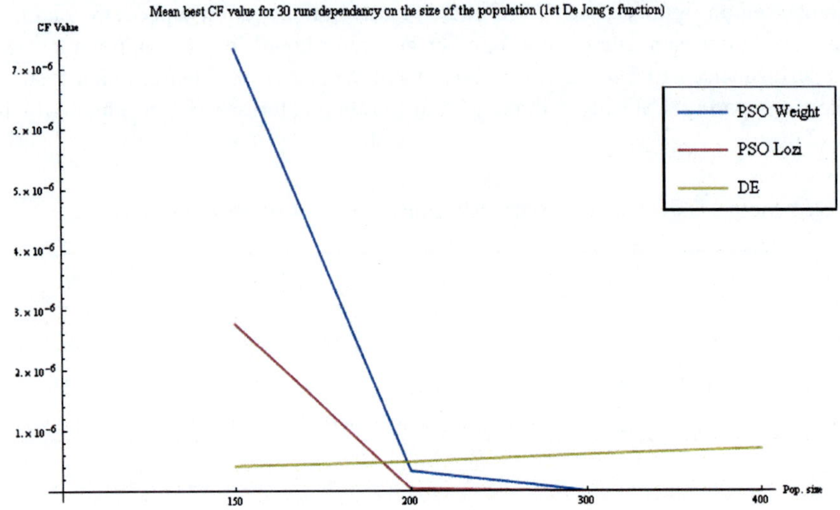

Fig. 4 PSO Weight, PSO Lozi and DE – influence of population size – 1st De Jong´s function

7 Conclusion

In this paper, the enhanced PSO algorithm with Lozi map chaos number generator was proposed and investigated. Four different test functions were used to show the performance of the proposed algorithm in comparison with classic non-chaotic version. The results obtained in this initial study seem to be promising and may indicate that chaotic Lozi map based random number generator can improve the performance of PSO algorithm. Furthermore the seemingly worse performance of chaotic PSO against DE in higher dimension was investigated. From this investigation it follows, that the population size has very significant influence on both algorithms. With increasing of size of the population DE is outperformed by PSO even though both algorithms perform the same amount of CF evaluations.

Future experiments should concentrate on tuning the control parameters of PSO and investigation on the performance of this algorithm on variety of problems. Overall, this study brought findings that chaotic Lozi map may improve the performance of PSO algorithm.

The primary aim of this work was not to develop a new type of random number generator, which should pass many statistical tests, but to try to use and test the implementation of natural chaotic dynamics into evolutionary algorithm as a chaotic random number generator.

Acknowledgements. This work was supported by European Regional Development Fund under the project CEBIA-Tech No. CZ.1.05/2.1.00/03.0089, by Internal Grant Agency of Tomas Bata University under the project No. IGA/FAI/2012/042 and by Internal Grant Agency of Tomas Bata University under the project No. IGA/FAI/2012/037.

References

1. Kennedy, J., Eberhart, R.: Particle Swarm Optimization. In: Proceedings of IEEE International Conference on Neural Networks, vol. IV, pp. 1942–1948 (1995)
2. Dorigo, M.: Ant Colony Optimization and Swarm Intelligence, Springer (2006)
3. Eberhart, R., Kennedy, J.: Swarm Intelligence. The Morgan Kaufmann Series in Artificial Intelligence. Morgan Kaufmann (2001)
4. Storn, R., Price, R.: Differential evolution—a simple and efficient heuristic for global optimization over continuous spaces. Journal of Global Optimization 11, 341–359 (1997)
5. Goldberg, D.E.: Genetic Algorithms in Search Optimization and Machine Learning. Addison Wesley, p. 41 (1989) ISBN 0201157675
6. Davendra, D., Zelinka, I., Senkerik, R.: Chaos driven evolutionary algorithms for the task of PID control. Computers & Mathematics with Applications 60(4), 1088–1104 (2010) ISSN 0898-1221
7. Araujo, E., Coelho, L.: Particle swarm approaches using Lozi map chaotic sequences to fuzzy modelling of an experimental thermal-vacuum system. Applied Soft Computing 8(4), 1354–1364 (2008)
8. Nickabadi, A., Ebadzadeh, M.M., Safabakhsh, R.: A novel particle swarm optimization algorithm with adaptive inertia weight. Applied Soft Computing 11(4), 3658–3670 (2011) ISSN 1568-4946
9. Sprott, J.C.: Chaos and Time-Series Analysis. Oxford University Press (2003)

Custom Winding Ratio Analysis of Evolutionary Optimized Audio Transformer

Martin Pospisilik[1], Lukas Kouril[1], Milan Adamek[1],
Ivan Zelinka[2], and Roman Jasek[1]

[1] Tomas Bata University in Zlin, Faculty of Applied Informatics, Nam. T.G.M. 5555, Zlin, Czech Republic
{pospisilik,kouril}@fai.utb.cz

[2] Department of Computer Science, Faculty of Electrical Engineering and Computer Science, VŠB-TUO 17. listopadu 15 708 33, Ostrava-Poruba, Czech Republic

Abstract. This paper is intended to present a trend analysis resulting from different secondary to primary windings ratio of evolutionary optimized audio transformer. Evolutionary optimization of audio transformers is a perspective approach to the wideband audio transformers design by employing methods of artificial intelligence algorithms. This novel approach to audio transformers designing has a notable implication for problematics related to audio transformers complex design process. The authors are building on their previous research by trying to determine the trends which can be observed by selecting different secondary to primary windings ratio of evolutionary designed audio transformer. These trends can be useful at further optimization of the designing algorithms.

1 Introduction

At designing of audio electrical circuits the transformers are somehow overlooked because of their complex problems in design and construction, relatively high price and bulky dimensions. However, sometimes this is a pity because the transformers can increase the noise parameters of the circuit significantly which can result in matching the desired parameters with lower total price of the whole construction. Many times the transformers are omitted only because the designers are not sure how to treat with the amount of equations to be evaluated in order the design of the proper transformer could be realised. Therefore the authors of this paper aimed to creating a set of algorithms that should help the designers with the transformer design issues.

Previously, the authors published several papers on this topic. The equivalent circuit of an audio transformer that is suitable for simulations is described in [1], the mechanical issues determining the transformer manufacturability are described in [2] while the methods of applying the optimizing algorithms at audio transformers designing can be found in [3] and [4]. In [3] also a complex set of equations describing the transformer behaviour is included.

In this paper the authors describe how the particular parameters of the transformer vary depending on the required secondary to primary winding ratio N.

I. Zelinka et al. (Eds.): Nostradamus: Mod. Meth. of Prediction, Modeling, AISC 192, pp. 179–190.
springerlink.com

Single primary and double secondary winding step-up transformers are considered in this paper as this configuration is utilizable at microphone preamplifiers with symmetrical circuitry which is the field in which the application of transformers is still reasonable. This transformer configuration is also widely spread in vacuum valve amplifiers that currently undergo a certain renaissance.

2 Problem Definition

The secondary to primary windings turns ratio N is probably the most known parameter of the transformers as it specifies the transformation ratio. However, in the field of audio transformers many more parameters have to be considered. Briefly, the significant parameters of the transformer are secondary to primary winding ratio N [-], number of primary winding turns n_1 [-], number of secondary winding turns n_2 [-], windings to total resistances ratio m [%] (utilized as a correction factor, see below), secondary to primary winding resistance ratio r [-] (considering noise optimisation it should be close to 1), primary winding resistance R_p [Ω], secondary winding resistance R_s [Ω], core magnetizing inductance L_p [H], main resonant frequency f_r [Hz], maximum core induction B [T].

Because mechanical issues are an essential part of the transformer design, physical dimensions of the core, the wires and the former (bobbin) must be considered. The parameters of the transformer depends on how the winding is wound, how many turns fit in one layer, of how many layers the winding consists as well as on the ferromagnetic characteristics of the core and many more phenomenons. Because the core dimensions are normalised as well as the thicknesses of the wires and their lacquer insulation layers (see [2]), a considerable discontinuities are brought into the calculations. From the mechanical issues, let us briefly list the most important ones. These are primary winding wire cross section S_1 [m^2], secondary winding wire cross section S_2 [m^2], primary winding wire diameter considering the insulation lacquer d_{out1} [m], secondary winding wire diameter considering the insulation lacquer d_{out2} [m], thickness of the former (decreases the space for the winding), core material cross section S_m [m^2], core window width h (defines the width of the winding) [m], core window height t_w (defines the thickness of the winding) [m], average length of a magnetic line of force inside the core mass l_m [m], average length of a single current turn o [m], thickness of the isolation layer between separated windings t_{ps} and the number of such layers.

Secondary electrical parameters outgoing from the mechanical parameters are leakage inductance L_l [H], primary winding capacity C_p [F], secondary winding capacity C_s [F], primary to secondary winding coupling capacity C_{ps} [F], secondary winding to core capacity C_{sc} [F] (considering the secondary winding is wound as the first).

The electrical parameters mentioned above influence the transfer function of the transformer mainly at high frequencies as the capacities, being spread across the whole winding and thus dependent on its geometrical constitution, resulting together with the leakage inductance in a series of resonances. Usually, when the transformer is loaded, only one resonant frequency prevails, being treated as the fr frequency.

Usually, the transformer is treated as a part of a complex circuit so the following electrical parameters must also be taken into account. These are signal source impedance Z_G [Ω] or signal source resistance R_G [Ω], load impedance L_L [Ω] or load resistance R_L [Ω], primary winding resistance R_p [Ω], secondary winding resistance R_s [Ω].

Together with other above mentioned parameters the secondary parameters that describe the behaviour of the transformer connected in a particular circuit can be expressed as low frequency roll-off f_{min} [Hz], high frequency roll-off or the highest applicable frequency f_{max} [Hz] which may comply to the resonant frequency f_r [Hz], maximum input voltage u_{inmax}(considering the maximum allowed core induction B) or maximum total harmonic distortion (THD) if the magnetization curve of the core is known as well as the core magnetizing induction, signal attenuation Att [Hz] based on the attenuation caused by the own winding resistance.

The list of here described parameters is definitely not exhaustive but in most designs these parameters will be enough to specify and verify. Moreover, there are various relationships among all the parameters as described in [1], [2] or [3].

For the complexity of the calculations the authors decided to utilize evolutionary algorithms. In [3] it has been proven that it is possible to solve the transformer designing task with the aid of a Differential evolution [5], [6]. In [4] it has been also proven that Self-Organizing Migrating Algorithm [6], [7], [8] can be used for audio transformers designing. In the consequent research the authors dealt with the problem of proper specification of the boundary conditions to be specified for the evolutionary algorithm. The present practice consisted in specifying the ideal and the boundary (tolerable) parameters. According to [3], [4] these were f_{min} minimal transferred frequency in [Hz] desired value, f_{min_max} minimal transferred frequency in [Hz] the highest (worst) value that can be tolerated, m the optimal relative resistance of the windings, m_{max} the maximum relative resistance of the windings that can be tolerated, r the primary to secondary winding resistance ratio, r_{min} the minimal acceptable primary to secondary winding resistance ratio, r_{max} the maximal acceptable primary to secondary winding resistance ratio, B the optimal maximum core induction in [T] to be reached at the frequency f_{pmin}, B_{max} the maximum tolerable core induction in [T] to be reached at the frequency f_{pmin}, att optimal attenuation of the transformer in [dB], att_{max} maximal tolerable attenuation of the transformer in [dB], t_0 core window height according to the type of the core (see [2]), t_w total winding thickness.

The results achieved in [3], [4] were considered as satisfactory but there still remained some headroom for improving the optimization in the way it worked more efficiently which was the main motivation for the improvement of the approach the authors applied in their previous work.

3 Approach Improvement

The signal attenuation *Att* arises from the losses caused by the ineligible resistance of the wires in the transformer winding and is partially compensated by increasing the number of secondary winding turns by m [%] (see the parameter m above). In fact this affects the theoretical transformation ratio N and changes the impedance ratios that are considered in the circuit employing the transformer with the specified N ratio. The higher the N ratio is, the higher are the losses resulting in the attenuation *Att* that counteract the voltage gain reached by the increased N ratio. Generally, the real voltage gain A obtained by the transformer can be expressed by the following equation:

$$A = 20 \cdot log_{10}(N) - Att \text{ [dB]} \tag{1}$$

Another issue consists in the transformer signal bandwidth. According to O'Mear the bandwidth M of the transformer manufactured with the same technology on the same core depends at the N factor as follows:

$$M = \frac{1}{\sqrt{1 + \frac{(\frac{1}{N} - N)^2}{2}}} \tag{2}$$

This brings the danger of decreasing of the achievable bandwidth when the N ratio is increased.

Considering the facts mentioned above there arises a question: "How high the N ratio can be in order the required parameters were met and the transformer was still manufacturable?". In practical design tasks this question will probably be converted to: "What N ratio is really convenient to be required?"

Therefore the authors decided to process the previously published transformer design [3] for different N factors. The N factor was changed from $N = 7.5$ to $N = 25$ and the trends of the parameters changes were observed. The transformer is equipped with a single primary winding and a double secondary winding.

3.1 *Customized Optimization by Differential Evolution*

As in [3] the evolutionary algorithm selected was Differential evolution [5], [6] again. This method of artificial intelligence is inspired by evolutionary theory by Darwin. The difference is an environment where the evolution is proceeded. The environment is represented *in silico* when Differential evolution is considered. Thus individuals, mutation and cross-over process have a form of vectors or mathematical expressions. Similarly to evolution *in situ* there are originated new individuals. These represents temporary solutions of the problems being optimized. The structure of individuals is specified by specimen which states how many parameters the temporary solution has and range of these parameters. The goal of optimization is to originate new generation of individuals on the basis of individual quality. It means that it is necessary to evaluate each individual and express its quality. In the nature the quality of individual is represented as e.g. ability to survive on the basis of strength etc. In the case of evolution realized *in silico* the individual quality is calculated by cost function which specifies conditions each solution has to be met. Individuals are also processed by mutation and cross-over similarly to nature evolution. New generation is formed by individuals with higher cost value which expresses individual quality than their ancestors. Then the optimization process is repeated thus new generations are subsequently originated until sufficient temporary solution is founded or number of generation stated is exceeded.

The optimization which is proceeded by Differential evolution is comprised of following steps:

1. Random initialization of individuals contained in first generation in according to specimen.
2. Starting optimization.
3. Subsequent selection of each individual x_r of current generation.
4. Random selection of other three individuals x_{r1}, x_{r2}, x_{r3}.
5. Creation of the differential vector as a subtraction of x_{r1} and x_{r2}.
6. Creation of the weighted differential vector as a mutation.
7. Creation of the noise vector (3) by addition x_{r3} to the weighted differential vector

$$v = x^G_{r3,j} \cdot (x^G_{r1,j} - x^G_{r2,j}), \tag{3}$$

   where $j = 1, ..., D$, D - dimension of problem.
8. Creation of the test vector (4)

$$x^{test}_{i,j} = \begin{cases} x^G_{r3,j} + F \cdot (x^G_{r1,j} - x^G_{r2,j}) & \text{if } rand_j[0,1] \leq CR \vee j = j_{rand} \\ x^G_{r,j} & \text{else} \end{cases} \tag{4}$$

   where $i = 1, ..., NP$, NP - number of population.
9. Evaluation of the current individual x_r and the test vector by cost function.
10. Better-evaluated individual or the test vector is passed to the next generation (5).

$$x^{G+1}_i = \begin{cases} x^{test}_r & \text{if } f_{cost}(x^{test}) \leq f_{cost}(x^G_i) \\ x^G_i & \text{else} \end{cases} \tag{5}$$

Differential evolution has several parameters which have to be set before starting optimization process. These are number of population (NP), mutation constant (F), cross-over value (CR) and number of generations (G). Above mentioned parameters were set as NP = 1000, F = 0.8, CR = 0.4, G = 10 000. Specimen was expressed as can be seen in [3], [4].

The optimisation task was simplified because m, m_{max}, att, att_{max} conditions were not used any more (compare with [3], [4]) thus conditions contained in cost function are following:

- $f_{min} \leq f_{min,max}$,
- $r \leq r_{max}$,
- $r \geq r_{min}$,

- $B \leq B_{max}$,
- $t_0 \geq t_w$,
- $0.8 \cdot t_0 \leq t_w$,

where above parameters are explained in Tab. 1 and Tab. 3.

Observing of these parameters would be rather inconvenient because it would slow down the design processing and moreover it could counteract the trends the authors wanted to show. The parameters of the transformer to be designed for the different N ratios are specified in the following table:

Table 1 Requirements on the transformer to be designed

Parameter	Values		
	Optimal	Minimal	Maximal
R_G	$R_G = 500\Omega$		
R_L	$R_L = 2 \times 200$ kΩ		
C_L	$C_L = 50$ pF		
N	Changed from 7.5 to 25 by 0.5 step		
f_{min}	$f_{min} = 15$ Hz		$f_{minmax} = 25$ Hz
f_{pmin}			$f_{pmin} = f_{minmax}$
U_{imax}			$U_{imax} = 2$V
S_{min}		$S_{min} = 7 \cdot 10^{-4}$mm^2	
t_{ps}	$t_{psid} = 100\mu$m	$t_{ps} = \{0, 100, 200, 300, 400\}\mu$m	
r	$r_{id} = 1$	$r_{min} = 0.25$	$r_{max} = 4$
B			$B_{max} = 0.3$T
d_{in1}			
d_{in2}			
d_{out1}		see Table 2	
d_{out2}			
d_{S1}			
d_{S2}			

As stated above, the N parameter was changed from 7.5 to 25 with the step of 0.25. As in [3], the EI38/20 core has been chosen considering the mechanical parameters enlisted in Tab. 3.

The results were exported into graphs and trends were attempted to be found. Moreover, electrical parameters of the selected transformers were fully simulated (for $N = 7.5$, $N = 10$, $N = 15$, $N = 20$, $N = 25$). At the simulations, the model presented in [1] was employed.

Table 2 List of applicable wires

Nominal diameter d_{in} [mm]	Outer diameter d_{out} [mm]	Nominal wire cross-section [mm^2]
0.03	0.05	0.007
0.04	0.06	0.0013
0.05	0.07	0.0020
0.056	0.078	0.0025
0.063	0.088	0.0031
0.071	0.095	0.0039
0.08	0.105	0.0050
0.09	0.118	0.0063
0.1	0.128	0.0078
0.112	0.15	0.0099
0.125	0.165	0.0122
0.132	0.172	0.0136
0.14	0.18	0.0153
0.15	0.19	0.0176
0.16	0.2	0.0200
0.17	0.216	0.0226
0.18	0.227	0.0253
0.19	0.238	0.0282
0.2	0.25	0.0314

Table 3 Mechanical parameters of EI38/20 core

Parameter	Value
Core mass cross-section	$S = 2.4$ cm^2
Total height of the coil section	$t_0 = 6.5$ mm
Coil former material thickness	$t_f = 0.5$ mm
Average length of a single turn	$o = 84$ mm
Average length of a magnetic line of force inside the core mass	$l = 71.5$ mm
Core window width	$h_w = 19$ mm
Estimated air gap caused bz the inaccuracy of the transformer sheets cutting	$l_{air} = 0.1$ mm
Core mass relative permeability	$\mu_r = 1.000$

4 Results

Two different results sets were obtained. For all the results generated by the Differential evolution algorithm the graphs of dependence of the following parameters on the N ratio were assembled really achieved (effective) gain A [dB], resonant frequency f_r [kHz], lowest transferred frequency f_{min} [Hz], secondary to primary winding resistance ratio r [-].

For the selected N ratios the transfer function of the appropriate transformer connected in the specified circuit was examined.

4.1 Observed Trends

First of all the impact of the increasing attenuation Att on the total voltage gain A was observed. The results obtained by the Differential evolution are shown in Fig. 1. As obvious from Fig. 1, due to increased attenuation, for higher N ratio the really achieved voltage gain does not rise as one could expect. This is caused by the fact that for high N too many secondary winding turns are needed and moreover thinner wire must be used in order the winding fitted into the core window. Due to these issues the winding resistance increases rapidly and, being comparable to the load resistance, it creates a voltage divider. Generally it could be said that for the required

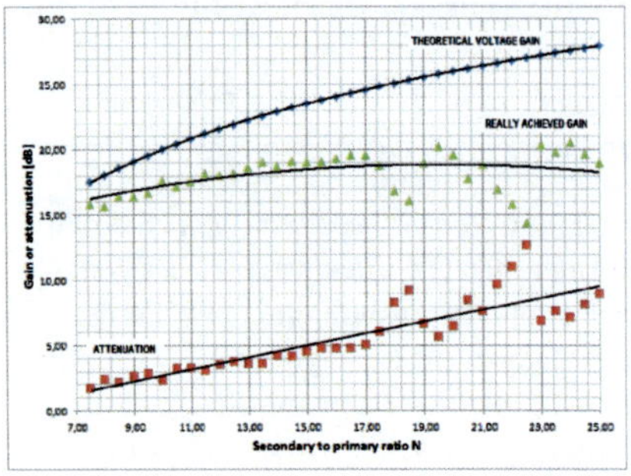

Fig. 1 Theoretical and really achieved gain dependence on the N ratio

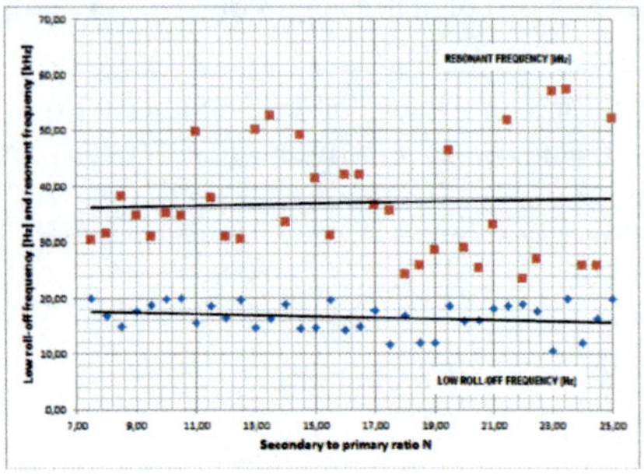

Fig. 2 The lowest transferred frequency and the resonant frequency of the transformer dependence on the N ratio

parameters the effective voltage gain of the transformer cannot exceed the level of approximately 20 dB.

Unfortunately a significant dispersion of the resulting values has been observed, especially at higher N. This is probably caused by two issues: Firstly, the Differential evolution searches for all the results that comply with the requirements. There are several parameter combinations leading to different but always satisfactory results. Moreover, the evolution of each of the solutions (one solution per one N ratio was obtained) reaches the solution from "different side". For these reasons the parameters combination may vary within the framework of the specified boundaries, still reaching the solution that can be considered as the satisfactory one. The second reason for the values dispersion consists in the fact that several parameters are chosen from the tables, thus discontinuous. Moreover, as stated above, there exists more than one combination of the selected parameters. All these factors can be considered as the source of the values dispersion.

With regard to the above mentioned dispersions, the obtained parameters f_{min} and f_r (the lowest transferred and the resonant frequency) practically do not possess any trend as the values dispersion is too high. The results for these parameters are shown in Fig. 2.

According to Fig. 2 it can be stated that the resonant frequency and the lowest transferred frequency (at which the 3 dB roll-off is considered) do not change significantly according to N. It seems that this ascertainment denies the O'Mears equation for the achievable bandwidth (2) but it must be taken into account that with increased N the attenuation Att also raised up which results in increased damping of the circuit. This effect will be obvious especially at the lowest frequencies where the 3 dB roll-off is given by the combination of magnetizing inductance L_p, the source resistance R_G, winding resistances and the load resistance R_L. Damping of the transfer via the transformer at the centre frequencies results in decrease of the f_{min}. At high frequencies the situation is more complicated. As there are more turns at the secondary winding split into more layers, the winding capacity is decreased (see [3] for the winding capacity estimation equations). This results in the higher resonant frequency.

Generally it is not possible to extract any trend from the data depicted in Fig. 2. It seems that rather to search for trends, it is more convenient to run the computation more times for more *N* ratios that are close each to other and select the best achieved result.

Whereas it seems that the change of the *N* parameter does not directly influence the transformed signal bandwidth (which is obviously not true, see below), the noise optimisation can be affected by this parameter quite a lot. In Fig. 3 there is the dependence of the r parameter on the *N* parameter depicted. The *r* parameter expresses the ratio of the secondary to the primary winding resistance considering the secondary winding resistance is recalculated to the primary winding according to the transformation ratio *N*. For noise-optimized design it is generally expected that the resistances at the primary and the secondary winding shall be balanced and therefore the *r* parameter shall be close to 1. Unfortunately this requirement is not achievable for higher *N*.

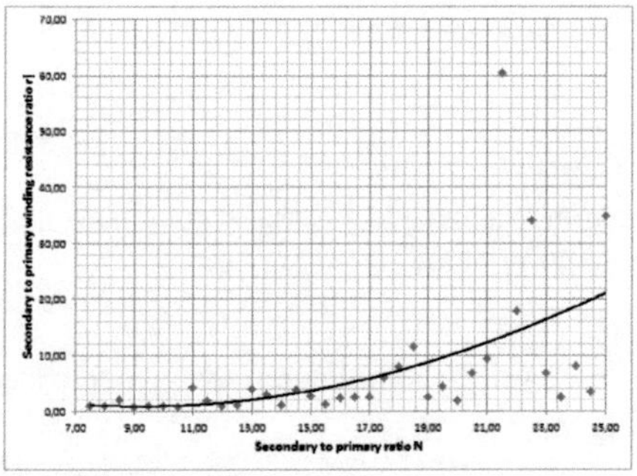

Fig. 3 The *r* parameter dependence on the *N* ratio

According to Fig. 3 there is still a chance to achieve the favourable *r* at high *N* ratios but the probability of this finding is decreased. If the noise-optimised design is required, more runs of the computation shall be processed with slightly different *N* parameter in order the most favourable result could be selected.

4.2 Simulations

The transfer function simulations were processed according to the model published in [1] for $N = 7.5, N = 10, N = 12.5$ and $N = 15$. By this verification it was revealed that for *N* greater than approximately 13 the solutions proposed by the Differential evolution do not make sense.

For high *N* the bandwidth of the transformed signal was really limited (which corresponds to previously known (2)). The high resonant frequency does not tell anything about the highest transferred frequency as the influence of the capacities prevails. Together with the source resistance the winding of the transformer create a considerable integrator. In order to achieve high N the evolutionary algorithm must limit the number of the primary winding turns in order the whole winding fitted into the core window. This results in limited magnetizing inductance L_p causing increased core magnetizing induction B. The distortion caused by the transformer will possibly rise and the low-frequency cut-off will rise definitely.

Based on the simulation results, the dependences of the magnetizing inductance L_p and the induction in the core B were observed. The trends can be seen at Fig. 4 and Fig. 5.

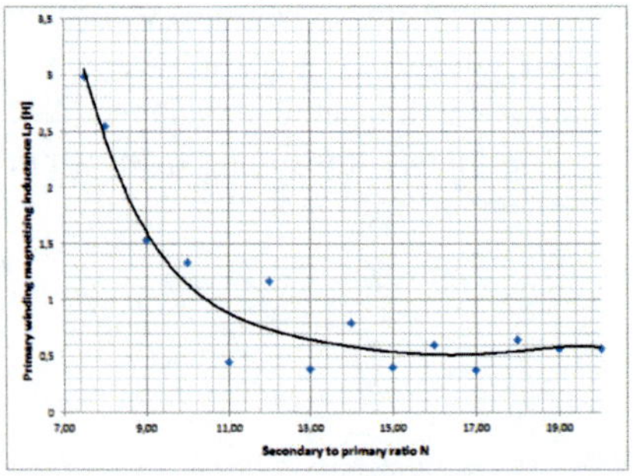

Fig. 4 The primary winding magnetizing inductance L_p dependence on the N ratio

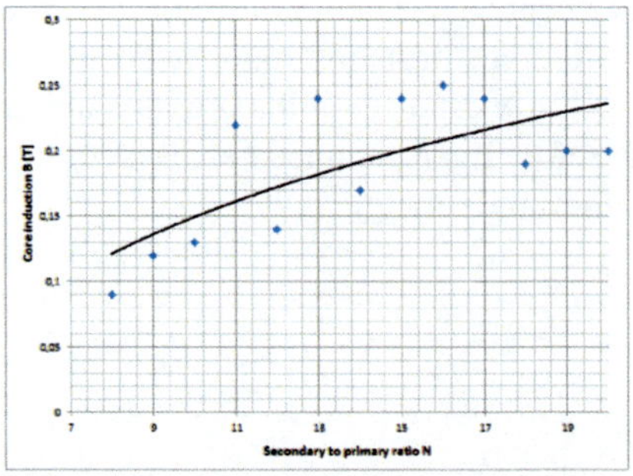

Fig. 5 Core induction B dependence on the N ratio

The frequency responses for the simulated transformer, processed in Maple software, are depicted in Fig. 6.

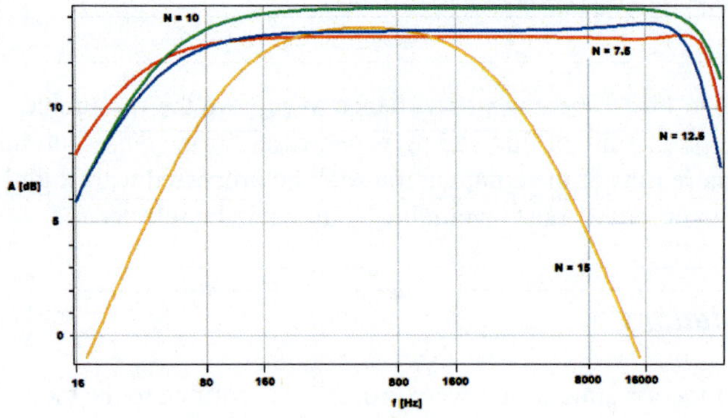

Fig. 6 Simulated frequency response of the transformers for selected N factors

According to the simulations depicted in Fig. 6 it was proven that in the case of the designed transformer the increasing of the N factor above approximately 10 brings no benefits. For $N > 10$ the effective gain is not increased correspondingly according the effects described above (see Fig. 1). Moreover, the bandwidth is really limited, according to O'Mear's research (see (2)).

5 Conclusion

In the research described in this paper the authors tried to find possibilities of improving their algorithm for simple audio transformers designing by means of the Differential evolution that was previously published in [3], [4]. Unfortunately certain surprising facts arisen.

According to the expectations the computations showed that there is not unlimited headroom for increasing the transforming ratio N. It was proven that the transformer parameters even cannot be improved by increasing the transforming ratio N together with the increased damping.

Moreover it was also showed that as the complexity of the task rises together with the rising transforming ratio because less utilisable results are achievable, the probability of an error increases. For high N ratios the resulting parameter dispersion risen significantly. Moreover some results were not complying with all the pre-set restrictions.

From the results obtained in this research the following facts are obvious: It is worth applying the effective gain computation (1) in the algorithm. For higher N the results are unreliable and must be always verified by simulation. The accuracy of the computation must be increased.

Acknowledgement. This paper is supported by the Internal Grant Agency at TBU in Zlin, project No. IGA/FAI/2012/056 and IGA/FAI/2012/053 and by the European Regional Development Fund under the project CEBIA-Tech No. CZ.1.05/2.1.00/03.0089.

References

1. Pospisilik, M., Adamek, M.: Audio Transformers Simulation. In: Proceedings of the 16th WSEAS Multiconference, Kos, Greece (Accepted for publication, in print, 2012)
2. Pospisilik, M., Adamek, M.: Determining the parameter manufacturability. In: Proceedings of the 16th WSEAS Multiconference, Kos, Greece (Accepted for publication, in print 2012)
3. Kouril, L., Pospisilik, M., Adamek, M., Jasek, R.: Designing an Audio Transformer by Means of Evolutionary Algorithms. In: Proceedings of the 5th WSEAS World Congress on Applied Computing Conference, Faro, Portugal, pp. 133–138 (2012) ISBN 978-1-61804-089-3
4. Pospisilik, M., Kouril, L., Adamek, M., Zelinka, I., Jasek, R.: SOMA-Based Audio Transformers Optimization. In: Proceedings of the 18th International Conference on Soft Computing MENDEL 2012, Brno, Czech Republic, pp. 326–331 (2012) ISBN 978-80-214-4540-6. ISSN 1803-3814

5. Lampinen, J., Zelinka, I.: Mechanical Engineering Design Optimization by Differential Evolution. In: New Ideas of Optimization, 1st edn., McGraw-Hill, London (1999) ISBN 007-709506-5
6. Zelinka, I., Oplatkova, Z., Seda, M., Osmera, P., Vcelar, F.: Evolucni vypocetni techniky - principy a aplikace. BEN - technicka literatura, Praha (2008) ISBN 80-7300-218-3
7. Zelinka, I.: SOMA - Self-Organizing Migrating Algorithm. In: Onwubolu, G., Babu, B.V. (eds.) New Optimization Techniques in Engineering, pp. 3–540. Springer, Heidelberg (2004)
8. Varacha, P.: Innovative Strategy of SOMA Control Parameter Setting. In: Proceedings of the 12th WSEAS International Conference on Neural Networks, Fuzzy Systems, Evolutionary Computing & Automation. WSEAS Press, Timisoara (2011) ISBN 978-960-474-292-9

Evolutionary Synthesis of Control Rules by Means of Analytic Programming for the Purpose of High Order Oscillations Stabilization of Evolutionary Synthesized Chaotic System

Roman Senkerik[1], Zuzana Oplatkova[1], and Ivan Zelinka[2]

[1] Tomas Bata University in Zlin, Faculty of Applied Informatics, T.G. Masaryka 5555, 760 01 Zlin, Czech Republic.
{senkerik,oplatkova}@fai.utb.cz
[2] VŠB-Technical University of Ostrava, Faculty of Electrical Engineering and Computer Science, Department of Computer Science, 17. listopadu 15, 708 33 Ostrava-Poruba, Czech Republic.
ivan.zelinka@vsb.cz

Abstract. In this paper, it is presented a utilization of tool for symbolic regression, which is analytic programming, for the purpose of the synthesis of a new control law. This synthesized chaotic controller secures the stabilization of high periodic orbit – oscillations between several values of selected one-dimensional discrete chaotic system, which is artificially evolutionary synthesized system. The paper consists of the descriptions of analytic programming as well as chaotic system, used evolutionary techniques and the cost function. For experimentation, Self-Organizing Migrating Algorithm (SOMA) and Differential evolution (DE) were used.

1 Introduction

During the recent years, usage of new intelligent systems in engineering, technology, modeling, computing and simulations has attracted the attention of researchers worldwide. The most current methods are mostly based on soft computing, which is a discipline tightly bound to computers, representing a set of methods of special algorithms, belonging to the artificial intelligence paradigm. The most popular of these methods are neural networks, evolutionary algorithms, fuzzy logic, and genetic programming. Presently, evolutionary algorithms are known as a powerful set of tools for almost any difficult and complex optimization problem.

The interest about the interconnection between evolutionary techniques and control of chaotic systems is spread daily. First steps were done in [1], [2], where

I. Zelinka et al. (Eds.): Nostradamus: Mod. Meth. of Prediction, Modeling, AISC 192, pp. 191–201.
springerlink.com

the control law was based on Pyragas method: Extended delay feedback control – ETDAS [3]. These papers were concerned to tune several parameters inside the control technique for chaotic system. Compared to previous research, this paper shows a possibility how to generate the whole control law (not only to optimize several parameters) for the purpose of stabilization of a chaotic system. The synthesis of control is inspired by the Pyragas's delayed feedback control technique [4], [5]. This method is very advantageous for the evolutionary computation, due to its amount of easy accessible control parameters, which can be easily tuned by means of evolutionary algorithms (EA).

Instead of EA utilization, analytic programming (AP) is used in this research. AP is a superstructure of EAs and is used for synthesis of analytic solution according to the required behavior. Control law from the proposed system can be viewed as a symbolic structure, which can be synthesized according to the requirements for the stabilization of the chaotic system. The advantage is that it is not necessary to have some "preliminary" control law and to estimate its parameters only. This system will generate the whole structure of the law even with suitable parameter values.

This work is focused on the expansion of AP application for synthesis of a whole control law instead of parameters tuning for existing and commonly used method control law to stabilize desired Unstable Periodic Orbits (UPO) of chaotic systems.

This work is an extension of previous research [6 - 8] focused on the stabilization of simple p-1 orbit, which is stable state and p-2 orbit. In general, this research is concerned to stabilize p-4 UPO – high periodic orbit (oscillations between four values).

Firstly, selected evolutionary algorithms are described, and then a problem design is proposed. The next sections are focused on the description of used cost function, analytic programming and evolutionary algorithms. Results and conclusion follow afterwards.

2 Selected Chaotic System

The chosen example of chaotic systems was an evolutionary synthesized chaotic system presented in [9]. The paper [10] discusses the use of evolutionary algorithms for the optimization of control of synthesized chaotic systems by means of AP, but only from the point of view of optimal tuning of ETDAS control technique parameters. In this research, a complete synthesis of control rule structure including all parameters is presented. One of the motivations was also, that previous research showed that some of synthesized systems are barely controllable. One of them (1) was chosen for the experiments.

$$x_{n+1} = \frac{A\left(2A - 2x_n^{\,2} - 3x_n\left(A - x_n + Ax_n\right)\right)}{-A + x_n - x_n^{\,2}} \tag{1}$$

This system exhibits chaotic behavior for the control parameter A in the ranges <0.1, 0.13> and <0.8, 1.2> (see Figure 1).

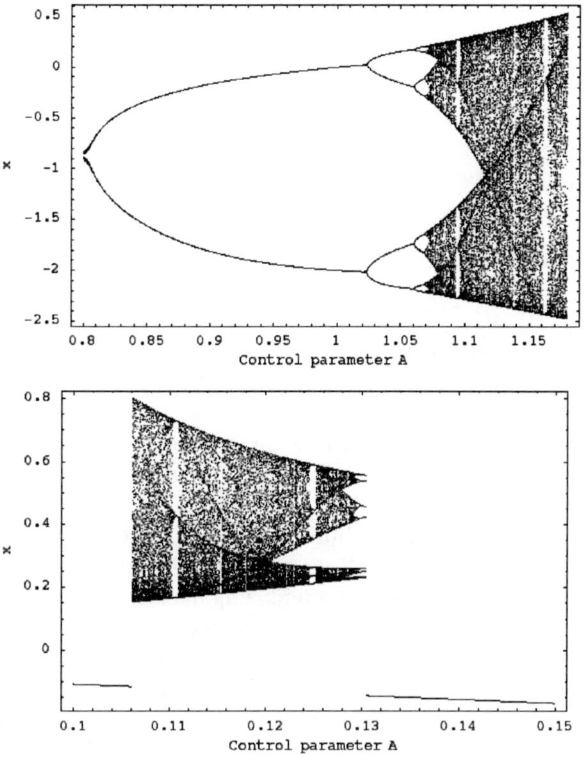

Fig. 1 Bifurcation diagram for A=<0.8, 1.2> (upper), Bifurcation diagram for A=<0.1, 0.15> (lower)

3 Original Control Method - ETDAS

This work is focused on the explanation of application of AP for synthesis of a whole control law instead of demanding tuning of classical ETDAS method control law to stabilize chaotic system. In this research desired UPO is p-4 orbit, which represents oscillations between 4 values. ETDAS method was obviously an inspiration for preparation of sets of basic functions and operators for AP. The original control method – ETDAS has form (2).

$$F(t) = K\left[(1-R)S(t-\tau_d) - x(t)\right]$$
$$S(t) = x(t) + RS(t-\tau_d) \tag{2}$$

Where: K and R are adjustable constants, F is the perturbation; S is given by a delay equation utilizing previous states of the system and τ_d is a time delay. The original control method – ETDAS in the discrete form has the form (3).

$$F_n = K[(1-R)S_{n-m} - x_n]$$
$$S_n = x_n + RS_{n-m} \tag{3}$$

Where: m is the period of m-periodic orbit to be stabilized. The perturbation F_n in equations (3) may have arbitrarily large value, which can cause diverging of the system. Therefore, F_n should have a value between $-F_{max}$, F_{max}. In this work a suitable F_{max} value was taken from the previous research.

4 Cost Function for Chaotic System Stabilization

Proposal for the cost function comes from the simplest Cost Function (CF). The idea was to minimize the area created by the difference between the required state and the real system output on the whole simulation interval – τ_i.

But another universal cost function had to be used for stabilizing of extremely sensitive chaotic system and having the possibility of adding penalization rules. It was synthesized from the simple CF and other terms were added. In this case, it is not possible to use the simple rule of minimizing the area created by the difference between the required and actual state on the whole simulation interval – τ_i, due to many serious reasons, for example: including of initial chaotic transient into the final CF value or degrading of the possible best solution by phase shift of higher periodic orbit, which represents the oscillations between several values.

This CF is in general based on searching for desired stabilized periodic orbit and thereafter calculation of the difference between desired and found actual periodic orbit on the short time interval - τ_s (40 iterations) from the point, where the first min. value of difference between desired and actual system output is found. Such a design of CF should secure the successful stabilization of either p-1 orbit (stable state) or higher periodic orbit anywise phase shifted. The CF_{Basic} has the form (4).

$$CF_{Basic} = pen_1 + \sum_{t=\tau 1}^{\tau 2} |TS_t - AS_t|, \tag{4}$$

where:

TS - target state, AS - actual state

τ_1 - the first min value of difference between TS and AS, τ_2 – the end of optimization interval ($\tau_1 + \tau_s$)

$pen_1 = 0$ if $\tau_i - \tau_2 \geq \tau_s$; $pen_1 = 10*(\tau_i - \tau_2)$ if $\tau_i - \tau_2 < \tau_s$ (i.e. late stabilization).

5 Analytic Programming

This tool was used for the synthesis of used chaotic system and for synthesis of control rule within presented research. Basic principles of the AP were developed in 2001 [11], [12]. Until that time only genetic programming (GP) and grammatical evolution (GE) had existed. GP uses genetic algorithms while AP can be used with any evolutionary algorithm, independently on individual representation.

The core of AP is based on a special set of mathematical objects and operations. The set of mathematical objects is set of functions, operators and so-called terminals (as well as in GP), which are usually constants or independent variables. This set of variables is usually mixed together and consists of functions with different number of arguments. Because of a variability of the content of this set, it is called here "general functional set" – GFS. The structure of GFS is created by subsets of functions according to the number of their arguments. For example GFS_{all} is a set of all functions, operators and terminals, GFS_{3arg} is a subset containing functions with only three arguments, GFS_{0arg} represents only terminals, etc. The subset structure presence in GFS is vitally important for AP. It is used to avoid synthesis of pathological programs, i.e. programs containing functions without arguments, etc. The content of GFS is dependent only on the user. Various functions and terminals can be mixed together [11], [12].

The second part of the AP core is a sequence of mathematical operations, which are used for the program synthesis. These operations are used to transform an individual of a population into a suitable program. Mathematically stated, it is a mapping from an individual domain into a program domain. This mapping consists of two main parts. The first part is called discrete set handling (DSH) [11] – [14] and the second one stands for security procedures which do not allow synthesizing pathological programs. The method of DSH, when used, allows handling arbitrary objects including nonnumeric objects like linguistic terms {hot, cold, dark…}, logic terms (True, False) or other user defined functions. In the AP DSH is used to map an individual into GFS and together with security procedures creates the above mentioned mapping which transforms arbitrary individual into a program.

AP needs some evolutionary algorithm that consists of population of individuals for its run. Individuals in the population consist of integer parameters, i.e. an individual is an integer index pointing into GFS. The individual contains numbers which are indices into GFS. The detailed description is represented in [12].

AP exists in 3 versions – basic without constant estimation, AP_{nf} – estimation by means of nonlinear fitting package in *Mathematica* environment and AP_{meta} – constant estimation by means of another evolutionary algorithms; meta means meta-evolution.

6 Used Evolutionary Algorithms

This research used two evolutionary algorithms: Self-Organizing Migrating Algorithm (SOMA) [15] and Differential Evolution (DE) [16], [17]. Future simulations expect a usage of soft computing GAHC algorithm (modification of HC12) [18] and a CUDA implementation of HC12 algorithm [19] and DE.

6.1 Self-organizing Migrating Algorithm (SOMA)

Self-Organizing Migrating Algorithm is a stochastic optimization algorithm that is modeled on the basis of social behavior of cooperating individuals [15]. It was chosen because it has been proven that the algorithm has the ability to converge towards the global optimum [15] and due to the successful applications together with AP [20], [21].

6.2 Differential EVOLUTION (DE)

DE is a population-based optimization method that works on real-number-coded individuals, [16], [17]. DE is quite robust, fast, and effective, with global optimization ability. It does not require the objective function to be differentiable, and it works well even with noisy and time-dependent objective functions.

7 Results

As described in section about Analytic Programming, AP requires some EA for its run. In this research, AP_{meta} version was used. Meta-evolutionary approach means usage of one main evolutionary algorithm for AP process and second algorithm for coefficient estimation, thus to find optimal values of constants in the evolutionary synthesized control law.

SOMA algorithm was used for main AP process and DE was used in the second evolutionary process. Settings of EA parameters for both processes were based on performed numerous experiments with chaotic systems and simulations with AP_{meta} (See Table 1 and Table 2).

Table 1 SOMA settings for AP

PathLength	3
Step	0.11
PRT	0.1
PopSize	50
Migrations	4
Max. CF Evaluations (CFE)	5345

Table 2 DE settings for meta-evolution

PopSize	40
F	0.8
Cr	0.8
Generations	150
Max. CF Evaluations (CFE)	6000

Due to the recursive attributes of delay equation *S* utilizing previous states of the system in discrete ETDAS (3), the data set for AP had to be expanded and cover longer system output history, thus to imitate inspiring control method for the successful synthesis of control law securing the stabilization of chaotic system. The Analytic Programming used following settings:

Basic set of elementary functions for AP:
GFS2arg= +, -, /, *, ^
GFS0arg= $data_{n-9}$ to $data_n$, K

All simulations were successful and have given new synthesized control law, which was able to either fully stabilize the system at required behavior (p-4 orbit) within short simulation interval of 200 iterations or at least secure the temporary stabilization. Total number of cost function evaluations for AP was 5345, for the second EA it was 6000, together 32.07 millions per each simulation. All experiments were 50 times repeated and performed in the Wolfram Mathematica environment. One experiment (simulation) took approx. 72 hours.

The novelty of this approach represents the synthesis of feedback control law F_n (5) (perturbation) for the artificial synthesized chaotic system inspired by original ETDAS control method.

$$x_{n+1} = \frac{A\left(2A - 2x_n^{\ 2} - 3x_n\left(A - x_n + Ax_n\right)\right)}{-A + x_n - x_n^{\ 2}} + F_n \tag{5}$$

Following two presented simulation results represent the best example of synthesized control law as well as the example of temporary stabilized system. Based on the mathematical analysis, the real p-4 UPO for unperturbed synthesized chaotic system has following values: $x_1 = -0.43$, $x_2 = -2.17$, $x_3 = 0.25$, $x_4 = -1.52$.

Description of the two selected simulation results covers direct output from AP – synthesized control law without coefficients estimated; further the notation with simplification after estimation by means of second algorithm DE and corresponding CF value.

7.1 Example 1

The first example of a new synthesized feedback control law F_n (perturbation) for the controlled synthesized system (1) inspired by original ETDAS control method (3) has the form (6) – direct output from AP and form (7) – with estimated coefficients by means of the second EA.

$$F_n = -\frac{x_n}{(K_2 - K_3)\left(\frac{x_n}{K_4} + K_1 - K_5 + K_6 + x_n\right)} \tag{6}$$

$$F_n = -\frac{0.00691794 x_n}{0.977563 x_n - 0.284243} \tag{7}$$

Simulation depicted in Figure 2 lends weight to the argument, that AP is able to synthesize a new control law securing quick and precise stabilization. The CF V value was 0.460952 for 40 measured iterations, which means that average error between actual and required system output was 0.01152 per iteration.

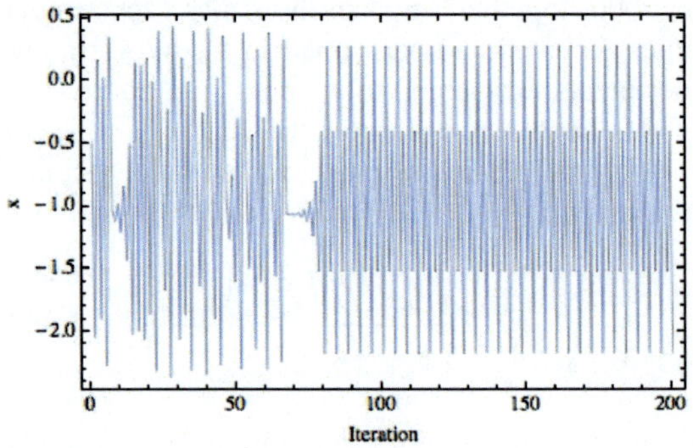

Fig. 2 Simulation results - the first example

7.2 *Example 2*

The second example of a new synthesized feedback control law has the form (8), which represents direct output from AP and form (9) with estimated coefficients.

$$F_n = -\frac{x_{n-8}}{K_1 K_2} \tag{8}$$

$$F_n = 0.000372493 x_{n-8} \tag{9}$$

Simulation output representing very quick but only temporary stabilization is depicted in Figure 3. The CF Value was 0.4639 (average error 0.01159 per iteration). These numerical results are only slightly different from the previous case, but confirm the extreme sensitivity of chaotic systems. More about this phenomenon is written in the conclusion.

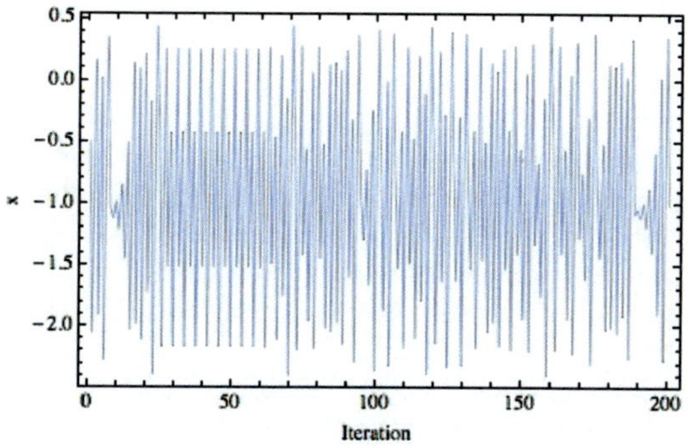

Fig. 3 Simulation results - the second example

8 Conclusions

This paper deals with a synthesis of the control laws by means of AP for stabilization of four selected chaotic system. The system of interest was the example of evolutionary synthesized chaotic system. In this presented approach, the analytic programming was used instead of tuning of parameters for existing control technique by means of EA's as in the previous research.

Presented results reinforce the argument that AP is able to solve this kind of difficult problems and to produce a new synthesized control law in a symbolic way securing desired behavior of chaotic system and stabilization.

Two simulation examples show two different results: not very quick but precise stabilization, as well as very fast and only temporary stabilization, although the numerical difference between these solutions was very low. Also the complexity of notation of chaotic controller is very different. This phenomenon is caused by the design of CF, which was borrowed from the previous research focused on the simpler cases, which were stabilization of stable state and p-2 orbit, and it has given satisfactory results. This CF brings too much of chaotic dynamics into the evolutionary process and searching for the best solutions. Nevertheless this fact lends weight to the argument, that AP is a powerful symbolic regression tool, which is able to strictly and precisely follow the rules given by cost function and synthesizes any symbolic formula, in the case of this research – the feedback controller for chaotic system.The question of more precise and robust stabilization for wide range of UPOs will be included into future research together with development of better cost functions, different AP data sets, and performing of numerous simulations to obtain more results and produce better statistics, thus to confirm the robustness of this approach.

Acknowledgments. This work was supported by European Regional Development Fund under the project CEBIA-Tech No. CZ.1.05/2.1.00/03.0089 and the project IT4Innovations Centre of Excellence No. CZ.1.05/1.1.00/02.0070.

References

[1] Zelinka, I., Senkerik, R., Navratil, E.: Investigation on evolutionary optimization of chaos control. Chaos, Solitons & Fractals 40(1), 111–129 (2009)

[2] Senkerik, R., Zelinka, I., Davendra, D., Oplatkova, Z.: Utilization of SOMA and differential evolution for robust stabilization of chaotic Logistic equation. Computers & Mathematics with Applications 60(4), 1026–1037 (2010)

[3] Pyragas, K.: Control of chaos via extended delay feedback. Physics Letters A 206, 323–330 (1995)

[4] Just, W.: Principles of Time Delayed Feedback Control. In: Schuster, H.G. (ed.) Handbook of Chaos Control. Wiley-Vch (1999)

[5] Pyragas, K.: Continuous control of chaos by self-controlling feedback. Physics Letters A 170, 421–428 (1992)

[6] Senkerik, R., Oplatkova, Z., Zelinka, I., Davendra, D.: Synthesis of feedback controller for three selected chaotic systems by means of evolutionary techniques: Analytic programming. Mathematical and Computer Modelling (2010), doi: 10.1016/j.mcm.2011.05.030

[7] Senkerik, R., Oplatkova, Z., Zelinka, I., Davendra, D., Jasek, R.: Evolutionary Synthesis of Control Law for Higher Periodic Orbits of Chaotic Logistic Equation. In: Proceedings of 25th European Conference on Modelling and Simulation. European, pp. 452–458 (2011)

[8] Oplatkova, Z., Senkerik, R., Belaskova, S., Zelinka, I.: Synthesis of Control Rule for Synthesized Chaotic System by means of Evolutionary Techniques. In: Proceedings of Mendel 2010, pp. 91–98 (2010)

[9] Zelinka, I., Guanrong, C., Celikovsky, S.: Chaos Synthesis by Means of Evolutionary algorithms. International Journal of Bifurcation and Chaos 18(4), 911–942 (2008)

[10] Senkerik, R., Zelinka, I., Oplatkova, Z.: Optimal Control of Evolutionary Synthesized Chaotic System, VUT Brno. In: Proceedings of 15th International Conference on Soft Computing MENDEL 2009, pp. 220–227 (2009)

[11] Zelinka, I., Oplatkova, Z., Nolle, L.: Boolean Symmetry Function Synthesis by Means of Arbitrary Evolutionary Algorithms-Comparative Study. International Journal of Simulation Systems, Science and Technology 9, 44–56 (2005)

[12] Zelinka, I., Davendra, D., Senkerik, R., Jasek, R., Oplatkova, Z.: Analytical Programming - a Novel Approach for Evolutionary Synthesis of Symbolic Structures. In: Kita, E. (ed.) Evolutionary Algorithms. InTech (2011)

[13] Lampinen, J., Zelinka, I.: New Ideas in Optimization – Mechanical Engineering Design Optimization by Differential Evolution, vol. 1, 20 p. McGraw-Hill, London (1999) ISBN 007-709506-5

[14] Oplatkova, Z., Zelinka, I.: Investigation on Evolutionary Synthesis of Movement Commands. Modelling and Simulation in Engineering 2009 (2009)

[15] Zelinka, I.: SOMA – Self Organizing Migrating Algorithm. In: Babu, B.V., Onwubolu, G. (eds.) New Optimization Techniques in Engineering. Springer (2004) ISBN 3-540-20167X

[16] Price, K.: An Introduction to Differential Evolution. In: Corne, D., Dorigo, M., Glover, F. (eds.) New Ideas in Optimization, pp. 79–108. McGraw-Hill, London (1999)

[17] Price, K., Storn, R.: Differential evolution homepage (2001), `http://www.icsi.berkeley.edu/~storn/code.html` (accessed May 15, 2012)

[18] Matousek, R.: GAHC: Improved GA with HC station. In: Proceedings of WCECS 2007, pp. 915–920 (2010) ISBN: 978 - 988 - 98671 - 6 - 4
[19] Matousek, R.: HC12: The Principle of CUDA Implementation. In: Proceedings of MENDEL 2010, pp. 303–308 (2010) ISBN: 978 - 80 - 214 - 4120 - 0
[20] Varacha, P., Jasek, R.: ANN Synthesis for an Agglomeration Heating Power Consumption Approximation. In: Recent Researches in Automatic Control, pp. 239–244. WSEAS Press, Montreux (1804) ISBN 978-1-61804-004-6
[21] Varacha, P., Zelinka, I.: Distributed Self-Organizing Migrating Algorithm Application and Evolutionary Scanning. In: Proceedings of the 22nd European Conference on Modelling and Simulation ECMS, pp. 201–206 (2008) ISBN 0-9553018-5-8

Application of Self-Organizing Migrating Algorithm in Five-Dimensional Chaotic Synchronization Systems via Active-Passive Decomposition

Thanh Dung Nguyen[1], T.T. Dieu Phan[2], and Ivan Zelinka[3]

[1] Faculty of Applied Informatics, Tomas Bata University in Zlin, Nad Stranemi 4511, 76005 Zlin, Czech Republic
nguyen.utb12@yahoo.com

[2] Faculty of Technology, Tomas Bata University in Zlin, Namesti T. G. Masaryka 275, 76272 Zlin, Czech Republic
phan@ft.utb.cz

[3] Faculty of Electrical Engineering and Computing Science, Technical University of Ostrava, Tr. 17. Listopadu 15, Ostrava, Czech Republic
ivan.zelinka@ieee.org

Abstract. This paper aims to present the combination of chaotic signal and evolutionary algorithm to estimate the unknown parameters in five-dimensional chaotic synchronization system via the active-passive decomposition method. The self-organizing migrating algorithm was used to estimate the unknown parameters. Based on the results from evolutionary algorithm, two identical chaotic systems were synchronized.

1 Introduction

Chaos theory is one of the most important achievements in nonlinear system research. Chaos dynamics are deterministic but extremely sensitive to initial conditions. The high unpredictability of chaotic signal is the most attractive feature of chaos based on secure communication. Several types of synchronization have been considered in communication systems, such as Pecora and Carroll (PC) method [7], Ott, Grebogi and Yorke (OGY) method [2], feedback approach [1],etc. In practice, some or all of the system's parameters are unknown. So that, many of proposed solutions focused on synchronization-based on the methods of parameter estimation [9], [5], [6], [10]. In [6], the parameters of a given dynamic model were estimated by minimizing the average synchronization error using a scalar time series, etc. However, most researches about chaos synchronization concern the synchronization of low dimensional systems. Recently, Roy presented a five-dimensional system by adding two additional variables into the three-dimensional Lorenz system [8]. Basic properties of the five-dimensional system have been analyzed by means of the Lyapunov exponents and bifurcation diagrams. Their study showed

I. Zelinka et al. (Eds.): Nostradamus: Mod. Meth. of Prediction, Modeling, AISC 192, pp. 203–213.
springerlink.com

that the system could generate various complex chaotic attractors when the system parameters were changed.

A new class of stochastic optimization algorithm called self-organizing migrating algorithm (SOMA) was proposed in literature [3], [11]. It was demonstrated that SOMA has ability to escape the traps in local optimal and easily to achieve the global optimal. Therefore, SOMA has attracted much attentions and wide applications in different fields mainly in various continuous optimization problems.

Motivated by the aforementioned studies, this paper aims to present the combination of chaotic signal with SOMA to estimate the unknown parameters in 5D chaotic synchronization system via active passive decomposition method (APD method). Based on the results of SOMA, the estimated parameters were used to synchronize two 5D chaotic systems.

2 Problem Formulation

2.1 Active-Passive Decomposition Method

L. Kocarev and U. Parlitz [4] proposed a general drive response scheme named as Active-Passive Decomposition (APD method). The basic idea of the active-passive synchronization approach consisted in a decomposition of a given chaotic system into an active and a passive part where different copies of the passive part synchronize when driven by the same active component. In the following, they explained the basic concept and terminology of the active passive decomposition. Consider an autonomous n-dimensional dynamical system, which is chaotic as

$$\dot{u} = g(u) \tag{1}$$

The system is rewritten as a non-autonomous system:

$$\dot{x} = f(x, s) \tag{2}$$

where x is a new state vector corresponding to u and s is some vector valued function of time given by

$$s = h(x) \tag{3}$$

The pair of functions f and h constitute a decomposition of the original vector field g, and are chosen such that any system.

$$\dot{y} = f(y, s) \tag{4}$$

given by the same vector field f, the same driving signal s, but different variables y, synchronizes with the original system. Here, x constitutes the active system while y is the passive one. Consider the difference of these two systems, the synchronization of the pair of identical systems (2) and (4) occur if the dynamical system describes the evolution of the difference $|x - y| \rightarrow 0$ as $t \rightarrow \infty$.

2.2 The Parameter Estimation

When estimating the parameters, suppose that the structure of the system is known in advance, the transmitter (drive) system is set with original parameters and the parameter in receiver (response) system is unknown. Therefore, the problem of parameter estimation can be formulated as the following optimization problem:

$$\mathrm{CF} = \sqrt{\frac{1}{m}\sum_{t=1}^{m}|x-y|^2} \tag{5}$$

where m denotes length of time used for parameter estimation, the parameter can be estimated by minimum the cost function CF (5).

Because of the irregular dynamic behavior of chaotic systems, the parameter estimation of chaotic systems is a multidimensional continuous optimization problem, the parameters are not easy to obtain. There are often multiple variables in the problem and multiple local optimums in the landscape of cost function, so traditional optimization methods are easy to trap in local optima and difficult to achieve the global optimal parameters. Therefore, SOMA was chosen because it has been demonstrated that SOMA has the ability to converge toward the global optimum.

3 Self-Organizing Migration Algorithm

Self-Organizing Migration Algorithm (SOMA)- one of the evolutionary algorithms was chosen. It imitates nature process of wildlife migration. The method was established in 1999, developed by professor Ivan Zelinka at the University of Tomas Bata, Zlin. SOMA is a stochastic optimization algorithm that is modeled on the social behavior of cooperating individuals [3], [11]. The approach is similar to that of genetic algorithms (GA), although it is based on the idea of a series of "migrations" by a fixed set of individuals, rather than the development of successive generations. It can be applied to any cost-minimization problem with a bounded parameter space, and is robust to local minimum. SOMA works on a population of candidate solutions in loops called migration loops. The population is initialized randomly distributed over the search space at the beginning of the search. In each loop, the population is evaluated and the solution with the highest fitness becomes the leader L. Apart from the leader, in one migration loop, all individuals will traverse the input space in the direction of the leader. Mutation, the random perturbation of individuals, is an important operation for evolutionary strategies (ES). It ensures the diversity amongst the individuals and also provides the means to restore lost information in a population. Mutation is different in SOMA compared with other ES strategies. SOMA uses a parameter called perturbation of migration (PRT) to achieve perturbation. The effect of this parameter on SOMA is the same as mutation parameter has for GA. The novelty of this approach is that the PRT vector is created before an individual starts its journey over the search space. The PRT vector defines the final movement of an active individual in search space. The randomly generated binary perturbation vector controls allowed dimensions for an individual. If an element of the perturbation vector is set to zero, then the individual is not allowed to change its position in the corresponding dimension. An individual will travel a certain distance (called the Path Length) towards the leader in n steps of defined length. If the Path Length is chosen to be greater than one, then the individual will overshot the leader. This path is perturbed randomly.

There are many of SOMA variations which are differentiated by way of migration. In our case, SOMA-All-To-All variation has been chosen, in which all individuals move towards the other individuals. This strategy often needs less cost function evaluations to reach the global optimum than the other strategies.

More detailed description of SOMA can be found in [3], [11], [12].

4 Simulation and Result

4.1 Synchronization of 5D Lorenz Chaotic System

In this section, APD technique is applied to achieve the synchronization between two identical chaotic systems. The mathematical description of 5D system is as follows [8]:

$$U = \begin{cases} \dot{x} = a(y-x) \\ \dot{y} = rx - y - zx \\ \dot{z} = xy - bz + 2uw \\ \dot{u} = -cau + 2(a/c)w \\ \dot{w} = -2uz + 2ru - cw \end{cases} \tag{6}$$

where x, y, z, u and w are the state variables, and a, b, c and r are the positive real constants. The 5D system exhibits a chaotic attractor for a=10, b=8/3, c=2 and r=24.75 as shown in Fig. 1. We use a subscript '1' for the signals in the response, the synchronization has been done by variables y. The response system U_1 is described by the following equations:

$$U_1 = \begin{cases} \dot{x}_1 = a(s_{(t)} - x_1) \\ \dot{y}_1 = rx_1 - y_1 - z_1x_1 \\ \dot{z}_1 = x_1y_1 - bz_1 + 2u_1w_1 \\ \dot{u}_1 = -cau_1 + 2(a/c)w_1 \\ \dot{w}_1 = -2u_1z_1 + 2ru_1 - cw_1 \end{cases} \tag{7}$$

where a, b, c and r are unknown parameters in response system, $s_t = y$ is the transmitted signal. Subtracting system (6) from (7) yields the error dynamical system between two system e(t) were used to create a cost function CF representing the root mean square error (RMSE) of synchronization between U and U_1:

$$\mathrm{CF} = \sqrt{\frac{1}{m}\sum_{t=1}^{m} |U(t) - U_1(t)|^2} \tag{8}$$

SOMA is used to find a suitable parameter a, b, c and r such that the cost function CF can be approached to minimum point. The minimum value of CF guarantees for the best solution with suitable parameters, two chaotic systems are synchronized.

4.2 Parameter Setup

In our simulations, the initial states of the drive system (6) and the response system (7) are taken as $x(0) = 1, y(0) = 2, z(0) = 1, u(0) = 1, w(0) = 1$ and $x_1(0) = -1, y_1(0) = -2, z_1(0) = -1, u_1(0) = -1, w_1(0) = -1$, respectively. Hence the error system has the initial values $e_x(0) = 2, e_y(0) = 4, e_z(0) = 2, e_u(0) = 2$ and $e_w(0) = 2$. SOMA was used to solve the systems, which the control parameters setting are given in Table 1. Simulations were implemented using Mathematica and executed on HP 8200 core i3-2120 3.3Ghz, 12GB personal computer.

4.3 Experimental Results

4.3.1 Case Study 1: simulation on One-Dimensional Parameter Estimation

In this case, one-dimensional parameter estimation is considered. That means three parameters are known in advance with the original value; one parameter 'a' is unknown and need to be estimated. The initial guesses are in the range [5, 15] for 'a', SOMA has found the best results of CF=0.555664. Both the worst and the best values of the cost function approached minimum value quickly after 3 migrations as shown in Fig. 2(a). SOMA has found the optimum value of a=10.0000 as

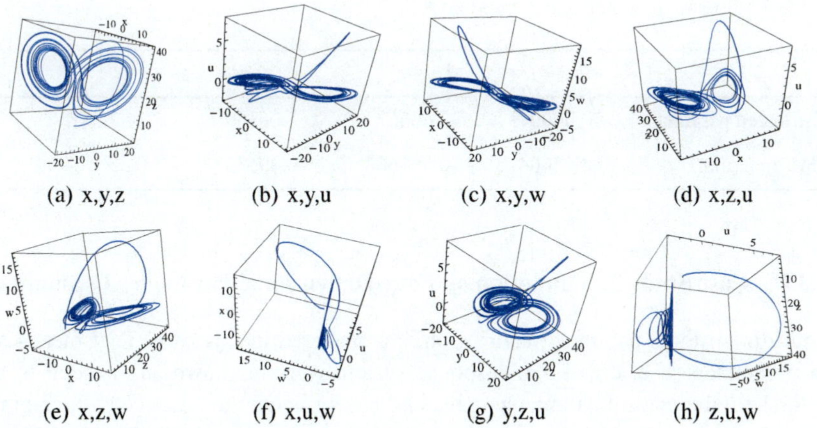

(a) x,y,z (b) x,y,u (c) x,y,w (d) x,z,u

(e) x,z,w (f) x,u,w (g) y,z,u (h) z,u,w

Fig. 1 Views of the chaotic attractor of 5D Lorenz system shown in Fig. 2(b). Similar for the others, the initial guesses are in the range for $b \in [0,5], c \in [0,5]$ *and* $r \in [20,30]$; the estimated parameters b,c,r were also found by SOMA. The estimated parameters and minimum CF values were presented in Table 2. It can be seen that the best results (estimated values) obtained by SOMA are almost the same original values.

Table 1 SOMA parameter setting

Parameter	Population size	Migrations	Step	Path length	Perturbation	Minimal diversity
Value	20	20	0.11	3	0.1	-1

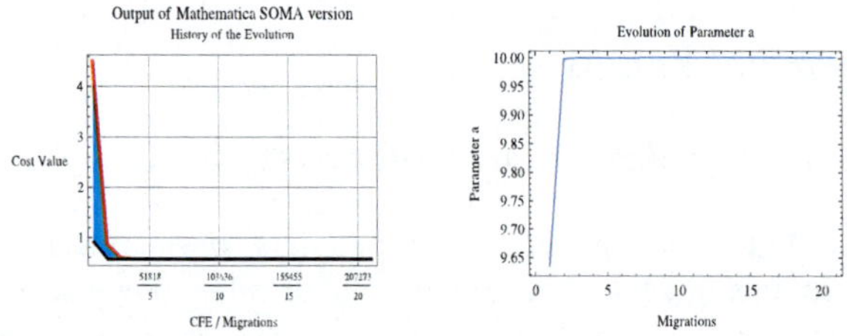

(a) (b)

Fig. 2 (a). CF_a evolution by SOMA and (b). Evolution history of a

Table 2 1D Estimated parameters by SOMA

	a	b	c	r
Estimated parameters	10.0000	2.66667	2.06159	24.7500
Cost function	0.555664	0.555664	0.496329	0.555664

4.3.2 Case Study 2: Simulation on Four-Dimensional Parameter Estimation

Four-dimensional parameter estimation is considered in this case. That means all of parameters a, b, c and r of response system were unknown and needed to be estimated, the control parameters a,b,c and r were known in advance with original value in drive system. The initial guesses are in the range for $a \in [5,15], b \in [0,5], c \in [0,5], r \in [20,30]$. Because of the sensitive of chaotic system, the cost function CF is so complex and has a lot of local optimums. But after 5 migrations, SOMA has found the best results of CF=0.496329, the best values of the cost function quickly approached minimum value. Both the worst and the best values of cost function approached minimum gradually after 8 migrations as shown in Fig. 3(c). SOMA had found the optimum value of a, b, c and r as shown in Table 3.

Table 3 4D Estimated parameters by SOMA

Estimated parameters				Cost function
a	b	c	r	CF
10.0000	2.66667	2.06159	24.75	0. 496329

In case 1 or 2 parameters among of a, b, c and r in response system were known in advance with the original value; the others (3 or 2) were also found with the optimum value by SOMA as shown in Table 4, 5. Both the worst and the best values of cost function quickly approach minimum after 8 migrations as shown in Fig. 3(a), Fig. 3(b).

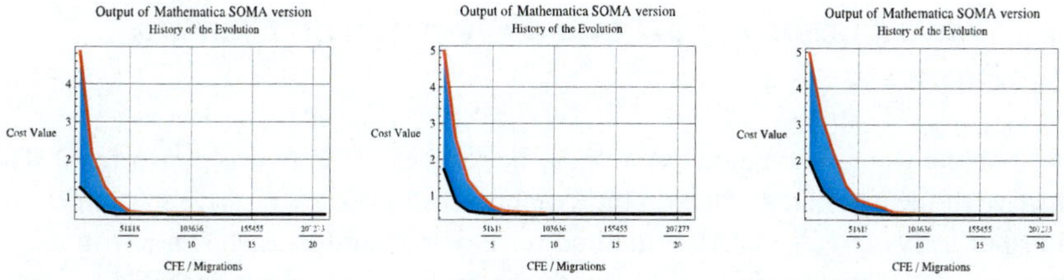

(a) CF_{ab} (b) CF_{abc} (c) CF_{abcr}

Fig. 3 CF_{ab},CF_{abc} and CF_{abcr} evolution by SOMA

As shown in Table 2- 5, the estimated parameters always have the similar values with original values. However, we can easily recognize the effect of parameter 'c' on system. The cost-functions were always found CF=0.555664 with 'c=2' (such as in case study $CF_a, CF_{ab}, CF_{abr}, ...$), while cost-functions CF approach to area CF=0.49632x in case 'c' were estimated by SOMA (such as $CF_c, CF_{ac}, CF_{abc}, \cdots$). The cost-functions CF=0.49632x are smaller than CF=0.555664. So that, the final estimated values were chosen: a=10.00000, b=2.66667, c=2.06159 and r=24.75 to ensure that the synchronization error approaches to minimum. Thus, the actual parameters were fully identified. The values of cost function always approach to optimum values, the estimated parameters obtained by SOMA and original parameters have the similar values. So, it is demonstrated that SOMA is effective to estimate parameters for 5D-chaos synchronization system.

Table 4 2D Estimated parameters by SOMA

	Estimated parameters		Cost function
a, b	10.0000	2.66667	0.555664
a, c	10.0000	2.06159	0.496329
a, r	10.0000	24.7500	0.555664
b, c	2.66667	2.06159	0.496329
b, r	2.66667	24.7500	0.555664
c, r	2.06159	24.7500	0.496329

Table 5 3D Estimated parameters by SOMA

	Estimated parameters			Cost function
a, b, c	10.0000	2.66667	2.06159	0.496328
a, b, r	10.0000	2.66667	24.7500	0.555664
a, c, r	10.0000	2.06159	24.7500	0.496328
b, c, r	2.66667	2.06159	24.7500	0.496329

4.4 Synchronization of 5D Chaotic System with Estimated Parameter

Based on the values estimated by SOMA, the response system was constructed. The effective of the estimated value on the synchronization errors of drive system U and on response system U_1 via APD method were demonstrated as shown in Fig. 5. As shown in Fig. 4, the synchronization between drive system (dash line) and response system (red line) do not exist in the phase space of chaotic attractor. When APD was applied with the estimated parameters, Fig. 5 displays that the phase space of chaotic attractor between drive system and response system approached and synchronized although they were started under different initial values.

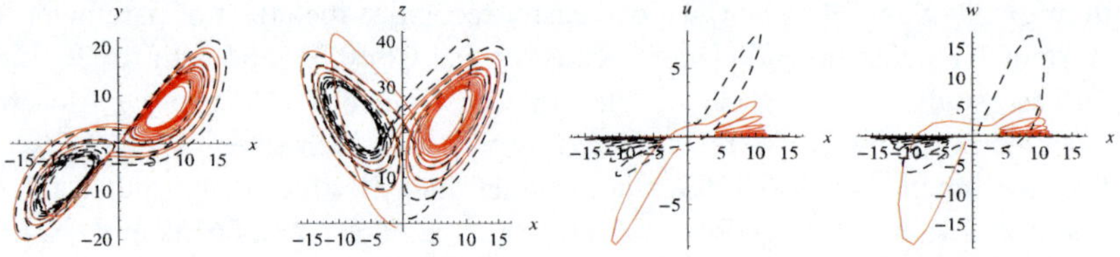

(a) x-y (b) x-z (c) x-u (d) x-w

Fig. 4 Projection onto 2D plane of the 5D Lorenz attractor: Non-synchronization

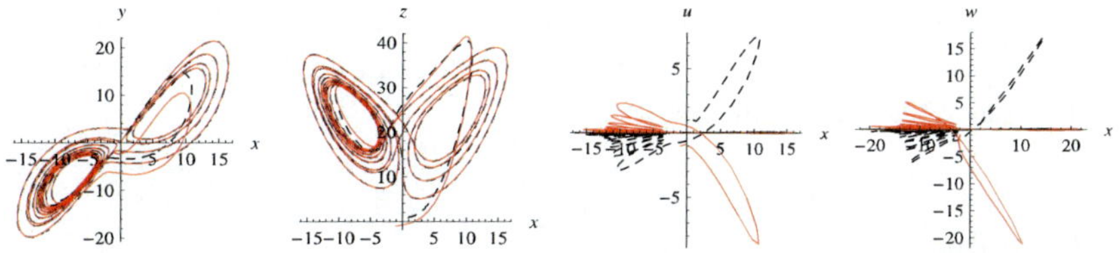

(a) x-y (b) x-z (c) x-u (d) x-w

Fig. 5 Projection onto 2D plane of the 5D Lorenz attractor: Synchronization

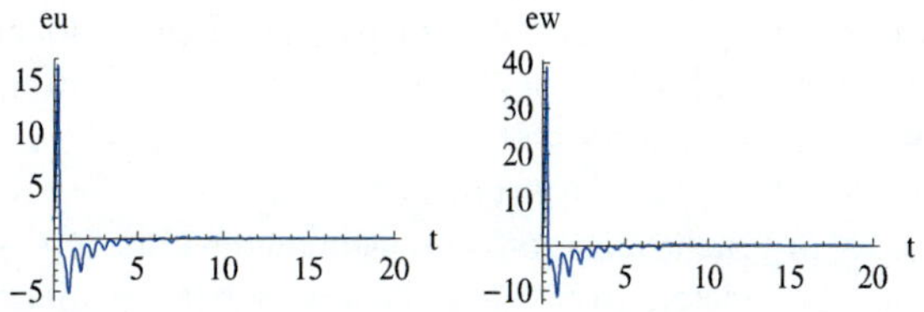

(a) (b)

Fig. 6 Synchronization error of variable u and w

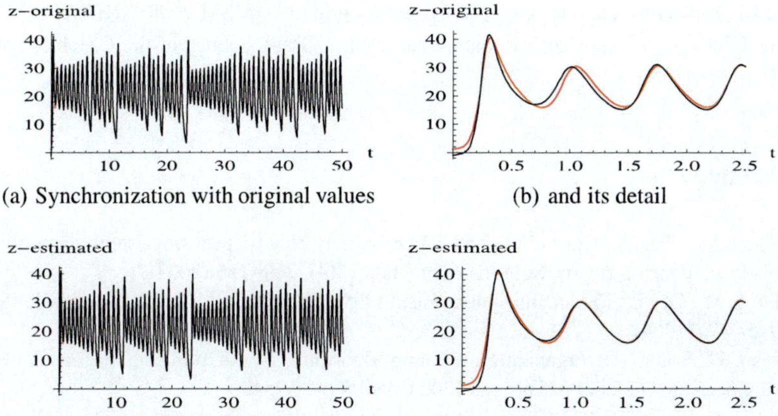

(a) Synchronization with original values (b) and its detail

(c) Synchronization with estimated values (d) and its detail

Fig. 7 Compare the synchronization quality between original with estimated values

It seems that there are not synchronization between drive system and response system of variable 'u' and 'w' in Fig. 5(c), Fig. 5(d). But Fig. 6 showed that the synchronization error of e_u and e_w approach to 0 after $t \geq 10$. That means the synchronization was achieved when both variables 'u' and 'w' approached 0 with

$t \geq 10$. This is a characteristic of these 5D chaotic systems as shown in Fig. 1. Therefore, it can be stated that the estimated values and APD method are effective to synchronize for two 5D-chaotic systems.

5 Conclusions

In this paper, the APD method was applied to synchronize two identical 5D-chaotic systems. Parameter estimation for chaotic synchronization system was formulated as a multidimensional optimization problem. SOMA was used to estimate the optimum values for unknown parameters, and it easily escaped the local optimum trap and achieved the global optimum parameters. Based on the estimated parameter from SOMA, two chaotic systems were synchronized. It is difficult to recognize what differs Fig. 7(a) from Fig. 7(c), but in its detail as shown in Fig. 7(b) and Fig. 7(d), the quality of synchronization with estimated parameters is clearly higher than that of original parameters from $t \geq 1$. That means the quality of communication system is increased with the estimated parameters.

As the future subject, it can be extended to other synchronization method to find the more suitable, which method is much stronger of synchronization.

Acknowledgements. This work was supported by grant No. MSM 7088352101 of the Ministry of Education of the Czech Republic and by the Grant Agency of the Czech Republic GACR 102/09/1680.

References

1. Chen, M., Han, Z.: Controlling and synchronizing chaotic genesio system via nonlinear feedback control. Chaos, Solitons & Fractals 17(4), 709–716 (2003)
2. Ding, M., Ott, E.: Enhancing synchronism of chaotic systems. Physical Review E 49(2), 945–948 (1994)
3. Ivan, Z.: Soma–self organizing migrating algorithm. New Optimization Techniques in Engineering, ch. 7, 33 p. Springer (2004) ISBN pp. 3–540
4. Kocarev, L., Parlitz, U.: General approach for chaotic synchronization with applications to communication. Physical Review Letters 74(25), 5028–5031 (1995)
5. Nguyen, T., Zelinka, I.: Using method of artificial intelligence to estimate parameters of chaotic synchronization system. In: Proceedings of 17th International Conference on Soft Computing, vol. 1, pp. 22–29 (2011)
6. Parlitz, U.: Estimating model parameters from time series by autosynchronization. Physical Review Letters 76(8), 1232–1235 (1996)
7. Pecora, L., Carroll, T.: Synchronization in chaotic systems. Physical Review Letters 64(8), 821–824 (1990)
8. Roy, D., Musielak, Z.: Generalized lorenz models and their routes to chaos. ii. energy-conserving horizontal mode truncations. Chaos, Solitons & Fractals 31(3), 747–756 (2007)

9. Shen, L., Wang, M.: Robust synchronization and parameter identification on a class of uncertain chaotic systems. Chaos, Solitons & Fractals 38(1), 106–111 (2008)
10. Wu, X., Hu, H., Zhang, B.: Parameter estimation only from the symbolic sequences generated by chaos system. Chaos, Solitons & Fractals 22(2), 359–366 (2004)
11. Zelinka, I.: Real-time deterministic chaos control by means of selected evolutionary techniques. Engineering Applications of Artificial Intelligence 22(2), 283–297 (2009)
12. Zelinka, I.: Self-organizing migrating algorithm (2011), `http://www.ft.utb.cz/people/zelinka/soma/`

Using Differential Evolution Algorithm in Six-Dimensional Chaotic Synchronization Systems

Thanh Dung Nguyen[1], T.T. Dieu Phan[2], and Ivan Zelinka[3]

[1] Faculty of Applied Informatics, Tomas Bata University in Zlin, Nad Stranemi 4511, 76005 Zlin, Czech Republic
nguyen.utb12@yahoo.com
[2] Faculty of Technology, Tomas Bata University in Zlin, Namesti T. G. Masaryka 275, 76272 Zlin, Czech Republic
phan@ft.utb.cz
[3] Faculty of Electrical Engineering and Computing Science, Technical University of Ostrava, Tr. 17. Listopadu 15, Ostrava, Czech Republic
ivan.zelinka@ieee.org

Abstract. This paper presents the use of artificial intelligence in the synchronization between two six-dimensional chaotic systems. Differential evolution algorithm is used to estimate the unknown parameters of six-dimensional chaotic synchronization system via Pecora and Carroll method. The parameters are estimated by minimizing the synchronization errors, and they are used to synchronize two systems. The optimal values ensure that the best quality of the synchronization is achieved.

1 Introduction

Chaotic systems and their applications in secure communications have received a great deal of attention since Pecora and Carroll proposed a method to synchronize two identical chaotic systems under different initial conditions [9]. The high unpredictability of chaotic signal is the most attractive feature of chaos based secure communication. Several secure communication techniques were proposed and analysed including chaos masking (CMS), chaotic shift keying (CSK), differential chaos shift keying (DCSK), chaotic frequency modulation, and in-phase/ anti phase synchronization, [3], [1], [2], etc. Most of the systems are based on synchronization of chaos between a transmitter and a receiver, which are linked by a transmission channel. However, because of the difference in the driving forces, the receiver chaotic system may not have exactly the same parameters as those of the transmitter system, synchronization of the two chaotic waveforms generated by the transmitter and the receiver is not complete. Therefore, the system performance generally depends on the quality of chaos synchronization. To ensure the quality of chaotic communication, minimization of synchronization errors is very important. In [11], the authors proposed an approach for constructing chaotically synchronizing systems by a suitable separation of these systems. The method is based on the stability criterion of

I. Zelinka et al. (Eds.): Nostradamus: Mod. Meth. of Prediction, Modeling, AISC 192, pp. 215–224.
springerlink.com

linear systems, and the synchronization of constructed systems is ensured by that all eigenvalues of the Jacobian matrix A have negative real parts. In [4], authors introduced a modified method for synchronization and cascading chaotic systems. It is based on generalizing the PC method in a systematic way to design a response system with same dimension as the drive and that yields the same result obtained within the original PC method with cascade of two subsystems. In [7], [8], the parameters of a given dynamic model were estimated by minimizing the average synchronization error using a scalar time series. However, most research about chaos synchronization concerns the synchronization of low (three-) dimensional systems. In [5], Kennamer constructed a six-dimensional (6D) system by adding three additional variables into the three-dimensional Lorenz system. This system has strange attractors that have properties very similar to the Lorenz strange attractor and their routes to chaos are via chaotic transients.

Besides, differential evolution algorithm (DE) has been proposed as an alternative genetic algorithm for unconstrained continuous optimization problems. In DE system, a population of solutions is initialized randomly, which is evolved to find optimal solutions through the mutation, crossover and selection of operation procedures. It uses a simple differential operator to create new candidate solutions and one-to-one competition scheme to select a new candidate, which works with real numbers in natural manner and avoids complicated generic search for operators in genetic algorithms. Therefore, DE has attracted much attention and wide applications in different fields mainly in various continuous optimization problems.

In this paper, artificial intelligence is used to find the best quality of synchronization between two 6D chaotic systems. Differential evolution algorithm is used to estimate the unknown parameters in 6D chaotic synchronization system via Pecora and Carroll method under different initial conditions. The parameters are estimated by minimizing the synchronization errors, and they are used to synchronize two 6D chaotic systems. The optimal values ensure that the best quality of the synchronization is achieved.

2 System Description

A six dimensional Lorenz system was constructed by Kennamer in his M.S thesis in 1995 [5]. The system have similar strange attractors with the 3D Lorenz system. In addition, the route to chaos via chaotic transients observed in the 6D Lorenz systems is the same as that identified for the 3D Lorenz system. The noteworthy points are the appear of parameter c, and the difference of parameter r between two systems. The 6D Lorenz system is described by following nonlinear differential equations:

$$X = \begin{cases} \dot{x}_1 = a(x_2 - x_1) \\ \dot{x}_2 = rx_1 - x_2 - x_1x_3 + x_3x_4 - 2x_4x_6 \\ \dot{x}_3 = x_1x_2 - bx_3 - x_1x_5 - x_2x_4 \\ \dot{x}_4 = -cax_4 + (a/c)x_5 \\ \dot{x}_5 = x_1x_3 - 2x_1x_6 + rx_4 - cx_5 \\ \dot{x}_6 = 2x_1x_5 + 2x_2x_4 - 4bx_6 \end{cases} \tag{1}$$

where $x_1 - x_6$ are the state variables, and a, b, c and r are the positive real constants. The system (1) has a chaotic attractor as shown in Fig. 1 when a = 10, b=8/3, c =3 and r=41.

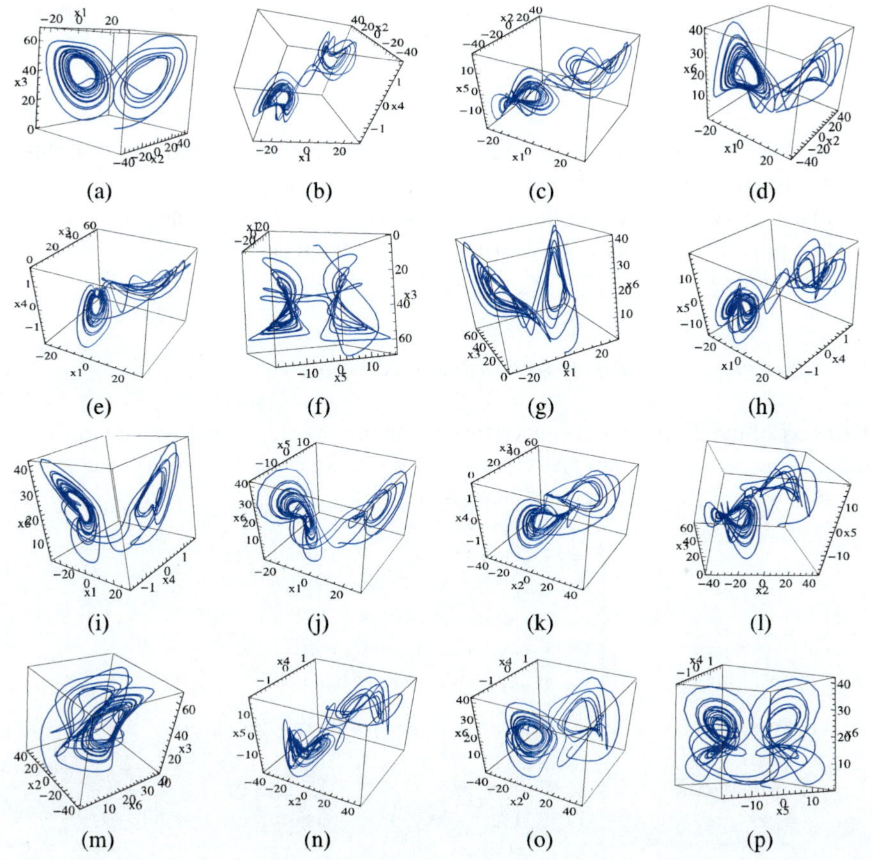

Fig. 1 Views of the chaotic attractor of 6D Lorenz system

3 Chaos Synchronization Using Pecora-Carroll Method

In 1989, Pecora and Carroll introduced a method for constructing synchronizing chaotic systems [9]. They showed that when a state variable from a chaotic system was input into a replica subsystem of the original one, both systems could be synchronized identically. They decomposed the dynamical system

$$\dot{u} = g(u) \tag{2}$$

into two subsystems,

$$\begin{aligned} \dot{v} &= g_v(v, w) \\ \dot{w} &= g_w(v, w) \end{aligned} \tag{3}$$

with $v = (u_1, ..., u_k)$ and $w = (u_{k+1}, ..., u_n)$, and considering one of the decomposed subsystems as the driving signal, say v, to be injected into the response system,

$$\dot{w}' = g_w(v, w') \tag{4}$$

that is given by the same vector field g_w, the same driving v, but different variables w' synchronizes with the original w subsystem.

Consider the difference of these two systems e = w'-w. The synchronization of the pair of identical systems (3) and (4) occurs if the dynamical system describing the evolution of the difference $|w' - w| \rightarrow 0$ as $t \rightarrow \infty$.

3.1 Synchronization of 6D Chaotic Systems

Using x_1 of the decomposed subsystems as the driving signal in drive system X. In this case, x_1 is injected into the response system Y. Since $x_1 = y_1$, we only consider the following drive and response subsystems:

$$X = \begin{cases} \dot{x}_2 = rx_1 - x_2 - x_1x_3 + x_3x_4 - 2x_4x_6 \\ \dot{x}_3 = x_1x_2 - bx_3 - x_1x_5 - x_2x_4 \\ \dot{x}_4 = -cax_4 + (a/c)x_5 \\ \dot{x}_5 = x_1x_3 - 2x_1x_6 + rx_4 - cx_5 \\ \dot{x}_6 = 2x_1x_5 + 2x_2x_4 - 4bx_6 \end{cases} \tag{5}$$

And the response system Y is described by the following equations:

$$Y = \begin{cases} \dot{y}_2 = ry_1 - y_2 - y_1y_3 + y_3y_4 - 2y_4y_6 \\ \dot{y}_3 = y_1y_2 - by_3 - y_1y_5 - y_2y_4 \\ \dot{y}_4 = -cay_4 + (a/c)y_5 \\ \dot{y}_5 = y_1y_3 - 2y_1y_6 + ry_4 - cy_5 \\ \dot{y}_6 = 2y_1y_5 + 2y_2y_4 - 4by_6 \end{cases} \tag{6}$$

where a, b, c and r are unknown parameters in response system.

Consider the difference of these two systems e = X-Y. The synchronization of the pair of identical systems (5) and (6) occurs if the dynamical system describing the evolution of the difference $|X - Y| \rightarrow 0$ as $t \rightarrow \infty$. Subtracting system (5) from system (6) yields the error dynamical system between two system e(t)=X(t)-Y(t) were used to create a cost function CF representing the root mean square error (RMSE) of synchronization between X and Y:

$$\text{CF} = \sqrt{\frac{1}{m}\sum_{t=1}^{m} |X(t) - Y(t)|^2} \tag{7}$$

The parameter estimation can be formulated as a multidimensional nonlinear problem to minimize the cost function CF. When estimating the parameters, suppose the structure of the system is known in advance, the transmitter (drive) system is set with original parameters and the parameter in receiver (response) system is unknown. Because of the irregular dynamic behavior of chaotic systems, the parameters are not easy to obtain. In addition, there are often multiple variables in the problem and multiple local optimums in the landscape of cost function, so traditional optimization methods are easy to trap in local optima and difficult to achieve the global optimal parameters. DE was chosen because it has been demonstrated that the algorithm has the ability to converge toward the global optimum. Therefore, DE are used to find a suitable parameter a, b, c and r such that the cost function CF can be asymptotical approach to minimum point. The minimum value of cost function guarantee of the best solution with suitable parameters. Systems are asymptotically synchronized.

3.2 *Differential Evolutionary Algorithms*

DE is the simple yet efficient population-based evolutionary algorithm introduced by Storn and Price in 1995 [10]. Over the last few years, DE has been investigated by many researches and proved very easy to be implemented with superior performance in many real optimization problems. In DE system, a population of solutions is initialized randomly, which is evolved to find optimal solutions through the mutation, crossover and selection of operation procedures. It uses a simple differential operator to create new candidate solutions and one-to-one competition scheme to select a new candidate, which works with real numbers in natural manner. The procedure of standard DE could be described:

Step 1: Choose control parameters: NP, F, Cr and the Generations (stopping criterion).

Step 2: Randomly initialize the population of DE with NP individuals in the search space.

Step 3: Evolve the system from a random initial state and obtain N sampling points of the state X at certain time points.

Step 4: Evaluate the objective function value: first, for each individual of DE, evolves the system from the same initial state as *Step 3* and obtains N sampling points of the state X at the same time points as *Step 3*, then calculate objective function value. Determine X_{best} in current generation.

Step 5: Perform mutation operation.

Step 6: Perform crossover operation to obtain crossover trial vectors.

Step 7: Calculate the objective function value of crossover trial vectors.

Step 8: Perform selection operation to generate individual for next generation.

Step 9: Determine the best individual of the current new generation with the best objective value. If the objective function value is better than the objective function value of X_{best}, then update X_{best} and its objective value with the value and objective value of the current best individual.

Step 10: Check if the stopping criterion is met, then output X_{best}; otherwise, go to *Step 4*.

Five strategy rules of DE are DERand1Bin, DERand2Bin, DEBest2Bin, DERand1DIter and DELocalToBest, which will be used in this study. More detailed description of DE can be found in [10], [6].

4 Simulation and Result

4.1 Parameter Setup

In our simulations, the initial states of the drive system (5) and the response system (6) are taken as $x_1(0) = 1, x_2(0) = 1, x_3(0) = 1, x_4(0) = 1, x_5(0) = 1, x_6(0) = 1$ and $y_2(0) = 0, y_3(0) = 0, y_4(0) = 0, y_5(0) = 0, y_6(0) = 0$, respectively. Hence the error system has the initial values $e_2(0) = 1, e_3(0) = 1, e_4(0) = 1, e_5(0) = 1$ and $e_6(0) = 1$. DE was used to solve the systems, which the control parameters setting are given in Table 1. Simulations were implemented using Mathematica programming language and executed on HP 8200 core i3-2120 3.3Ghz, 12GB personal computer.

Table 1 DE parameter setting

Parameter	Value
Population size NP	20
Crossover rate Cr	0.8
Mutation factor F	0.7
Generation number	200

4.2 Experimental Results

As mentioned above, the parameters a, b, c and r are known in advance with original value in drive system. In response system, they are unknown and needed to be estimated. The initial guesses are in the range for $a \in [5, 15]$, $b \in [0, 5]$, $c \in [0, 5]$ *and* $r \in [35, 45]$ and the control parameters of DE are set as Table 1. Because of the sensitive of chaotic system, the cost function CF is so complex and has a lot of local optimum. But DE has found the best results of CF as shown in Fig. 2(a), it showed the best values of the cost function gradually approached minimum value (CF=0.225836) after 55 generations. There are huge differences between the worst and the best in first 5 generations as can be seen in Fig. 2(b) and Fig. 2(a). But it quickly approached the best value after 10 generations, both the best and the worst approached the optimum value after 90 generations as shown in Fig. 2(c).

Fig. 3 displayed the history evolution of estimated parameters by DE. As can be seen in the history evolution of parameter 'a' in Fig. 3(a), there was the instability

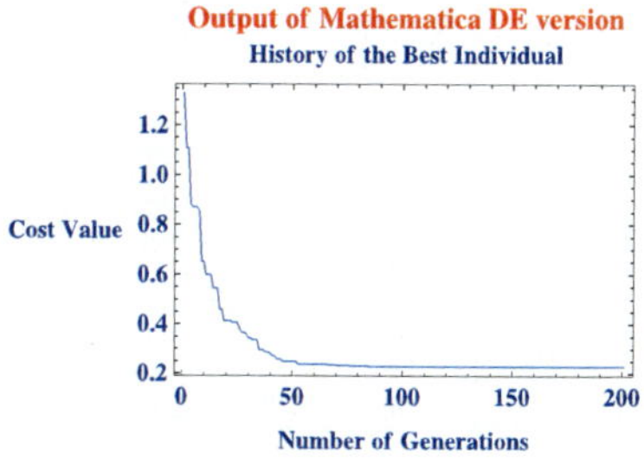

(a)

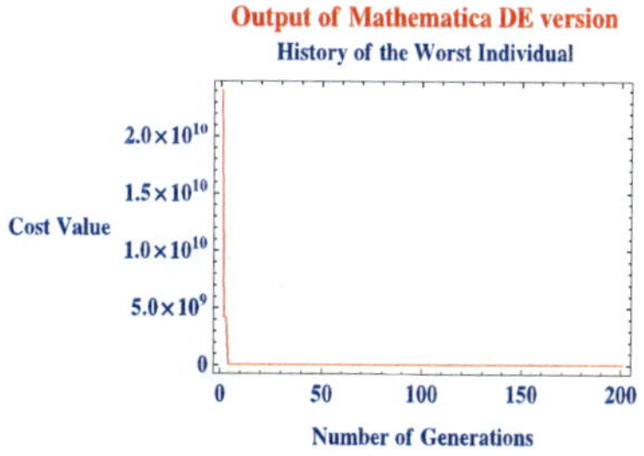

(b)

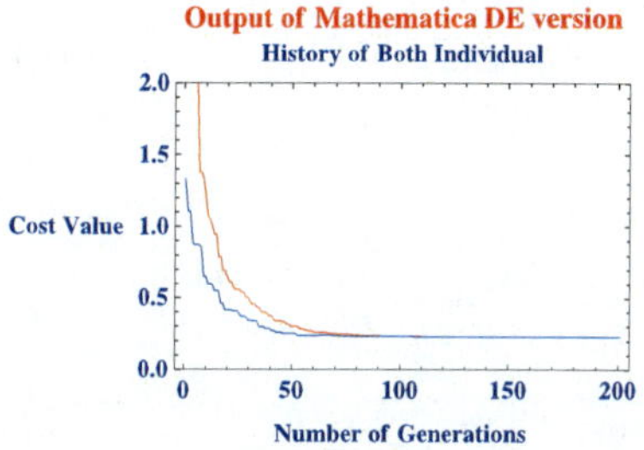

(c)

Fig. 2 CF_{abcr} evolution by DE

in the early stages, but after 50 generations, it reached the stability with a value of 10. Similar for the others, parameters b, c, r also achieved the optimum values after 50 generations as shown in Fig. 3. Therefore, it demonstrated that DE was success in finding the unknown parameters of response chaotic system.

As shown in Table 2, the estimated parameters were executed by five strategy rules of DE, and they got the similar values together. The values of cost function always approach to optimum values. There were not significant differences among the results from 5 simulations, the smallest value of CF is 0.225836. So that, the final estimated values were chosen: a= 10.0000, b= 2.66667, c= 3 and r= 41 to ensure that the synchronization error approaches to minimum. Thus, the parameters of response system were identified.

4.3 Synchronization of 6D Chaotic Systems with Estimated Parameters

Based on the estimated values, the response system was constructed. The effective of the estimated value on the synchronization errors of drive systems X and on response system Y via PC method were demonstrated as shown in Fig. 5.

As shown in Fig. 4, the synchronization between drive system and response system did not exist. Fig. 4(a) showed the drive state variable x_2 versus the response state variable y_2, the correlation did not occur in a single straight line, and the non-synchronized state looked so complex. Looking at the plot of x_3, x_4, x_5, x_6 versus

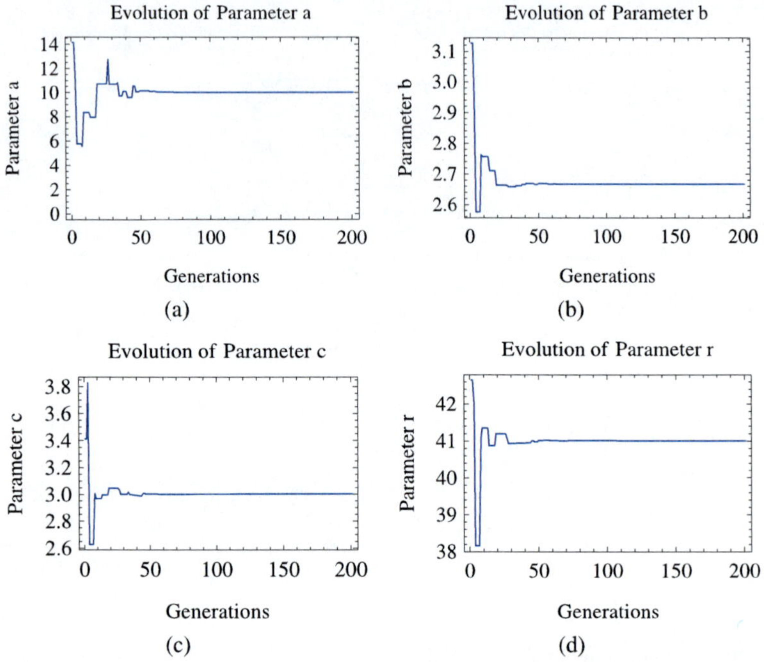

Fig. 3 Evolution history of parameters

Table 2 Estimated parameters by DE

	Estimated parameters				Cost function
	a	b	c	r	CF
DERand1Bin	10.000	2.66667	3	41	0.225836
DERand2Bin	10.000	2.66666	3	40.9999	0.225890
DEBest2Bin	10.000	2.66667	3	41	0.225836
DELocalToBest	10.000	2.66667	3	41	0.225836
DERand1DIter	9.98794	2.66668	2.99974	41.0014	0.227494

y_3, y_4, y_5, y_6 for the non-synchronized state in Fig. 4(b)-Fig. 4(e), the same results were easy recognized. It showed clear that oscillations in the drive and response systems were not synchronization.

Synchronized behavior, observed with estimated parameters and PC method, is shown in Fig. 5. The synchronization between the drive and the response can be easily observed. The behavior of a synchronization state appears as a sharp straight line $x_2 = y_2$ as shown in Fig. 5(a). The synchronization is confirmed by the stability of the diagonal in the correlation of variables x_3, x_4, x_5, x_6 versus y_3, y_4, y_5, y_6 as shown in Fig. 5(b)-Fig. 5(e). Thus the chaotic oscillations in drive and response system are synchronized.

Therefore, it demonstrated that the estimated values are effective to synchronize for two 6D-chaotic systems although they were started under different initial values.

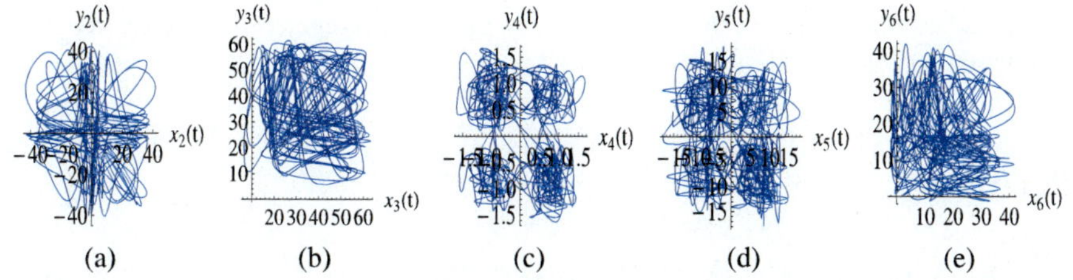

Fig. 4 Non-synchronization between drive and response system

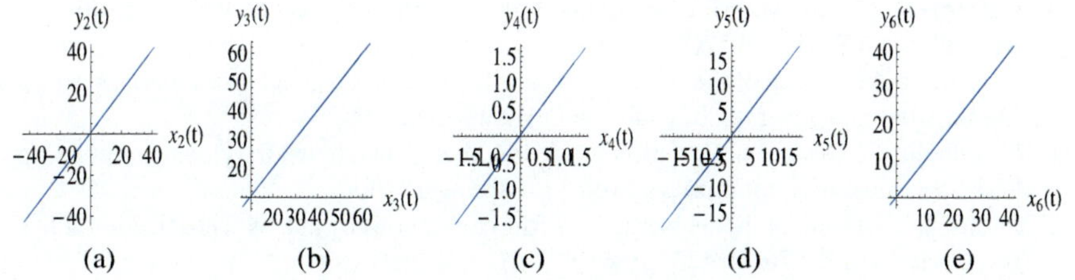

Fig. 5 Synchronization between drive and response system

5 Conclusions

The main aim of this paper is to show that an artificial intelligence was used as a method to find the best quality of 6D chaotic synchronization systems. The PC method was applied to synchronize two identical 6D chaotic systems. Parameter estimation for chaotic synchronization system was formulated as a multidimensional optimization problem. Because of the irregular dynamic behavior of chaotic systems, there is a lot of local optimum in the landscape of cost function. DE was used to estimate the optimum values for unknown parameters, and it easily escaped the local optimum trap and achieved the global optimum parameters. Based on the estimated parameter from DE, two chaotic systems were synchronized. The estimated parameters and the original parameters are the same; it indicated that the original parameters are optimal choice in this case.

As the future subject, it can be executed on other systems to compare the ability of this method on finding the optimum of chaotic synchronization systems. It can also be extended to the communication systems, which will consider the synchronization under effects of noise, fading, offset, etc.

Acknowledgements. This work was supported by grant No. MSM 7088352101 of the Ministry of Education of the Czech Republic and by the Grant Agency of the Czech Republic GACR 102/09/1680.

References

1. Cuomo, K., Oppenheim, A.: Circuit implementation of synchronized chaos with applications to communications. Physical Review Letters 71(1), 65–68 (1993)
2. Dung, N.: Chaos synchronization in none-ideal channel by differential evolution algorithm. In: 16th International Conference on Soft Computing Mendel 2010. Brno Univ. Technology Vut Press (2010)
3. Dung, N., Zelinka, I.: Chaos theory in secure communication. In: MENDEL 2009 (2009)
4. Güémez, J., Matías, M.: Modified method for synchronizing and cascading chaotic systems. Physical Review E 52(3), 2145–2148 (1995)
5. Kennamer, K.: Studies of the onset of chaos in the lorenz and generalized lorenz systems. Master's thesis, University of Alabama in Huntsville (1995)
6. Onwubolu, G., Davendra, D.: Differential evolution: A handbook for global permutation-based combinatorial optimization, vol. 175. Springer (2009)
7. Parlitz, U.: Estimating model parameters from time series by autosynchronization. Physical Review Letters 76(8), 1232–1235 (1996)
8. Parlitz, U., Junge, L., Kocarev, L.: Synchronization-based parameter estimation from time series. Physical Review E 54(6), 6253–6259 (1996)
9. Pecora, L., Carroll, T.: Synchronization in chaotic systems. Physical Review Letters 64(8), 821–824 (1990)
10. Storn, R., Price, K.: Differential evolution-a simple and efficient adaptive scheme for global optimization over continuous spaces. In: International Computer Science Institute-Publications-TR (1995)
11. Yu, H., Liu, Y.: Chaotic synchronization based on stability criterion of linear systems. Physics Letters A 314(4), 292–298 (2003)

Asynchronous Synthesis of a Neural Network Applied on Head Load Prediction

P. Vařacha

Tomas Bata University in Zlín, Faculty of Applied Informatics,
Nad Stráněmi 4511, Zlín, 760 05, Czech Republic
varacha@fai.utb.cz

Abstract. This paper introduces innovative method of an artificial neural network (ANN) optimization (synthesis) by means of Analytic Programming (AP). New asynchronous implementation of Self-Organizing Migration Algorithm (SOMA), which provides effective increase of AP computing potential, is introduced here for time as well as original strategy of communication between SOMA and AP that further contribute towards efficiency in search for optimal ANN solution. The whole ANN synthesis algorithm is applied on the real case of heating plant model identification. The heating plant is located in the town of Most, Czech Republic.

The method proves itself to be especially effective when formally identified non-neural parts of the heating plant model need to be made more accurate. Asynchronous distribution plays the key role here as the heating plant behavior data has to be acquired from a very large database and therefore learning of ANN may require a lot of computation time.

1 Introduction

Artificial neural networks (ANN) are parallel computation models comprised of densely interconnected adaptive processing units with extensive industrial usage and various applications on approximation, classification or problem prediction. However, ANN performance is significantly limited by its structure and related learning algorithm.

To overcome some disadvantages of commonly used methods of ANN design and learning, this paper introduces an innovative method of artificial neural network optimization (synthesis) by means of Analytic Programming (AP).

ANN is an established tool of real plant modeling. However, with increasing complexity of system in need of identification, limitations resulting from standard methods of ANN learning and structural optimization are more and more apparent. (Vařacha 2009) AP synthesis of ANN was developed to overcome such limitations.

AP creates an interface between individual and evolution algorithm (EA). The individual is represented as a vector of pointers (integers) to General Function Set

I. Zelinka et al. (Eds.): Nostradamus: Mod. Meth. of Prediction, Modeling, AISC 192, pp. 225–240.
springerlink.com

(GFS). GFS comprehends amount of base functions from which target function is synthesized. AP unique features provide for safe EA manipulation of individuals in accordance with optimal figure retrieval. (Zelinka 2002) If the figure contains any constants, these are fitted by separate run of EA in accordance with set cost function (CF).

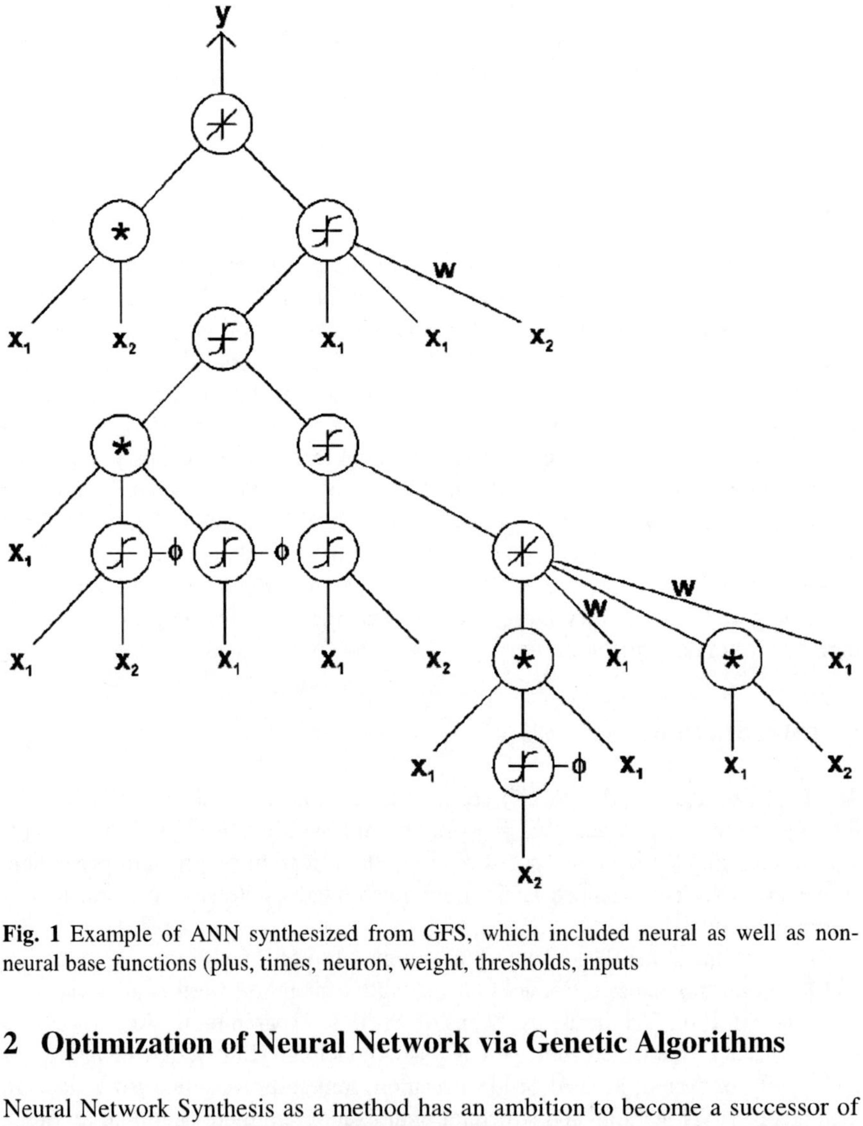

Fig. 1 Example of ANN synthesized from GFS, which included neural as well as non-neural base functions (plus, times, neuron, weight, thresholds, inputs

2 Optimization of Neural Network via Genetic Algorithms

Neural Network Synthesis as a method has an ambition to become a successor of older methods based of Genetic Algorithms (GA). This chapter summarized most important developments considering GA usage for ANN learning and structural optimization.

The development of evolutionary methods aiming to design the ANN structure and weight values experienced boom at the end of the millennium with the introduction of sufficiently fast computers into common scientific practice. A comprehensive survey considering history of evolutionary computation methods of designing the ANN structure can be found in (Vonk, Jain and Johnson 1997).

These methods can be used in the field of the ANN in several ways:

- to train the weights of the ANN
- to analyze the ANN
- to generate the architecture of the ANN
- to generate both the ANN's architecture and weights

The problem often encountered with GA is that they are quite slow in fine-tuning once they are close to a solution. Therefore, hybridization of GA and back propagation algorithm (Gisarion, et al. 2011) (BP), where BP is used to fine-tune a near-optimal solution found GA, has proven to be successful (Lohmann 1992). In (Lund, Parisi 1994) a GA is used to evolve the ecological ANN that can adapt to their changing environment. This is achieved by letting the fitness function, which in this case is seen as individual for every gene, to co-evolve with the weights of the ANN.

(Garis 1990) uses a method, which is based on the fully self-connected ANN modules. It is shown that by using this approach a network can be taught a task even though the time-dependent input changes so fast that the ANN never settles down.

In (Munro 1993) and (Hassoun 1995), GA is used in a fixed three layer feed forward ANN to find the optimal mapping from the input to a hidden layer. It is suggested that the hidden target space might have more optima than the weight space and that finding the optimum will therefore be easier.

(Montana 1991) and (Whistley, Starkweather and Bogard 1990) used chromosomes with real-valued genes instead of binary coded chromosomes. Satisfactory results are reported using a Genitor type Steady State Genetic Algorithm with relatively small population size of 50.

An alternative approach is to use GA, where the topology and weights are encoded as variable-length binary strings (Meniezzo 1990). In (Dasgupta, McGregor 1993) a structured GA is used that simultaneously optimizes the ANN topology and the values of weights.

In (Dasgupta, McGregor 1993) feed forward ANN are generated with GA, using a direct encoding scheme where every gene in a chromosome represents a connection between two neurons. This Approach is also known as *restrictive mating*. (Braun, Weisborod 1993).

Jacob and Rehder (Jacob, Rehder 1994) use a grammar-based genetic system, where the topology creation, neuron functionality and weight creation are split into three different modules, each using a separate GA. Similarly, (Turner, Jacobson and Taylor 2011) reports an approach, where modular ANN are generated using the direct encoding scheme.

(Angelia, Saunders and Pollack 1994) implemented a system based on evolutionary programming where ANN evolve using both parametric mutation and structural mutation and in (McDonnell, Waagen 1993) evolutionary programming is used where the initial network is a three-layered fully connected feed forward ANN and the evolutionary programming algorithm is used to prune the connection.

In (Nix, Vose 1992) a modular design approach is used, where a distinction is made between the structure, connectivity and weights optimization. Kitano (Kitano 1994) uses a GA-based matrix grammar approach with chromosome code grammar rewriting rules that can be used to build a connectivity matrix. Gruau (Gruau 1994) uses a graph grammar system called Cellular Encoding. The graph grammar rules work directly with neurons and their connections and include various kinds of cell divisions and connection pruning rules. Boers and Kuiper (Boers, Kuiper 1992) use a graph grammar system based on a class of fractals called L-system. The chromosomes used in the GA code the production rules in this grammar.

Genetic Programming offers an approach to the direct encoding scheme. The approach that consists of directly encoding ANN in the genetic tree structure used by GP is described in (Koza 1998).

According to (Koza 1999) the ANN topology as well as the values of the weights are defined within one structure and no distinction is made between learning of the ANN topology and its weights. In a similar way, the GE can be used to successfully design the ANN. Such an approach can be found for example in (Turner 2010).

In (Volná 2007), the transfer function has been shown to be an important part of architecture of ANN and have significant impact on ANN performance.

3 Analytic Programming

Main principle (core) of AP is based on discrete set handling (DSH) and is inspired by Genetic Algorithms. DSH shows itself as a universal interface between EA and symbolically solved problem. That is why AP can be used almost by any EA.

Briefly put, in AP, individuals consist of non-numerical expressions (operators, functions, …), which are within evolutionary process represented by their integer indexes. Such indexes then serve like pointers into the set of expressions and AP uses them to synthesize output function-program for Cost Function evaluation.

All simple functions and operators are in so-called General Function Set (GFS) divided into groups according to the number of arguments, which can be inserted during evolutionary process to create subsets GFS_{3arg}, GFS_{2arg}...GFS_{0arg}.

Table 1 Example of GFS and its subsets

GFS Degree	Contains
GFS_{all}	BetaRegularized, +, -, *, /, Power, Abs, Round, Sin, Cos, t, K, π, 1, 2
GFS_{3arg}	BetaRegularized
GFS_{2arg}	+, -, *, /, Power
GFS_{1arg}	Abs, Round, Sin, Cos
GFS_{0arg}	t, K, π, 1, 2

Example of AP translation from an individual to the Cost Function is given in Fig. 2:

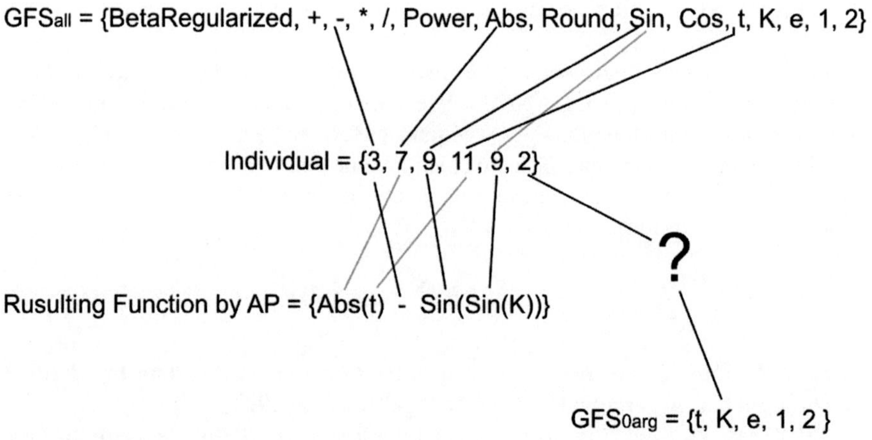

Fig. 2 Main principles of AP

The individual consists of 6 arguments (indices, pointers to GFS). The first index is 3, meaning that it is taken from the set of functions GFS_{all}. Function *minus* has two arguments; therefore indexes 7 and 9 are arguments of *minus*.

$$7 + 9 \tag{1}$$

Index 7 is then replaced by *Abs* and index 9 by *Sin*.

$$Abs + Sin \tag{2}$$

Abs and *Sin* are one-argument functions. Then, index 9 follows index 11, which is replaced by *t*.

$$Abs(t) + Sin \tag{3}$$

Sin is also one-argument function. Then, after index 11, the individual takes index 9, which is replaced by *Sin* and this *Sin* becomes argument of previous *Sin*.

$$Sin(Tan) + Sin(Sin(\tag{4}$$

The last index is 2, but in our case there is a function *Plus*. *Plus* needs two arguments to work properly. AP will not allow this, as there is not any other free pointer to be used as argument. Instead of *Plus*, AP will jump into subspace, in this case directly to GFS_{0arg}. In the GFS_{0arg} it finds second element, which is *K*. And by doing so, we get (5).

$$Sin(Tan(t)) + Cos(t) \tag{5}$$

Number of actually used pointers from an individual before the synthesized expression is closed is called *depth*. This example is based on relevant and previously published work in (Oplatková, Zelinka 2007).

4 Neural Network Synthesis

Several architectures of ANN exist, however, synthesis introduced in this chapter considers only feed-forward ANN, where network function $f(x)$ is defined as composition of other functions $g_i(x)$, which can further be defined as composition of other functions. This can be conveniently represented as a network structure with arrows depicting dependencies between variables. Widely used type of composition is the non-linear weighted sum,

$$f(x) = F(\sum_i w_i g_i(x)) \tag{6}$$

where F (commonly referred to as activation function) is a certain pre-defined function, such as hyperbolic tangent (see (8)). (Bischop 1995)

What has attracted the most interest in ANN is possibility of learning (see part 4.1). Given a specific task to solve and a class of functions F (in this case $\forall f \in F$ contains only sub functions from GFS), learning means using a set of observations to find $f^* \in F$ which solves the task in certain optimal sense. This entails definition of the Cost Function $C{:}F \rightarrow \Re$ such that for the optimal solution f^*, $C(f^*) \leq C(f)\ \forall f \in F$ (i.e. no solution costs less than the cost of the optimal solution). (Gurney 2007)

Many different ways of using EA for ANN optimization exist, nevertheless AP provides possibility to synthesize ANN with almost infinitely variable structure, complexity and scope. There is a very easy way of using AP for ANN synthesis. (Vařacha 2009) The most important part is to define items from which ANN will be composed. In this case GFS (7) contains only three items:

$$GFS_{all} = \{+,\ AN,\ K * x\} \tag{7}$$

Most important item (7) is Artificial Neuron (AN) (8) with weighted hyperbolic tangent as transfer function (9). Weight of output, steepness and biases are computed as *K* in AP (11).

$$GFS_1 = \{AN\} \tag{8}$$

$$AN(S) = w\frac{e^{2\lambda(S+\phi)}-1}{e^{2\lambda(S+\phi)}+1} \tag{9}$$

$$AN(S) = K_1\frac{e^{2K_2(S+K_3)}-1}{e^{2K_2(S+K_3)}+1} \tag{10}$$

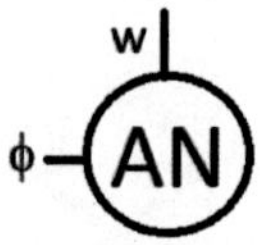

Fig. 3 Graphical example of AN

To allow more inputs into one ANN, simple plus operator (11) is used.

$$GFS_2 = \{+\} \tag{11}$$

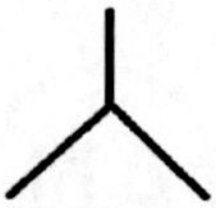

Fig. 4 Graphical example of plus operator

Finally, (12) represents weighted input data.

$$GFS_0 = K * x \tag{12}$$

Fig. 5 Graphical example of weighted input

Under such circumstances, translation of individual into ANN can be easily understood from Fig. 6.

The whole process is cyclical. Individuals provided by EA are translated into ANNs. ANNs are evaluated in accordance with training data set and their global errors are used to set fitness of these individuals. Consequently, a new generation is chosen and the whole process is repeated in next migration loop.

The introduced approach is not the only one possible. Different settings of GFS were successfully used to synthesize ANN performing classification as can be seen, for example, in Fig. 1 or in (Vařacha, Zelinka 2007).

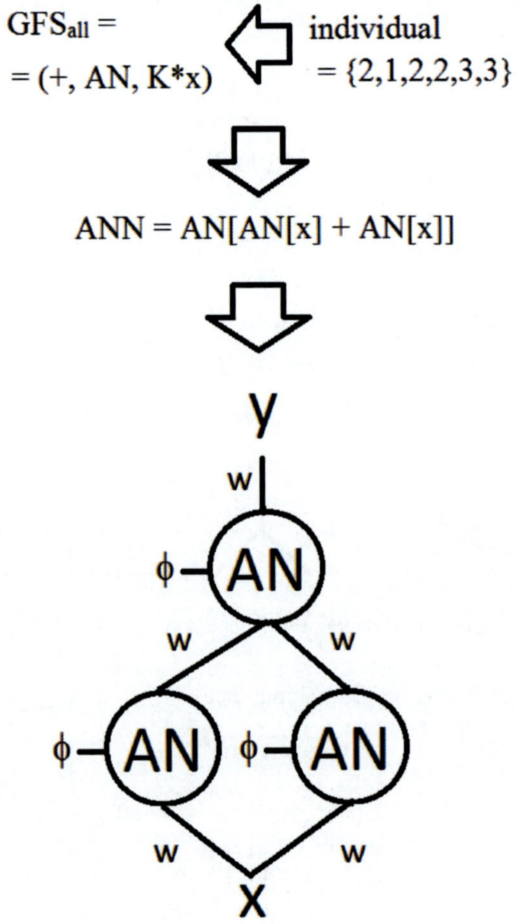

Fig. 6 Translation of individual into ANN

4.1 Reinforced Evolution

Reinforced Evolution is a common part of AP. (Oplatková, Zelinka 2006) If ANN of adequate quality cannot be obtained during AP run, AP puts the best ANN it finds as a sub ANN into GFS_0 and starts over. This arrangement considerably improves AP ability to find ANN with desirable parameters.

4.2 Constants Processing

Synthesized programs, formulas, or more importantly ANN, may also contain constants "K" that can be defined in GFS_0 or be a part of other functions included in GFS_{all}. When the program is synthesized, all Ks are indexed, so K_1, K_2, …, K_n, are obtained and then all K_n are estimated. Several versions of AP exist in accordance with K_n estimation. (Oplatková, Zelinka 2006)

In this case, asynchronous implementation of SOMA (inside another SOMA which operates AP) is used to estimate K_n. This is especially convenient for ANN synthesis. K_n can be interpreted as various weights and biases and their optimization done by SOMA as ANN learning. (Vařacha, Zelinka 2007)

5 Asynchronously Parallel Distribution

Described process can be successfully used for optimal ANN synthesis if GFS is loaded with elemental structural parts of ANN and constants represent thresholds and weights. (Vařacha, Zelinka 2007a) However, learning of such ANN - constants setting via EA can be extremely time-consuming operation. For example, let us have main EA, which runs AP and where every individual runs its own EA for ANN learning. This situation forced new innovative way of EA parallelism and distribution.

Self-Organizing Migration Algorithm (SOMA) is a perfect tool for AP based on communication between Leader and population of individuals (Šenkeřík, et al. 2011) (see Fig. 7), nevertheless, single individuals run on hyper surface of CF mostly independently. (Zelinka 2004) Every individual executes (usually) 27 evaluations of CF prior to next communication with Leader.

This SOMA quality provides highly effective asynchronous distribution between more available computational units. This asynchronous way of distribution thus enables effective synthesis of individuals with high but unequal demands on computation capacity. (Vařacha, Zelinka 2007b) This fact is especially important in case of ANN.

In comparison with standard SOMA, implementation of ANN synthesis created new strategy of mutation probability (PRT parameter) adjustment. Actual value of PRT is optimized through communication with AP in order to prevent ineffectively extensive destructions of mutated ANN.

Side effects of asynchronous distribution also introduce very interesting behavior of individuals when smaller (and computationally faster) individuals move on CF hyper surface faster. This causes integrated preference of less complicated ANN and therefore naturally prevents synthesis of an overgrown ANN. (Fig. 8)

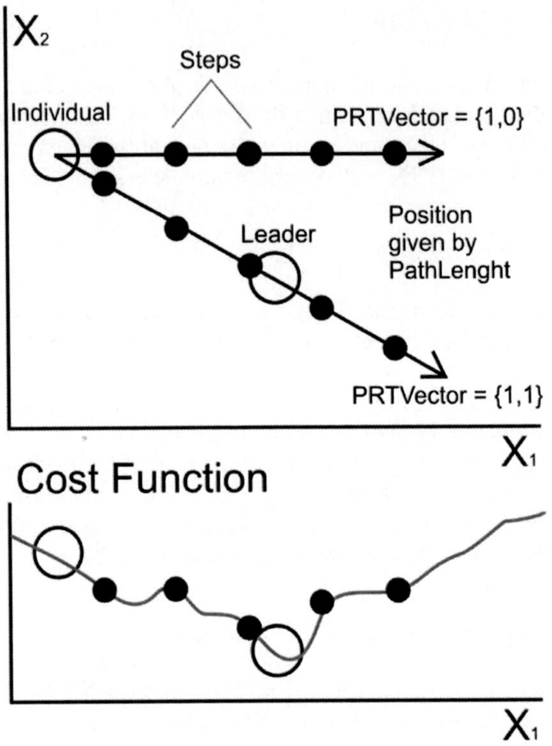

Fig. 7 Main principle of Self-Organizing Migration Algorithm

The method proves itself to be especially effective when formally identified non-neural parts of the heating plant model need to be made more accurate. Asynchronous distribution plays the key role here as the heating plant behavior data has to be acquired from a very large database and therefore learning of ANN may require a lot of computation time.

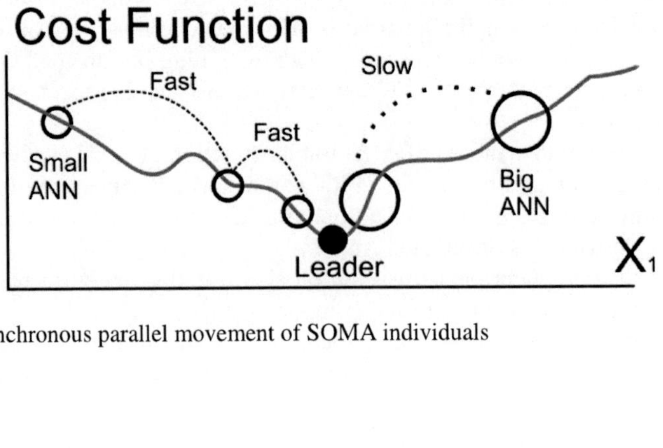

Fig. 8 Asynchronous parallel movement of SOMA individuals

6 Neural Network Synthesis for Head Load Prediction

The method of ANN synthesis described in the previous five parts is applied in order to optimise Head Load Approximation (HLA) function of the heating plant in Komořany (Czech Republic). The function is later used to predict head load of Most agglomeration in order to provide valuable information for heating plant control.

The heating plant uses (13) to approximate HLA by sum of time-dependent and temperature dependent components. (Chramcov, Balátě 2009)

$$f_p(t, \vartheta_{ex}) = f_{time}(t) + f_{temp}(\vartheta_{ex}) \tag{13}$$

Where

$f_{time}(t)$	is the time dependent, component,
t_0	is the time offset,
ϑ_{ex}	is the outdoor temperature,
$f_{temp}(\vartheta_{ex})$	is the outdoor temperature, dependent component.

The task of ANN synthesis here is to create ANN that provides HLA with data containing measured head load, time and external temperature. Data cover period from Nov 3, 2009 to Dec 31, 2009, which includes 1416 samples taken in one-hour steps. Formal HLA function (Král, et al. 2011) resulted in 4.28% Normal Mean Square Error (NRMSE) within provided data. Therefore, ANN with lower NRMSE is desirable.

Used data was normalized into <0; 1> interval and divided into training, validation and test sets. The whole experiment was conducted in accordance with rules proposed in (Prechelt 1994). Simple but effective GFS structure was used for ANN synthesis during this experiment:

$$GFS_{all} = \{+, AN, K * \vartheta_{ex}, K * t\} \tag{14}$$

The Cost Function was defined as follows:

$$\min_{ANN} \sum_t (P(t) - f_p((t - t_0) \bmod 24, \vartheta_{ex}, ANN)) \tag{15}$$

Where

ANN	is the vector of ANN structure, weight and biases,
P	is the measured value of head load

In case the best-synthesized ANN does not improve its Cost Function by at least 0,001%, breeding is stopped.

Setting of Asynchronous SOMA used as EA for AP can be seen in Table 2.

Table 2 Setting of SOMA used as EA for AP

Parameter	Value
Number of Individuals	48
Individual Parameters	100
Low	0
High	10
PathLength $_{rp}$	3
Step	0,11
PRT	1/ depth
Divergence	0.00001
Period	3

Table 3 shows setting of SOMA used for ANN learning:

Table 3 Setting of SOMA used to optimize Kn (ANN learning)

Parameter	Value
Number of Individuals	number of Kn * 0.5
Individual Parameters	(at least 10)
Low	100
High	-10
PathLength $_{rp}$	10
Step	3
PRT	0,11
Divergence	1 / number of Kn
Period	0.01

6.1 Results

In 100 cases AP was always able to synthesize ANN with NRMSD = 3.46%, however, final results vary in number of used AN. One example of typically obtained ANN structure is shown below:

$$\boldsymbol{ANN0} = \mathbf{AN}[t] + \vartheta_{ex} \tag{16}$$

$$\boldsymbol{ANN1} = \boldsymbol{ANN0} + \mathbf{AN}[t] \tag{17}$$

$$\boldsymbol{ANN2} = \boldsymbol{ANN1} + \mathbf{AN}[\mathbf{AN}[t] + t + \mathbf{AN}[\mathbf{AN}[t]]] \tag{18}$$

$$\boldsymbol{ANN3} = \mathbf{AN}[\mathbf{AN}[t]] + \boldsymbol{ANN2} \tag{19}$$

$$\boldsymbol{ANN4} = \boldsymbol{ANN3} + \mathbf{AN}[\mathbf{AN}[\vartheta_{ex}] + \mathbf{AN}[\vartheta_{ex}] + t] \tag{20}$$

In these cases, AP used 5 sub ANNs to form final ANN. ANNs (16) - (20) have non-trivial structure, nevertheless, they can be easily simplified, if necessary, by cutting later sub ANNs with positive influence on ANN computation speed.

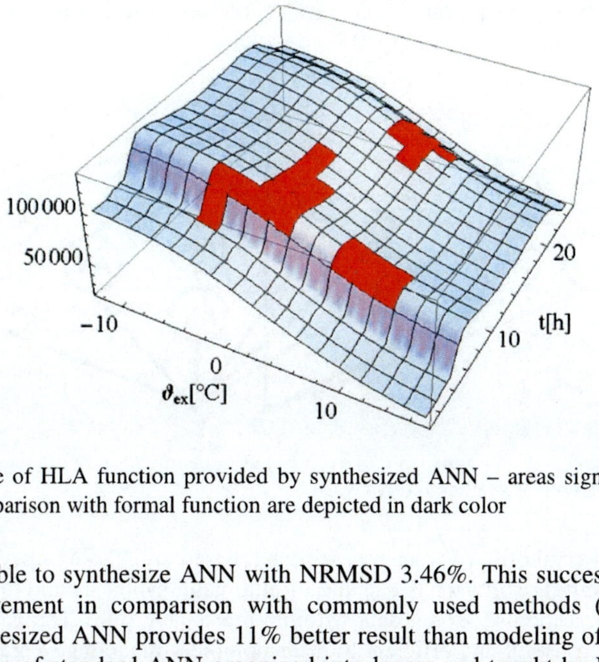

Fig. 9 Surface of HLA function provided by synthesized ANN – areas significantly corrected in comparison with formal function are depicted in dark color

AP was able to synthesize ANN with NRMSD 3.46%. This success represents 19% improvement in comparison with commonly used methods (Král, et al. 2011). Synthesized ANN provides 11% better result than modeling of HLA function with help of standard ANN organized into layers and taught by Back Propagation Algorithm and 7% better result than ANN optimized via Genetic Algorithms (Vařacha 2009).

Application of this method on the real case of the heating plant was possible only due to successful distribution method described in part 5.4. The algorithm was run on 24 AMD cores (Opteron 6128, G34, 2.0 GHz, 12MB L3, 80W, WOF) of Super Micro Server. Each core was occupied by two individuals of the algorithm. In this configuration, one algorithm's run took approximately 14 minutes, which resulted in the whole experiment lasting less than 24 hours.

Synthesized ANN can positively influence control quality in Komořany heating plant and may be integrated into wider expert system developed thought National Research Programme II (see Acknowledgement).

AP proves its ability to successfully synthesize ANN in dynamic and irregular environments.

7 Conclusion

The Neural Network Synthesis was developed on the basis of AP and SOMA algorithms. The method was successfully tested on the real life problems (Prechelt 1994), (Král et al. 2011) as well as on widely recognized benchmark functions (Zelinka 2004) with respect to the function approximation, prediction and classification problems. These results were published for example in (Vařacha 2009, 2011). An example of synthesized ANN structure working as successful head load predictor can be seen in Fig. 10.

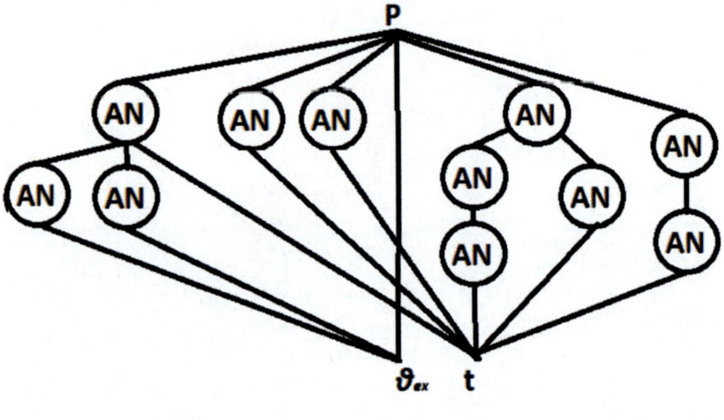

Fig. 10 An example of synthesized ANN

The ANN synthesis software was designed based on .NET Framework technology. The resulting software is capable of automatic synthesis and optimizing the ANN based on the user-given data within a reasonable time. The ANN synthesis proves to be a useful and efficient tool for nonlinear modeling (Vařacha 2011) in comparison with competing methods, e.g. (Tsulog et al. 2008).

Acknowledgments. This work was supported in part by European Regional Development Fund within the project CEBIA-Tech No. CZ.1.05/2.1.00/03.0089..

References

Garis, H.: Genetic Programming Building Nanobrains with Genetically Programmed Neural Network Modules. In: IEEE International Joint Conference on Neural Networks, New York, vol. 3, pp. 511–516 (1990)

Whitley, D., Starkweather, T., Bogard, C.: Genetic Algorithms and Neural Networks: Optimizing Connections and Connectivity. Parallel Computing 14, 347–361 (1990)

Montana, D.J.: Automated Parameter Tuning for Interpretation of Synthetic Images. In: Handbook of Genetic Algorithms, pp. 202–221 (1991)

Boers, E.J.W., Kuiper, H.: Biological Metaphors and Design of Modular Artificial Neural Networks, Technical report, Department of Computer Science and Experimental and Theoretical Psychology, Leiden University, The Netherlands (1992)

Lohmann, R.: Structure Evolution in Neural Systems, Dynamic, Genetic and Chaotic Programming, ch. 15 (1992)

Nix, A.E., Vose, M.D.: Modelling Genetic Algorithms with Markov Chains. Annals of Mathematics and Artificial Inteligence 5, 79–88 (1992)

Braun, H., Weisborod, J.: Evolving Neural Feedforward Networks. In: International Conference on Artificial Neural Nets and Genetic Algorithms (ANNGA 1993), Innsbruck, Austria, pp. 25–32 (1993)

Dasgupta, D., McGregor, D.R.: sGA: A Structured Genetic Algorithm, Technical Report: IKBS-11-93, Department of Computer Science, University of Strathclyde, Glasgow (1993)

McDonnell, J.R., Waagen, D.: Evolving Neural Network Connectivity. In: IEE International Conference on Neural Networks, San Francisco (1993)

Munro, P.W.: Genetic Search for Optimal Representations in Neural Networks. In: International Conference on Artificial Neural Nets and Genetic Algorithms (ANNGA 1993), Innsbruck, Austria, pp. 628–634 (1993)

Angelia, P.J., Saunders, G.M., Pollack, J.M.: An Evolutionary Algorithm that Constructs Recurrent Neural Networks. IEEE Transactions on Neural Networks 5 (1994)

Gruau, F.: Genetic Microprogramming of Neural Networks. Advances in Genetic Programming. MIT Press (1994)

Maniezzo, V.: Genetic Evolution of the Topology and Weight Distribution of Neural Networks. IEEE Transaction on Neural Networks 5 (1994)

Prechelt, L.: Proben1—A Set of Neural Network Benchmark Problems and Benchmarking Rules. Universität Karlsruhe, Germany (1994)

Lund, H.H., Parisi, D.: Simulations with an Evolvable Fitness Formula, Technical Report PCIA-1-94, C.N.R, Rome (1994)

Happel, B.L.M., Murrer, J.M.J.: Deign and Evolution of Modular Neural Network Architectures. Neural Networks 7, 985–1004 (1994)

Kitano, H.: Neurogenetic Learning: An Integrated Method of Designing and Training Neural Networks Using Genetic Algorithms. Physica D 75, 225–228 (1994)

Hassoun, M.H.: Fundamentals of Artificial Neural Networks. MIT Press (1995)

Gurney, K.: An Introduction to Neural Networks. Routledge, London (1997)

Vonk, E., Jain, L.C., Jjohnson, R.P.: Automatic Generation of Neural Network Architecture Using Evolutionary Computation. Advances in Fuzzy Systems – Applications and Theory, vol. 14. World Science (1997)

Koza, J.R.: Genetic Programming. MIT Press (1998)

Kkoza, J.R., et al.: Genetic Programming III; Darwinian Invention and problem Solving. Morgan Kaufmann Publisher (1999)

Zelinka, I.: Analytic Programming by means of Soma Algorithm. In: First International Conference on Intelligent Computing and Information Systems, Egypt, Cairo (2002)

Zelinka, I.: SOMA - Self Organizing Migrating Algoritm. In: Batu, B.V., Onwubolu, G. (eds.) New Optimization Techniques in Engineering, ch. 7, p. 33. Springer (2004)

Oplatková, Z., Zelinka, I.: Investigation on Artificial Ant using Analytic Programming. In: Genetic and Evolutionary Computation Conference. The Association for Computing Machinery, USA (2006)

Vařacha, P., Zelinka, I.: Synthesis of artificial neural networks by of evolutionary methods. In: Workshop ETID 2007 in DEXA 2007. IEEE Computer Society (2007a)

Vařacha, P., Zelinka, I.: Distributed Self-Organizing Migrating Algorithm (DISOMA). In: 8th International Carpathian Control Conference, Slovak Republic, Košice (2007b)

Volna, E.: Forming neural networks design through evolution. In: Artificial Neural Networks and Intelligent Information Processing, pp. 13–20 (2007) ISBN: 978-972-8865-86-3

Oplatková, Z., Zelinka, I.: Creating evolutionary algorithms by means of analytic programming - design of new cost function. In: European Council for Modelling and Simulation, ECMS 2007, pp. 271–276 (2007)

Tsoulog, I., Gavrilis, D., Glavas, E.: Neural network construction and training using grammatical evolution. Neurocomputing 72(1-3), 269–277 (2008)

Hu, X.: Applications of the general projection neural network in solving extended linear-quadratic programming problems with linear constraints. Neurocomputing 72 (2009)

Chramcov, B., Balátě, J.: Model-building for time series of heat demand. In: Proceedings of the 20th International DAAAM Symposium Intelligent Manufacturing and Automation: Focus on Theory, Practice and Education. DAAAM International Vienna, Vienna (2009)

Vařacha, P.: Impact of Weather Inputs on Heating Plant - Agglomeration Modeling. In: Proceedings of the 10th WSEAS Ing. Conf. on Neural Networks, pp. 159–162. WSEAS World Science and Engineering Academy and Science, Athens (2009)

Turner, S.D., Dudek, S.M., Ritchie, M.D.: Grammatical Evolution of Neural Networks for Discovering Epistasis among Quantitative Trait Loci. In: Pizzuti, C., Ritchie, M.D., Giacobini, M. (eds.) EvoBIO 2010. LNCS, vol. 6023, pp. 86–97. Springer, Heidelberg (2010)

Gisario, A., et al.: Springback control in sheet metal bending by laser-assisted bending: Experimental analysis, empirical and neural network modelling. Optic and Lasers in Engineering 49(12) (2011)

Turner, E., Jacobson, D.J., Taylo, J.W.: Genetic Architecture of a Reinforced, Postmating, Reproductive Isolation Barrier between Neurospora Species Indicates Evolution via Natural Selection. Plos Genetics 7(8) (2011)

Vařacha, P.: Neural network synthesis dealing with classification problem. In: Recent Researches in Automatic Control: Proceedings of the 13th WSEAS International Conference on Automatic Control, Modelling & Simulation (ACMOS), pp. 377–382. WSEAS Press, Lanzarote (2011)

Král, E., et al.: Usage of peak functions in heat load modeling of district heating system. In: Recent Researches in Automatic Control: Proceedings of the 13th WSEAS International Conference on Automatic Control, Modelling & Simulation (ACMOS), pp. 404–406. WSEAS Press, Lanzarote (2011)

Šenkeřík, R., Oplatková, Z., Zelinka, I., Davendra, D.: Synthesis of feedback controller for three selected chaotic systems by means of evolutionary techniques: Analytic programming. Mathematical and Computer Modelling (May 27, 2011) ISSN 0895-7177, doi:10.1016/j.mcm.2011.05.030

Prediction by Means of Elliott Waves Recognition

Eva Volna, Martin Kotyrba, and Robert Jarušek

University of Ostrava, 30. dubna 22, 70103 Ostrava, Czech Republic

Abstract. This article deals with prediction by means of Elliott waves recognition. The goal is to find and recognize important Elliott wave patterns which repeatedly appear in the market history for the purpose of prediction of subsequent trader's action. The pattern recognition approach is based on neural networks. The article is focused on reliability of Elliott wave patterns recognition made by developed algorithms which allows also causes the reduction of the calculation costs.

1 Introduction

The main topic of the article is to develop and optimize the pattern recognition algorithm in order to recognize Elliott wave patterns in time series for the purpose of prediction.

Elliott wave patterns are not exact, they are slightly different every time they appear. They can have different amplitude and different duration, albeit visually the same pattern can look differently despite being the same. These patterns also do not cover every time point in the series, but are optimized so that the developed classifier would be able to learn their key characteristics and accurately recognize them. Such optimized inputs also reduce calculation costs. One of important challenges is to recognize the input pattern reliably.

Recent studies show that Elliott wave patterns might implicate useful information for stock price forecasting. Currently, there are mainly two kinds of pattern recognition algorithms: an algorithm based on rule-matching [1] and an algorithm based on template-matching [3]. Nonetheless, both of these two categories have to design a specific rule or template for each pattern. However, both types of algorithms require participation of domain experts, and lack the ability to learn. For the last few decades, neural networks have shown to be a good candidate for solving problems with the market analysis. A typical illustration is the study conducted in [2], where a recognition algorithm for triangle patterns based upon a recurrent neural network was introduced. Elliott wave patterns can be classified into two categories: continuation patterns and reversal patterns. Continuation patterns indicate that the market price is going to keep its current movement trend; while reversal patterns indicate that the market price will move to the opposite trend. Elliott patterns can be seen as some sort of maps which helps us to orientate in certain situations and navigate us to profitable trades.

I. Zelinka et al. (Eds.): Nostradamus: Mod. Meth. of Prediction, Modeling, AISC 192, pp. 241–250.
springerlink.com

We focus on a prediction by means of Elliott waves recognition. The article proposes the Elliott waves pattern recognition approach based on a backpropagation neural network.

2 Elliott Wave-Pattern Recognition

Elliott wave theory is a form of market analysis based on the theory that market patterns repeat and unfold in cycles. Ralph Nelson Elliott developed this theory in the 1930s. Elliott argued that upward and downward market price action was based on mass psychology and always showed up in the same repetitive patterns. These patterns were divided into what Elliott called "waves." According to Elliott, crowd psychology moves from optimism to pessimism and back again and this is seen in the price movements of market trend which are identified in waves.

The Elliott wave theory assumes that market moves up in a series of five waves and down at a series of three waves. Elliott concluded that markets moves are not random but follow repetitive cycles driven by mass psychology. The repetitive cycles show up in waves. A wave is a movement in the market either upwards or down. Elliot discovered two basic types of wave patterns, impulse waves consisting of five waves and three smaller corrective waves. Impulse waves move in the direction of the main market trend. The corrective wave moves in the opposite direction of the main market trend.

Elliott labelled five impulse wave patterns to describe various stages of mass psychology. The waves are subdivided into five smaller waves [4]. Generally, waves three and five are with the trend and two and four corrections within the trend (Fig. 1).

- *Wave one* - its trend and sentiment is overwhelmingly negative but a few buyers emerge as the market starts to move up. This wave is marked by limited public participation
- *Wave two* corrects wave one and does not extend beyond the starting point of wave one. The news is negative but shows signs of improving and some early buyers look to take profits not convinced of the viability of the trend. In wave two, public participation has increased.
- *Wave three* is usually the most powerful wave. There's more positive news and prices begin to rise quickly. The public looks to get on board.
- *Wave four* is corrective and offers opportunity to buy the market on a pullback. Early buyers look to take profits and some who missed the early move look at the pullback to get into the market.
- *Wave five* is the last stage of the dominant trend. Mass psychology is universally positive reaching euphoric stage and the market becomes over priced. The public are heavy buyers of the market in this stage. The market is ripe for a trend change.

Elliott also concluded that markets move in a three wave corrective pattern. The corrective wave pattern is normally referred to as the ABC correction [4]. The

A wave is hard to identify. Wave B, prices reverse higher. In wave C, prices move lower. Generally wave A and C are trend corrections and wave B a countertrend move within the correction. These waves are known as corrective (Fig. 1).

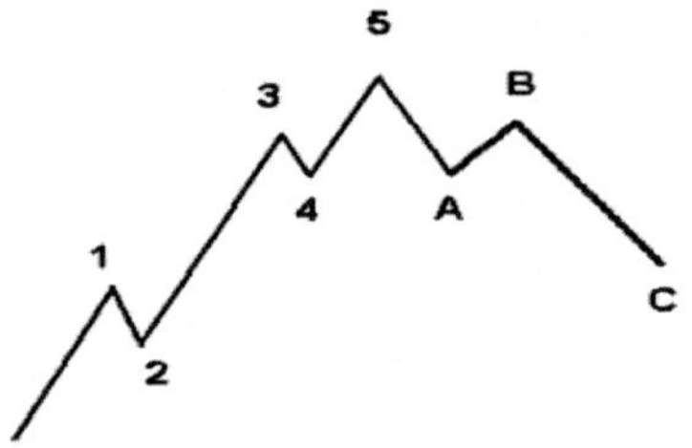

Fig. 1 The 5-3 wave pattern and ABC correction

One of the basic tenets of Elliott Wave theory is that market structure is fractal in character. Elliott Wave patterns that show up on long term charts are identical to, and will also show up on short term charts, albeit with sometimes more complex structures. Here is an example of a graph with waves within the broader wave structure (Fig. 2).

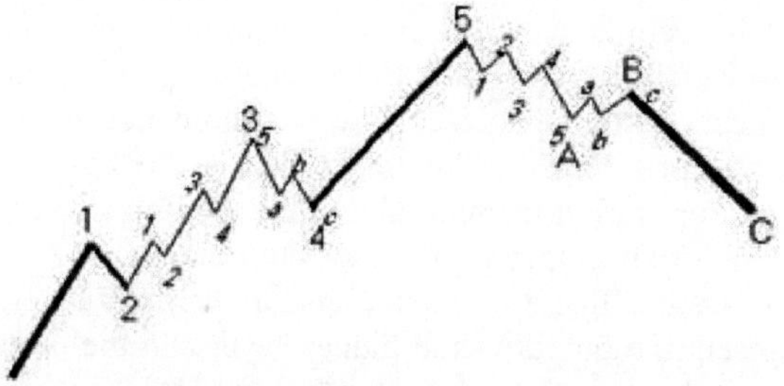

Fig. 2 The fractal character of Elliott wave patterns

Elliott wave is another technical tool that may be used to try to identify market trends and determine whether trends are about to change. Elliott wave can be used to generate short-term trading opportunities and analyze whether current market trends will continue. To apply Elliott wave to some analysis we need to identify which wave is being formed. The major waves determine the major trend of the market. The minor waves determine the minor trends in the market. Once we identify the main wave look to buy the market in the 1, 3 and 5 waves, and sell the market in waves to 2 and 4. In the corrective phase look to buy wave A and C and look to sell wave B. The most difficult part of Elliott wave analysis to correctly label the waves.

3 Backpropagation Neural Networks

A neural network is a parallel, distributed information processing structure consisting of processing elements (which can possess a local memory and can carry out localized information processing operations) interconnected together with unidirectional signal channels called connections. Each processing element has a single output connection which branches into as many collateral connections as desired (each carrying the same signal - the processing element output signal). The processing element output signal can be of any mathematical type desired. All of the processing that goes on within each processing element must be completely local: i.e., it must depend only upon the current values of the input signals arriving at the processing element via impinging connections and upon values stored in the processing element's a local memory.

The backpropagation neural network architecture is a hierarchical design consisting of fully interconnected layers or rows of processing units (with each unit itself comprised of several individual processing elements. Backpropagation belongs to the class of mapping neural network architectures and therefore the information processing function that it carries out is the approximation of a bounded mapping or function $f: A \subset R^n \to R^m$, from a compact subset $\boldsymbol{A}$ of n-dimensional Euclidean space to a bounded subset $f[\boldsymbol{A}]$ of m-dimensional Euclidean space, by means of training on examples (x_1, z_1), (x_2, z_2), . . . , (x_k, z_k). . . . It will always be assumed that such examples of a mapping f are generated by selecting $\boldsymbol{x}_k$ vectors randomly from $\boldsymbol{A}$ in accordance with a fixed probability density function $p(\boldsymbol{x})$. The operational use to which the network is to be put after training also assumed to involve random selections of input vectors $\boldsymbol{x}$ in accordance with $p(\boldsymbol{x})$. The backpropagation architecture described in this paper is the basic, classical version (Fig. 3). The backpropagation learning algorithm is composed of two procedures: (a) feed-forward and (b) back-propagation weight training.

Feed-forward. Assume that each input factor in the input layer is denoted by x_i, the y_j and z_k represent the output in the hidden layer and the output layer, respectively. And, the y_j and z_k can be expressed as follows (1):

$$y_j = f(X_j) = f\left(w_{oj} + \sum_{i=1}^{I} w_{ij} x_i\right) \text{ and } z_k = f(Y_k) = f\left(w_{ok} + \sum_{j=1}^{J} w_{ik} y_j\right) \quad (1)$$

where the w_{0j} and w_{0k} are the bias weights for setting threshold values, f is the activation function used in both hidden and output layers, and X_j and Y_k are the temporarily computing results before applying activation function f. In this study, a sigmoid function is selected as the activation function. Therefore, the actual outputs y_j and z_k in hidden and output layers, respectively, can be also written as:

$$y_j = f(X_j) = \frac{1}{1+e^{-X_j}} \text{ and } z_k = f(Y_k) = \frac{1}{1+e^{-Y_k}} \quad (2)$$

The activation function f introduces the non-linear effect to the network and maps the result of computation to a domain (0, 1). This sigmoid function is differentiable. The derivative of the sigmoid function in eq. (2) can be easily derived as: $f' = f(1+-f)$.

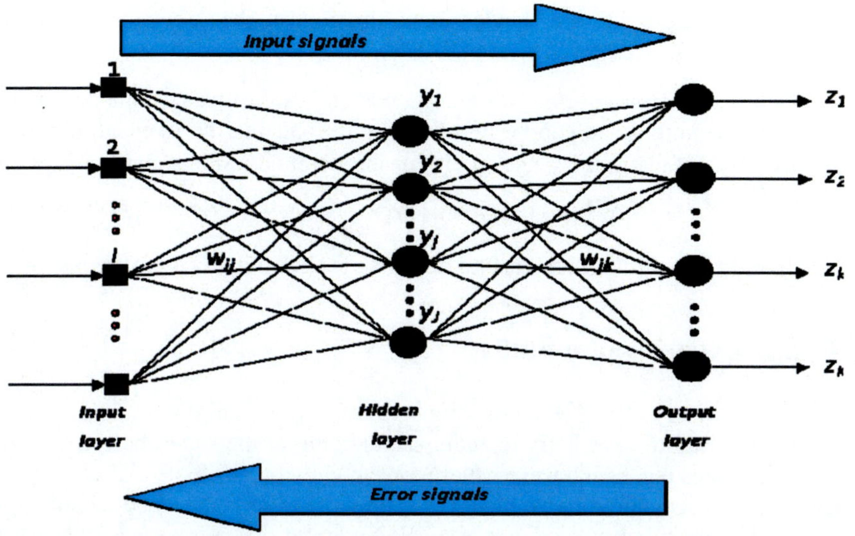

Fig. 3 A backpropagation network architecture

Back-propagation weight training. The error function is defined as [7] (3):

$$E = \frac{1}{2}\sum_{k=1}^{K} e_k^2 = \sum_{k=1}^{K} (t_k - z_k)^2 \tag{3}$$

where t_k is a predefined network output (or desired output or target value) and e_k is the error in each output node. The goal is to minimize E so that the weight in each link is accordingly adjusted and the final output can match the desired output. To get the weight adjustment, the gradient descent strategy is employed. In the link between hidden and output layers, computing the partial derivative of E with respect to the weight w_{jk} produces, as (4)

$$\frac{\partial E}{\partial w_{jk}} = -e_k f'^{(Y_k)y_j} = -\delta_k y_j \quad \text{where} \quad \delta_k = (t_k - z_k) f'(Y_k) \tag{4}$$

The weight adjustment in the link between hidden and output layers is computed by $\Delta w_{jk} = \alpha \times y_j \times \delta_k$, where α is the learning rate, a positive constant between 0 and 1. The new weight herein can be updated by the following $w_{jk}(n+1) = w_{jk}(n) + \Delta w_{jk}(n)$, where n is the number of iteration. Similarly, the error gradient in links between input and hidden layers can be obtained by taking the partial derivative with respect to w_{ij}, as (5)

$$\frac{\partial E}{\partial w_{ij}} = -\Delta_j x_j = f'(X_j)\sum_{k=1}^{K} \partial_k w_{jk} \tag{5}$$

The new weight in the hidden-input links can be now corrected as: $\Delta w_{ij} = \alpha \times x_i \times \Delta_j$ and $w_{ij}(n+1) = w_{ij}(n) + \Delta_j$.

Training the BP-networks with many samples is sometimes a time-consuming task. The learning speed can be improved by introducing the momentum term η [5]. Usually, η falls in the range $\langle 0,1 \rangle$. For the iteration n, the weight change Δw can be expressed as $\Delta w(n+1) = \eta \times w(n) + \alpha \times \frac{\partial E}{\partial w(n)}$. The back-propagation learning algorithm used in artificial neural networks is shown in many textbooks [6]

4 The Experimental Study

Waves are usually identified by looking back at historic price action. The hard part in applying Elliott wave is try to anticipate drawing of the waves before the market action takes place. The wave patterns are actually quite simple. All you need to know is markets tend to move in waves and market direction can be identified by identifying repetitive pattern of waves. Elliott wave theory says markets will in move five waves up to three waves down. There is no absolute time to be complete a cycle. In theory a wave could last for years.

4.1 The Adaptation Phase

It is necessary to remark that determination of training patterns is one of the key tasks that needed our attention. Improperly chosen patterns can lead to confusion of neural networks. A neural network "adapted" on incorrect patterns can give meaningless responses. The search for the patterns is a complicated process which is usually performed manually by the user.

During our experimental work, we made some study included Elliott waves pattern recognition. Our method is based on backpropagation neural network and is able to recognize Elliot wave structures in given time series. Artificial neural networks need for their adaptation training sets. In our experimental work, the training set consists of 12 different types of Elliott wave's samples, see Fig. 4. Input data are sequences included always n consecutive numbers, which are transformed into interval <0, 1> by the formula (6). Samples are adjusted for the needs of backpropagation networks with sigmoid activation function in this way.

$$x'_j = \frac{x_j - \min(x_i, \ldots, x_{i+n-1})}{\max(x_i, \ldots, x_{i+n-1}) - \min(x_i, \ldots, x_{i+n-1})}, \ (j = i, \ldots, i+n-1) \tag{6}$$

where x'_j is normalized output value of the j-th neuron ($j = i,\ldots,i+n$-1) and $(x_i, \ldots x_{i+n-1})$ are n-1 consecutive output values that specify sequences (patterns) from the training set (e.g. training pars of input and corresponding output vectors). Input vector contains 20 components. We used 20 input neurons in the order to each component of the input vector was accepted with the same weight, therefore the used training set contains together with each component its complement to value **"1"**, i.e. (x_1, x_2, x_3, x_4, x_5, x_6, x_7, x_8, x_9, x_{10}, 1-x_1, 1-x_2, 1-x_3, 1-x_4, 1-x_5, 1-x_6, 1-x_7, 1-x_8, 1-x_9, 1-x_{10}). Such a proposed input vectors representing patterns is a guarantee that equal emphasis is placed on each value, because backpropagation algorithm usually has a tendency to put less emphasis on inputs near zero. Output vector has got 12 components and each output unit represents one of 12 different types of Elliott wave samples. A neural network architecture is *20 - 16 - 12* (e.g. 20 units in the input layer, 16 units in the hidden layer, and 12 units in the output layer). The net is fully connected. Adaptation of the neural network starts with randomly generated weight values. We used the backpropagation method for the adaptation with the following parameters: first 5000 iterations has learning rate value 0.5, and for the next 2000 iterations has learning rate value 0.1, momentum is 0. The conducted experimental studies also showed that in each cycle of adaptation is to present an adequate network of training patterns mixed randomly to ensure their greater diversity, but also acts as a measure of system stability. Uniform system in a crisis usually collapses entirely, while in the diversion system through a crisis of its individual parts, but the whole remains functional. Condition of end of the adaptation algorithm was specified limit value overall network error, $E \leq 0.07$.

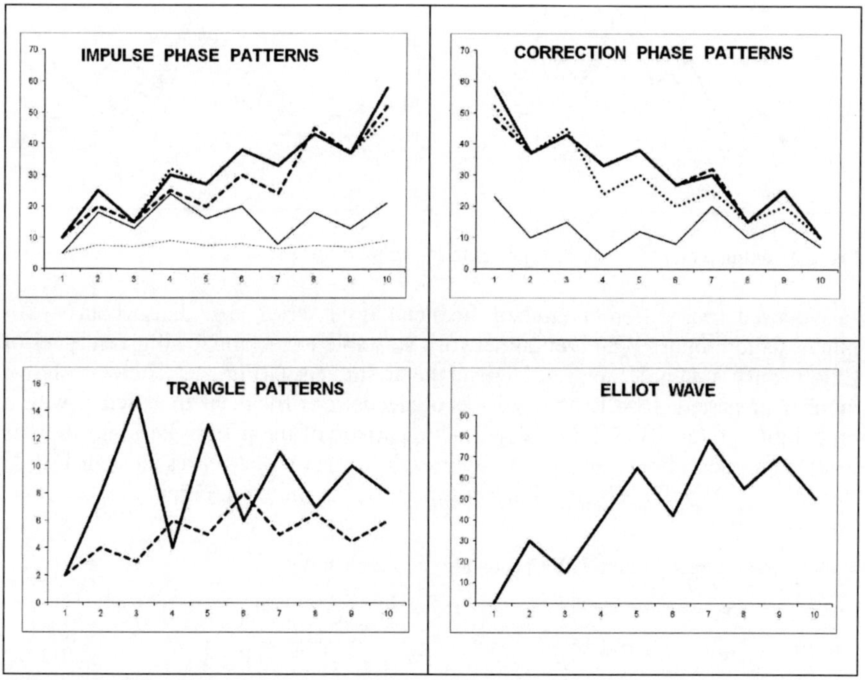

Fig. 4 Different types of Elliott wave's samples represented in the training set

4.2 The Test Phase

In order to test the efficiency of the method, we applied a database from the area of financial forecasting [8] that is a set of data that reflects the situation on the market. We used time series which shows the development of market values of EUR/USD, which reflect the exchange rate between EUR and USD.

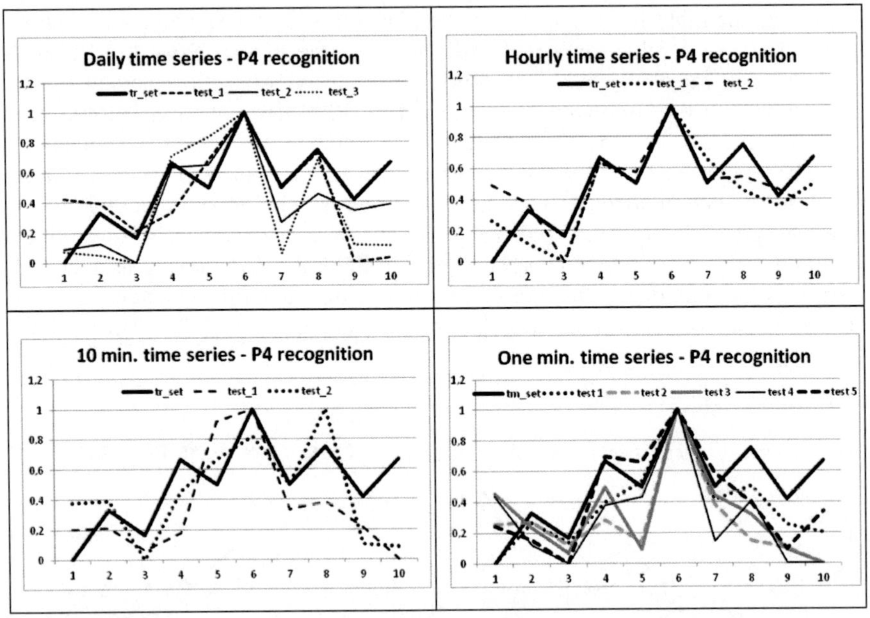

Fig. 5 Training pattern P4 and its representation in used test sets

We used four different kind of financial time series, e.g. daily, hourly, 10-minutes and minutes. Our Neural network was able to recognize all given types of Elliott wave's samples represented in the training set (Fig. 4). There is shown number of pattern P1-P12 that was recognized from financial time series with a probability grater than 0.9 in Table 1. Comparison of the pattern P4 looks, how is learned via neural network versus its present in test set is represented in Fig. 5. Pattern P4, which is recognized in test time series is shown in Fig.6.

Table 1 Pattern recognition with a probability grater than 0.9

time series	Patterns											
	P1	P2	P3	P4	P5	P6	P7	P8	P9	P1 0	P1 1	P12
Daily	0	19	8	24	13	0	28	5	21	25	35	11
Hourly	3	3	5	39	5	0	13	9	38	72	78	23
10 min.	6	25	4	44	1	1	14	10	33	59	95	4
One min.	5	31	1	39	0	0	4	21	35	62	87	2

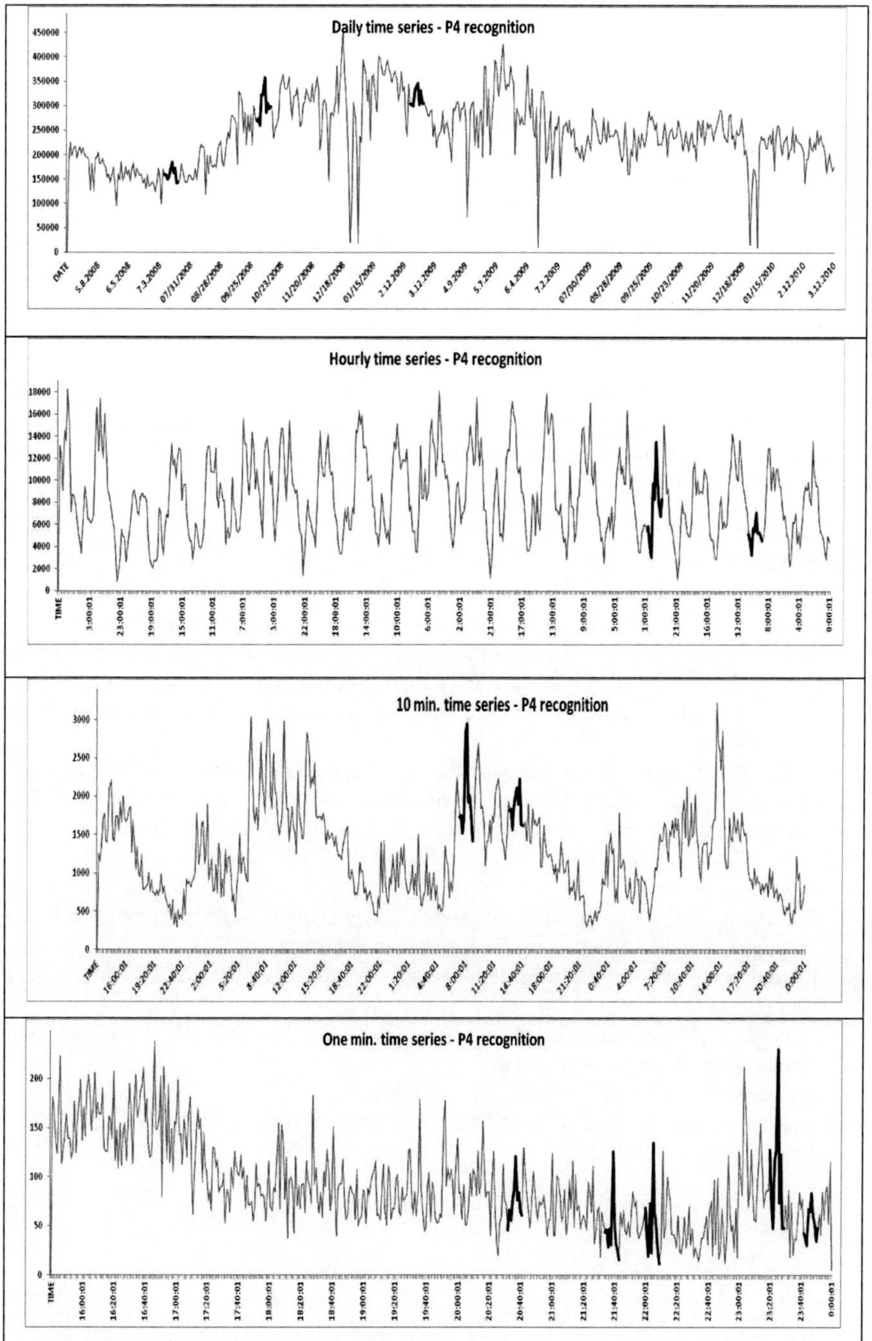

Fig. 6 P4 recognized patterns that occur in financial time series

5 Conclusion

In this paper, a short introduction into the field of Elliott waves recognition using backpropagation neural network has been given. According to the results of experimental studies, it can be stated that Elliot waves patterns were successfully extract in given time series in a varied time scale and recognize using suggested method, how as can be seen from figures in result section. It might result in better mapping of the time series behavior for better prediction. Elliott waves recognition allows time series trend prediction, as follows. If we recognize impulse phase of Elliott wave, trend prediction is downwards. If we recognize correction phase of Elliott wave or triangle patterns, trend prediction is upwards

Acknowledgments. The research described here has been financially supported by University of Ostrava grant SGS2/PRF/2012. Any opinions, findings and conclusions or recommendations expressed in this material are those of the authors and do not necessarily reflect the views of the sponsors.

References

[1] Anand, S., Chin, W.N., Khoo, S.C.: Chart Patterns on Price History. In: Proc. of ACM SIGPLAN Int. Conf. on Functional Programming, Florence, Italy, pp. 134–145 (2001)

[2] Kamijo, K., Tanigawa, T.: Stock Price Pattern Recognition: A Recurrent Neural Network Approach. In: Proc. of the Int. Joint Conf. on Neural Networks, vol. 1, pp. 215–221 (1990)

[3] Leigh, W., Modani, N., Hightower, R.: A Computational Implementation of Stock Charting: Abrupt Volume Increase As Signal for Movement in New York Stock Exchange Composite Index. Decision Support Systems 37(4), 515–530 (2004)

[4] Poser, S.: Applying Elliott Wave Theory Profitably. Wiley (2003) ISBN-10: 0471420077

[5] Rumelhart, D.E., Hinton, G.E., William, R.J.: Learning Representations by Back-Propagation Errors. Nature, London 323, 533–536 (1986)

[6] Russell, S., Norvig, P.: Artificial Intelligence—A Modern Approach, 2nd edn. Prentice Hall (2003)

[7] Widrow, B., Hoff, M.E.: Adaptive switching circuits. In: 1960 IRE WESCON Convention Record, pp. 96–104 (1960)

[8] A database from the area of financial forecasting, `http://www.google.com/finance/forex?=nas:eur/usd:daq` (accessed April 10, 2012)

Pattern Recognition Algorithm Optimization

Eva Volna, Michal Janosek, Martin Kotyrba, and Vaclav Kocian

University of Ostrava, 30. dubna 22, 70103 Ostrava, Czech Republic

Abstract. In this article, a short introduction into the field of pattern recognition in time series has been given. Our goal is to find and recognize important patterns which repeatedly appear in the market history. We focus on reliability of recognition made by the proposed algorithms with optimized patterns based on artificial neural networks. The performed experimental study confirmed that for the given class of tasks can be acceptable a simple Hebb classifier with a proposed modification that has been designed, tested, and used for the active mode of Hebb rule. Finally, we present comparison results of trading based on both recommendations: using proposed Hebb neural network implementation, and human expert.

1 Introduction

The main topic of the article is to optimize the pattern recognition algorithm in order to recognize the input pattern in real time with maximum reliability. These input patterns do not cover every time point in the series, but are optimized to be suitable candidates in experimental tasks so that the developed classifiers would be able to learn key characteristics of these patterns and accurately recognize them. Such optimized inputs reduce the calculation costs. One of important challenges is to recognize the input pattern reliably. Patterns in the market are not exact, they are slightly different every time they appear. They can have different amplitude and different duration, albeit visually the same pattern can look differently despite being the same. There are several problems concurrent with major ones as well as several optimization algorithms.

The article is arranged as follows: in *Section 2* describes features of pattern recognition systems. The pattern recognition algorithms based on neural networks are discussed in *section 3*. *Section 4* covers the experimental results and describes the feature extraction process in order to optimize the patterns used as inputs into our experiments. We have used data from X-Trade Brokers co. [12] which is a set of data and queries that reflect the needs of financial analysts who study patterns in stock market data. A conclusion is given in *section 5*. We focus on reliability of recognition made by the described algorithms with optimized patterns based on the reduction of the calculation costs. All results from the developed classifiers will be compared to each other.

I. Zelinka et al. (Eds.): Nostradamus: Mod. Meth. of Prediction, Modeling, AISC 192, pp. 251–260.
springerlink.com

2 Pattern Recognition Systems

Identification problems involving time-series data (or waveforms) constitute a subset of pattern recognition applications that is of particular interest because of the large number of domains that involve such data. The recognition of structural shapes plays a central role in distinguishing time-series analysis. Sometimes just one structural form (a bump, an abrupt peak or a sinusoidal component), is enough to identify a specific phenomenon. There is not a general rule to describe the structure – or structure combinations – of various phenomena, so specific knowledge about their characteristics has to be taken into account. Therefore, domain knowledge has to be added to the structural information.

The interest is focused on searching for specific patterns within waveforms [4]. The algorithms used in pattern recognition systems are commonly divided into two tasks, as shown in Figure 1. The description task transforms data collected from the environment into features (primitives).

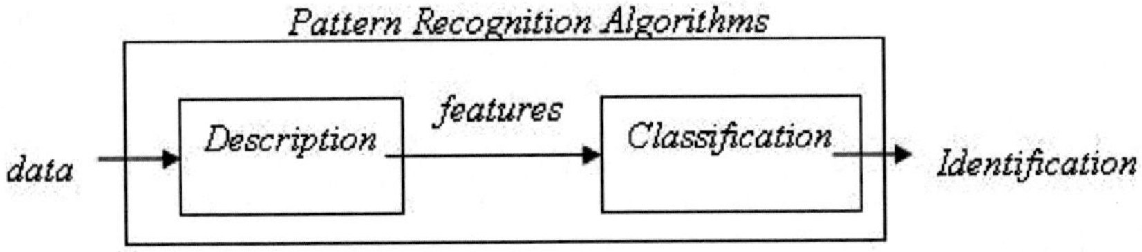

Fig. 1 Tasks in the pattern recognition systems

The classification task arrives at an identification of patterns based on the features provided by the description task. There is no general solution for extracting structural features from data. The selection of primitives by which the patterns of interest are going to be described depends upon the type of data and the associated application. The features are generally designed making use of the experience and intuition of the designer.

Image data are usually two-dimensional and the intuitive concept of "pattern" also corresponds to the two-dimensional shapes. This way allows showing a progress of a scalar variable. In this article, we consider experiments only over two-dimensional data (time series). Extending the principles of multidimensional vectors (random processes) will be the subject of our future projects. The feature extraction process is performed in order to optimize the patterns used as inputs into experiments. We propose approaches relying on neural networks, but in contrast to [8], the patterns do not cover every time point in the series. Here, a proposed segmentation is perceived as an optimization process to transform the original time series into the corresponding features. This sequence of features, instead of the whole time series, is designated as inputs into our experimental tasks. Such optimized inputs are able to reduce the calculation costs.

3 Pattern Recognition Algorithms

We can imagine a pattern as some object with same or similar properties. There are many ways how to recognize and sort them [3]. An input value can be any data regardless its origin as text, audio, image or any other data. Since we are working on computers, the input data and all patterns can be represented in a binary form without the loss of generality.

In order to build a better model to recognize patterns effectively and efficiently, this study aims at investigating the recognition performance that utilizes classification methods based on neural networks. Moreover, patterns in the market are analyzed and optimized in order to be suitable candidates as inputs in experimental tasks. Developed classifiers based upon neural networks should effectively learn the key characteristics of the patterns and accurately recognize them.

3.1 Artificial Neural Networks

An Artificial Neural Network (ANN) is an information processing paradigm that is inspired by the way biological nervous systems, such as the brain, process information. The key element of this paradigm is the structure of the information processing system. It is composed of a large number of highly interconnected processing elements (neurones) working in unison to solve specific problems. ANNs, like people, learn by example. An ANN is configured for a specific application, such as pattern recognition or data classification, through a learning process. Learning in biological systems involves adjustments to the synaptic connections that exist between the neurones. This is true of ANNs as well. It receives input from some other units, or perhaps from an external source [5].

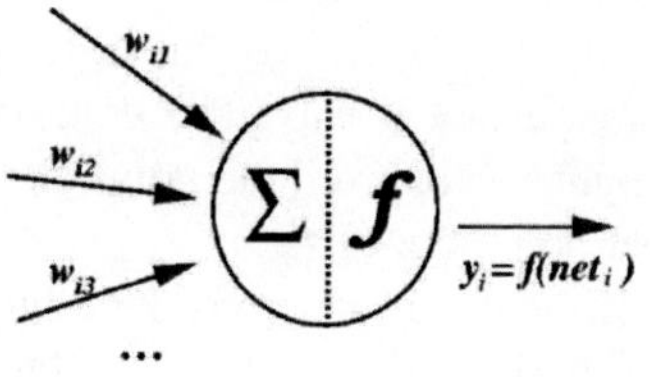

Fig. 2 A simple artificial neuron

Each input has an associated weight ***w***, which can be modified so as to model synaptic learning. The unit computes some function *f* of the weighted sum of its inputs (eq. 1):

$$y_i = f(net_i) \tag{1}$$

Its output, in turn, can serve as input to other units. The weighted sum (eq. 2) is called the net input to unit *i*, often written net_i. Note that w_{ij} refers to the weight from unit *j* to unit *i*. The function *f* is the unit's activation function.

$$net_i = \sum_j w_{ij} y_j \tag{2}$$

3.2 The Hebb Rule

The Hebb rule [6] determines the change in the weight connection from u_i to u_j by $\Delta w_{ij} = r * y_i * y_j$, where r is the learning rate and y_i, y_j represent the activations of u_i and u_j respectively. Thus, if both u_i and u_j are activated the weight of the connection from u_i to u_j should be increased.

Examples can be given of input/output associations which can be learned by a two-layer Hebb rule pattern associator. In fact, it can be proved that if the set of input patterns used in training are mutually orthogonal, the association can be learned by a two-layer pattern associator using Hebbian learning. However, if the set of input patterns are not mutually orthogonal, interference may occur and the network may not be able to learn associations. This limitation of Hebbian learning can be overcome by using the delta rule [1].

3.3 The Delta Rule

The delta rule [2], also called the Least Mean Square (LMS) method, is one of the most commonly used learning rules. For a given input vector, the output vector is compared to the correct answer. If the difference is zero, no learning takes place; otherwise, the weights are adjusted to reduce this difference. The change in weight from u_i to u_j is given by: $\Delta w_{ij} = r * y_i * e_j$, where r is the learning rate, y_i represents the activation of u_i and e_j is the difference between the expected output and the actual output of u_j. If the set of input patterns form a linearly independent set then arbitrary associations can be learned using the delta rule.

This learning rule not only moves the weight vector nearer to the ideal weight vector, it does so in the most efficient way. The delta rule implements a gradient descent by moving the weight vector from the point on the surface of the paraboloid down toward the lowest point, the vertex.

Generalizing the ideas of the delta rule, consider a hierarchical network with an input layer, an output layer and a number of hidden layers. We consider only the case where there is one hidden layer. The network is presented with input signals which produce output signals that act as input to the middle layer. Output signals from the middle layer in turn act as input to the output layer to produce the final output vector. This vector is compared to the desired output vector. Since both the output and the desired output vectors are known, we can calculate differences between both outputs and get an error of neural network. The error is backpropagated from the output layer through the middle layer to the units, which are responsible for generating that output. The delta rule can be used to adjust all the weights. More details are presented in [7].

4 Experimental Results

The search for the patterns is a complicated process which is usually performed manually by the user. In order to test the efficiency of the pattern recognition system we applied data from X-Trade Brokers [12] that is a set of data that reflect the situation on the market. We used time series which shows the development of market values of EURUSD, which reflect the exchange rate between EUR and USD.

4.1 Used Dataset

The testing time scale was four months from April to June 2011 on a 5-minute chart. That means that every 5 minutes a new record is created in the table. We use for experiments 17697 records totally that are indicated in table 1.

Table 1 The experiment's data sample

Date	Time	Open	High	Low	Close	V		C1	C2	C3
2011.04.01	00:00	1.4172	1.4172	1.4166	1.4166	10	I			
2011.04.01	00:10	1.4167	1.4169	1.4167	1.4168	12	I			
-	-	-	-	-	-	-	-	-	-	-
2011.04.28	18:30	1.4808	1.4812	1.4802	1.4805	28	I	27	28	
2011.04.28	18:35	1.4803	1.4804	1.4799	1.4801	24	I	27	28	
2011.04.28	18:40	1.4802	1.4804	1.4798	1.4800	32	I		28	
2011.04.28	18:45	1.4801	1.4802	1.4791	1.4792	28	I		28	
2011.04.28	18:50	1.4791	1.4794	1.4789	1.4792	30	I		28	
-	-	-	-	-	-	-	-	-	-	-
2011.04.30	23:50	1.4500	1.4500	1.4493	1.4495	13	I			

Columns description:
Date - corresponding date
Time - start time of the 5 minutes interval
Open - open value of the time interval
High - maximum value of the time interval
Low - minimum value of the time interval
Close - close value of the time interval
Volume - trade volume in the time interval

We choose the flip pattern for our experiments, see Figure 3. This pattern occurs in markets with both decreasing and increasing tendency. We can split this pattern to two subpatterns; a pattern anticipating price decrease (column C1 in table 1, Figure 4) and a pattern anticipating price increase (column C2 in table 1, Figure 5). Column C3 determines the pattern which doesn't belong either to C1 or C2 (column C3 in table 1, Figure 6).

Fig. 3 Flip pattern

In the following Figures 4-6 it is possible to see patterns' examples from each of the presented patterns' classes (C1, C2, C3). Vertical lines on the left and on the right side of the each chart represent marks of the beginning and the end of the pattern as it was marked by the user. Following figures represent some particular patterns in our dataset.

An example of the flip pattern characterizing the anticipated price decrease (Figure 4). This pattern emerged in the chart on 7 April 2011 from 07:00 till 9:55. It was the 6th pattern in our dataset, which is represented as a sample in the (table 1). The pattern was recorded to the table into the column C1 as a number 6 to the all rows representing the above-mentioned interval.

An example of the flip pattern characterizing the anticipated price increase (Figure 5). This pattern emerged in the chart on 14 April 2011 from 03:00 till 05:20. It was the 15th pattern in our dataset, which is represented as a sample in the (table 1). The pattern was recorded to the table into the column C2 as a number 15 to the all rows representing the above-mentioned interval.

An example of the pattern does not belong to any of presented class (Figure 6). This pattern emerged in the chart on 27 May 2011 from 13:45 till 15:10. It was the 44th pattern in our dataset, which is represented as a sample in the (table 1). The pattern was recorded to the table into the column C3 as a number 44 to the all rows representing the above-mentioned interval [11].

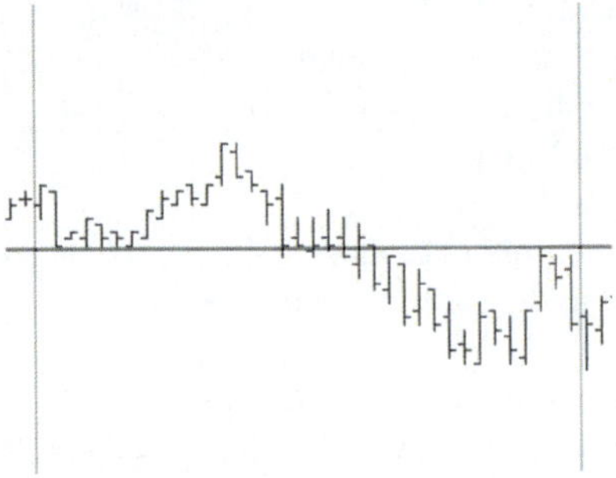

Fig. 4 Pattern - P6C1 (the pattern with id 6) - 2011.04.07 07:00 - 09:55

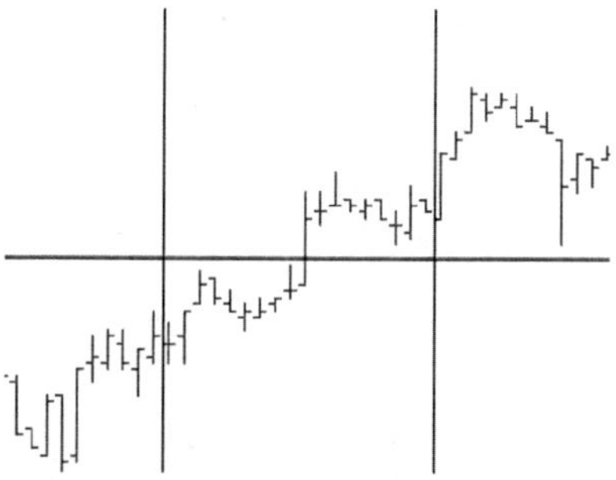

Fig. 5 Pattern P15C2 (the pattern with id 15) 011.04.14 03:00 - 05:20

Fig. 6 Pattern P44C3 (the pattern with id 44) 2011.05.27 13:45 - 15:10

4.2 Used Types of ANN

Hebb network is very simple, it contains minimum parameters and is adapted in one cycle. During our experimental study we worked with various neural networks, which used a method of a relevance evaluation of each input vector component.

Both of the tested algorithms (Hebb, backpropagation) tended to find almost all introduced patterns familiar. A new modification has been designed, tested and used for the active mode of Hebb rule. In a common operation, the output value of each output neuron is derived from its net value in a very simple way:

$$y_i = -1, \text{ if } net_i < 0, \; y_i = 1, \text{ if } net_i >= 0.$$

The problem there is that y_i takes the same output no matter how big its net_i (eq. 2) value is (apart from sign rules), topology you can see in table 2. In other words, we do not have any information about the certainty of the network result. Our network output function was replaced with simple equivalence: $y_i = net_i$.

Table 2 Hebb network configuration

Network topology	*Input layer*: x neurons, where x=number of bits in the input bitmap *Output layer*: 2 neurons *Interconnection*: fully connected
Type of I/O values:	bipolar

Setting up this type of network requires more testing. Finally, the following parameters were chosen for our experiments, see table 3:

Table 3 Backpropagation network configuration

Network topology	*Input layer*: x neurons, where x=number of bits in the input bitmap *Hidden layer*: 6 neurons *Output layer*: 3 neurons *Interconnection*: fully connected
Weight initialization algorithm:	Nguyen-Widrow
Weight initialization range:	(-0.5; +0.5)
Alpha - learning rate:	0.2
Maximal network error (difference between desired and computed network response)	0.01
Activation function:	Sigmoid (slope = 1.0)
Type of I/O values:	binary

There were no problems with learning patterns of all types. Usually it took some hundred cycles to learn them accurately. In several cases the network remained in a local minimum and therefore the learning process had to be restarted.

4.3 Trading Results

We present results of trading in the months April, May, and June 2011 based on both recommendations: using our Hebb neural network implementation, Figure 7a and human expert, Figure 7b. Results of the trading without commissions and with commissions are displayed in each graph. In the case of trading with commissions, value 0.0002 was subtracted from each transaction. The x axis indicates the number of trades during the given period, the y axis indicates the profit. As for the profit, the absolute value of the trading account is not important. More important is the trading equity itself.

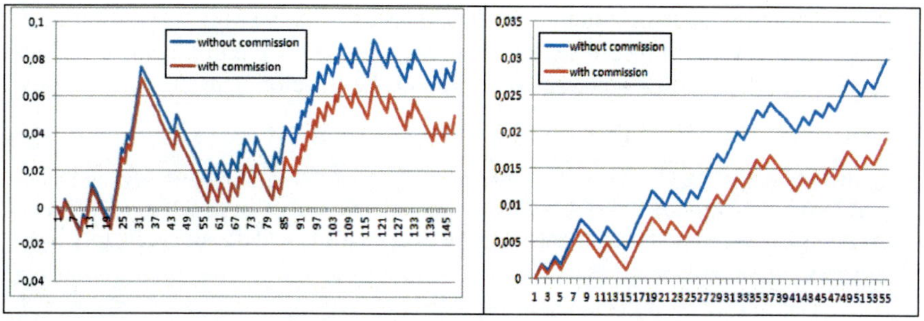

Fig. 7 Hebb algorithm managed trading and human managed trading

Figure 7 shows a graph with trading results based on the recommendations using outputs from our Hebb network. As we can see in the graph without commission, the result of trading settled just below 0.08. In a contrast to it, the result with commission dropped to below 0.06 for the given period. This is in particular due to the large number of trades.

Figure 7 also shows a graph with trading results performed based on the recommendations of the human expert. At first sight, there is a stable rise tendency and much lower count of trades. That was achieved by a strict selection of trades and limited business hours between 9:00 - 14:00 as well. Due to a lower number of trades the graph with commissions does not shows a rapid equity drop as in the previous case with Hebb rule.

Hebb network has surprised us because it was able to stay in positive account numbers in business without commissions. The disadvantage of this approach was stagnation after reaching the value 0.05 and also excessive overtrading. Due to high number of transactions it is not quite good from real trading because for each transaction we have to pay a commission fee. We would like to propose some method of control of the number of trades for adaptation algorithms in our future work.

5 Conclusion and Future Work

In this paper, a short introduction into the field of pattern recognition in time series has been given. We focus on reliability of recognition made by the described algorithms with optimized patterns based on artificial neural networks towards the reduction of the calculation costs. Two proposed approaches were used. Our experimental study confirmed that for the given class of tasks can be acceptable simple classifiers. The advantage of simple neural networks is its very easy implementation and quick adaptation. According to our results of experimental studies, it can be stated that pattern recognition in time series using chosen Hebb network was successful.

Our future research will be directed to the combination of several sub-branches, which we would like to improve the recognition of patterns and structures in market research (time series). We will continue our work with neural networks, business systems, and we would also like to focus on other soft-computing methods. The aim of our future research is to create a better environment for pattern recognition, which allows interactively predict the development of the stock market.

Acknowledgments. The research described here has been financially supported by University of Ostrava grant SGS02/PřF/2012. Any opinions, findings and conclusions or recommendations expressed in this material are those of the authors and do not necessarily reflect the views of the sponsors.

References

[1] Anand, S., Chin, W.N., Khoo, S.C.: Chart Patterns on Price History. In: Proc. of ACM SIGPLAN Int. Conf. on Functional Programming, Florence, Italy, pp. 134–145 (2001)

[2] Bulkowski, N.: Encyclopedia of Chart Patterns, 2nd edn. John Wiley and Sons (2005)

[3] Ciskowski, P., Zaton, M.: Neural Pattern Recognition with Self-organizing Maps for Efficient Processing of Forex Market Data Streams. AISC, pp. 307–314 (2010), doi:10.1007/978-3-642-13208-7_39

[4] Dormido-Canto, S., Farias, G., Vega, J., Dormido, R., Sánchez, J., Santos, M., et al.: Search and retrieval of plasma waveforms: structural pattern recognition approach. Review of Scientific Instruments 77, 10F514 (2006)

[5] Fausett, L.V.: Fundamentals of Neural Networks. Prentice-Hall, Inc., Englewood Cliffs (1994)

[6] Frost, A.J., Prechter, R.: Elliott wave principle: key to market behavior, 7th edn. New Classics Library, Gainesville (1998)

[7] Gershenson, C.: Design and Control of Self-organizing Systems. Dissertation. Vrije Universiteit Brussel (2007)

[8] Kamijo, K., Tanigawa, T.: Stock Price Pattern Recognition: A Recurrent Neural Network Approach. In: Proc. of the Int. Joint Conf. on Neural Networks, vol. 1, pp. 215–221 (1990)

[9] Kocian, V., Volna, E., Janosek, M., Kotyrba, M.: Optimizatinon of training sets for Hebbian-learningbased classifiers. In: Matoušek, R. (ed.) Proceedings of the 17th International Conference on Soft Computing, Mendel 2011, Brno, Czech Republic, pp. 185–190 (2011) ISBN 978-80-214-4302-0, ISSN 1803-3814

[10] Leigh, W., Modani, N., Hightower, R.: A Computational Implementation of Stock Charting: Abrupt Volume Increase As Signal for Movement in New York Stock Exchange Composite Index. Decision Support Systems 37(4), 515–530 (2004)

[11] Volna, E., Janosek, M., Kotyrba, M., Kocian, V., Oplatkova, Z.: Methodology for System Adaptation based on Characteristic patterns. In: Dutta, A. (ed.) Robotic Systems – Applications, Control and Programming, pp. 611–628. InTech, Croatia (2012) ISBN 978-953-307-941-7

[12] X-Trader Brokers, `http://xtb.cz` (accessed September 3, 2011)

Evolutionary Identification and Synthesis of Predictive Models

Ivan Zelinka[1], Lenka Skanderova[1], Mohammed Chadli[2], Tomas Brandejsky[3], and Roman Senkerik[4]

[1] VŠB-Technical University of Ostrava, Faculty of Electrical Engineering and Computer Science, Department of Computer Science, 17. listopadu 15, 708 33 Ostrava-Poruba, Czech Republic
{ivan.zelinka,lenka.skanderova.st}@vsb.cz
[2] University of Picardie Jules Verne, 7, Rue du Moulin Neuf, 80000, Amiens, France
mohammed.chadli@u-picardie.fr
[3] Faculty of Transportation Sciences, CTU in Prague
brandejsky@fd.cvut.cz
[4] Tomas Bata University in Zlin, Faculty of Applied Informatics, T.G. Masaryka 5555, 760 01 Zlin, Czech Republic
senkerik@fai.utb.cz

Abstract. In this contribution is demonstrated use of two evolutionary algorithms on parameter identification of selected predictive models. Both algorithms were used to indentify parameter of pre-selected ARMA models. At the end are discussed possibilities of use of synthesis of predictive models by means of methods of symbolic regression that has successfully been used on chaotic system identification by means of evolutionary algorithms on measured data.

1 Evolutionary Identification - Basic Ideas

For many years various methods for prediction of time series are used. These methods vary from classics to special methods like neural network and genetics algorithms. This contribution is focused on prediction by means of geometrical shapes (highly used on stock exchange) and its recognition and location in time series. For this purpose we can use professional software, which is usually expensive, or create own which can be demanding on time and programming skill. Another methods like neural networks can also be used.

In nowadays exist several algorithms, which are used for time series prediction. Time series in general sense can be produced by dynamical system, which can by vary from simple behavior to complicated one (see Fig. 2 a)). Many times we do not knew all inputs and output and thus our knowledge about given system is

I. Zelinka et al. (Eds.): Nostradamus: Mod. Meth. of Prediction, Modeling, AISC 192, pp. 261–272.
springerlink.com

uncompleted. In prediction we usually use wide class of algorithms which allow to us to do a good predictions based on knowledge of historical behavior of dynamical system which can vary from stable via quasi-periodic and catastrophic to chaotic (Fig. 2 a)). These algorithms are usually taken like "models" as for example AR, ARMA, ARIMA, etc. Some of them are self-adaptable" (neural networks) some of them no. For their using one have to found theirs parameters for example by Box-Jenkins methodology in case of AR models, etc.

In this contribution is shown how can be used evolutionary algorithms for retrieval of suitable model of type AR models like universal algorithms. Word "*universal*" means that these algorithms can be used not only for AR models, but for such "exotic" models like neural network (NN) where EA are able to learn NN [1, 2] or/and find its suitable-optimal structure for given time series. In this contribution are used two EA – Differential Evolution and Self-Organizing Migrating Algorithm [7]. Because DE and SOMA are explained in literature, (SOMA see this proceeding, DE [8] see also [3, 4], [5, 6]), attention is focused here on how DE and SOMA was used for retrieval of suitable predictive models.

2 Experiment Design and Used Evolutionary Algorithms

This research used two evolutionary algorithms: Self-Organizing Migrating Algorithm (SOMA) [7] and Differential Evolution (DE) [8], [9].

Self-Organizing Migrating Algorithm is a stochastic optimization algorithm that is modeled on the basis of social behavior of cooperating individuals [15]. It was chosen because it has been proven that the algorithm has the ability to converge towards the global optimum [7] and due to the successful applications together with AP [10], [11].

DE is a population-based optimization method that works on real-number-coded individuals, [8], [9]. DE is quite robust, fast, and effective, with global optimization ability. It does not require the objective function to be differentiable, and it works well even with noisy and time-dependent objective functions.

SOMA algorithm setting and DE setting is in Table 1 and Table 2.

Table 1 SOMA settings

PathLength	3
Step	0.11
PRT	0.1
PopSize	30
Migrations	10

Table 2 DE settings

PopSize	30
F	0.8
Cr	0.6
Generations	50

3 Evolutionary Identification of Predictive Models

Evolutionary algorithms here were used like searching algorithms, which were looking for the best model. Whole problem was defined like problem of optimization where cost function was defined so that its arguments were parameters (individual) of given kind of model and returned value (cost value) was (1). Thanks to this approach all possible models has created hyperplane on which each point represented one possible model (Fig. 1 and Fig. 2 a)). On such defined cost function (hyperplane) were used both algorithms (DE and SOMA) to find its minimum - the best model. All simulations were done on a few time series like for example solar activity and civilian unemployment (USA) 1950-1995 (Fig. 3 - Fig. 5). The cost function can be schematically depicted as here:

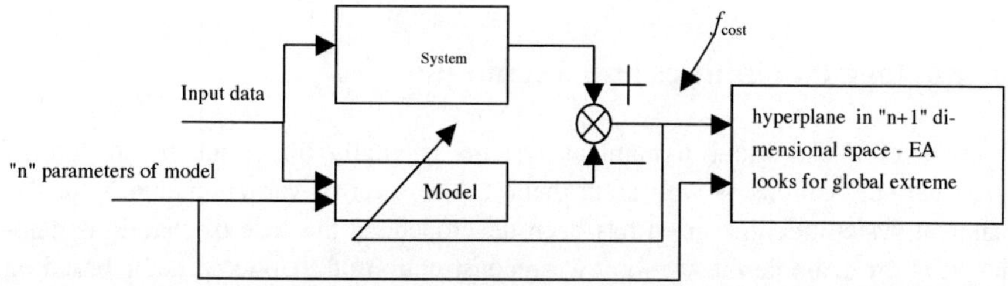

$$f_{cost} = \frac{\sum_{k=1}^{N} (y(k)_{original} - y(k)_{predicted})^2}{N} \tag{1}$$

$$y(k) = \frac{q^{-d} B(q^{-1})}{A(q^{-1})} u(k-1) + \frac{C(q^{-1})}{D(q^{-1}) A(q^{-1})} e(k) \tag{2}$$

$$y(k) = \frac{C(q^{-1})}{A(q^{-1})} e(k) \tag{3}$$

Generally such approach it is possible to use on any model like, fuzzy, neural network or ARIMAX model (2), [12], but in this contribution evolutionary identification was used on ARMA model (3), [13]. On Fig. 3 is result of evolutionary identification (EI) based on DE, which try to find only coefficients for ARMA model of second order for prediction one step ahead for Solar activity data. Model, which was found on time series in range 1860-1995, was then used on whole time series 1700-1995 with good result (Fig. 3).

In the next simulation was DE looking for not only coefficients but for their position ($a_1y(k-1)$, $a_5y(k-5)$,...). DE has tried to find the best combination of coefficients and their position in last 50 data points from given one Fig. 4 a) again with good results. Last simulation with DE was used on different time series, result one can see on Fig. 4 b).

For two last simulations was used SOMA. On Fig. 5 is visible how was successful on solar activity in comparing with simulations with DE (Fig. 3). Last simulation was focused on third kind of simulation - especially on identifying of suitable model, which was then compared with original model in [14]. Result is on Fig. 5 b).

$$Err = \sum_{i=1}^{N} \left| y_{i_{th}\ original} - y_{i_{th}\ \mathrm{mod}\,el} \right| \quad N \Rightarrow Dimension\ of\ data\ vector \qquad (4)$$

4 Further Possibilities and Techniques

Identification of various dynamical systems is vitally important in the case of practical applications as well as in theory. A rich set of various methods for dynamical system identification has been developed. In the case of chaotic dynamics, it is for example the well-known reconstruction of chaotic attractor based on research of Takens [29] who has shown that, after the transients have died out, one can reconstruct the trajectory on the attractor from the measurement of a single component. Because whole trajectory contains too much information, series of papers by [15], [16], [17] is introduced to show a set of averaged coordinate invariant numbers (generalized dimensions, entropies, and scaling indices) by which different strange attractors can be distinguished. Method presented here is based on evolutionary algorithms (EAs), see [18], and allows reconstruction not only of chaotic attractors as a geometrical objects, but also their mathematical description. All those techniques belong to the class of genetic programming techniques; see [19], [20]. Generally, when it is used on data fitting, these techniques are called symbolic regression (SR).

The term symbolic regression (SR) represents a process, in which measured data is fitted by a suitable mathematical formula such as x2 + C, sin(x)+1/ex, etc., Mathematically, this process is quite well known and can be used when data of an unknown process is obtained. Historically SR has been in the preview of manual manipulation, however during the recent past, a large inroad has been made through the use of computers. Generally, there are two well-known methods, which can be used for SR by means of computers. The first one is called genetic programming or GP, [19, 20] and the other is grammatical evolution, [21, 22].

The idea as to how to solve various problems using SR by means of EA was introduced by John Koza, who used genetic algorithms (GA) for GP. Genetic programming is basically a symbolic regression, which is done by the use of evolutionary algorithms, instead of a human brain. The ability to solve very difficult problems is now well established, and hence, GP today performs so well that it can be applied, e.g. to synthesize highly sophisticated electronic circuits [23].

In the last decade of the 20th century, C. Ryan developed a novel method for SR, called grammatical evolution (GE). Grammatical evolution can be regarded as an unfolding of GP due to some common principles, which are the same for both algorithms. One important characteristic of GE is that it can be implemented in any arbitrary computer language compared with GP, which is usually done (in its canonical form) in LISP. In contrast to other evolutionary algorithms, GE was used only with a few search strategies, for example with a binary representation of the populations in [24]. Another interesting investigation using symbolic regression was carried out by [25] working on Artificial Immune Systems or/and systems, which are not using tree structures like linear genetic programming.

Put simply, evolutionary algorithm simulates Darwinian evolution of individuals (solutions of given problem) on a computer and are used to estimate-optimize numerical values of defined cost function. Methods of GP are able to synthesize in an evolutionary way complex structures like electronic circuits, mathematical formulas etc. from basic set of symbolic (nonnumeric) elements.

Another alternative methods of symbolic regression is in [28]. This method can be used as well as GP or GE for predictive model synthesis.

In paper [26] and [27], analytic programming (AP) is applied, for the identification of selected chaotic system. Identification is not done on the "level" of strange attractor reconstruction, but it produces a symbolic-mathematical description of the identified system.

5 Results

All important and selected results are depicted in Fig. 2-4 and referred in [26] and [27]. SOMA as well as DE has demonstrated good performance in model parameter identification and also in its synthesis, as depicted in [26].

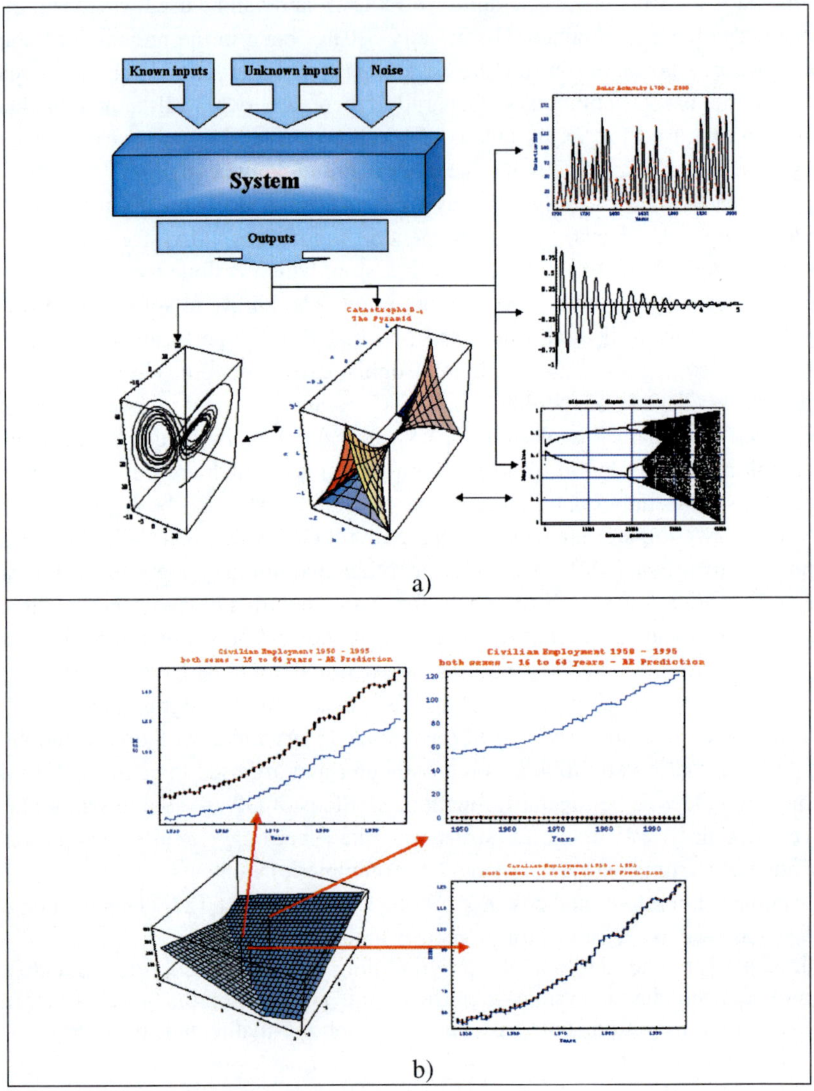

Fig. 1 a) Dynamical system and its possible behavior and b) Principle of evolutionary identification

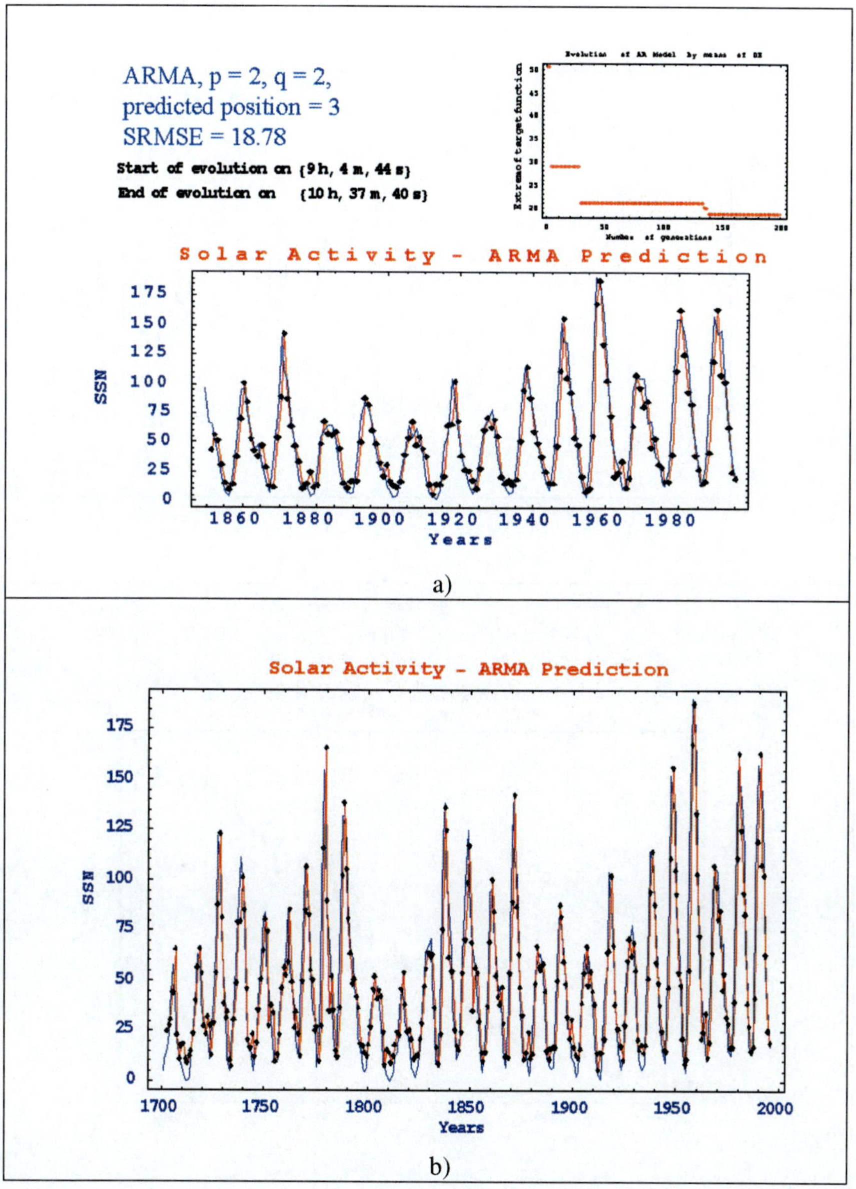

Fig. 2 a) First simulation b) Second simulation (model from the first one used on whole time series)

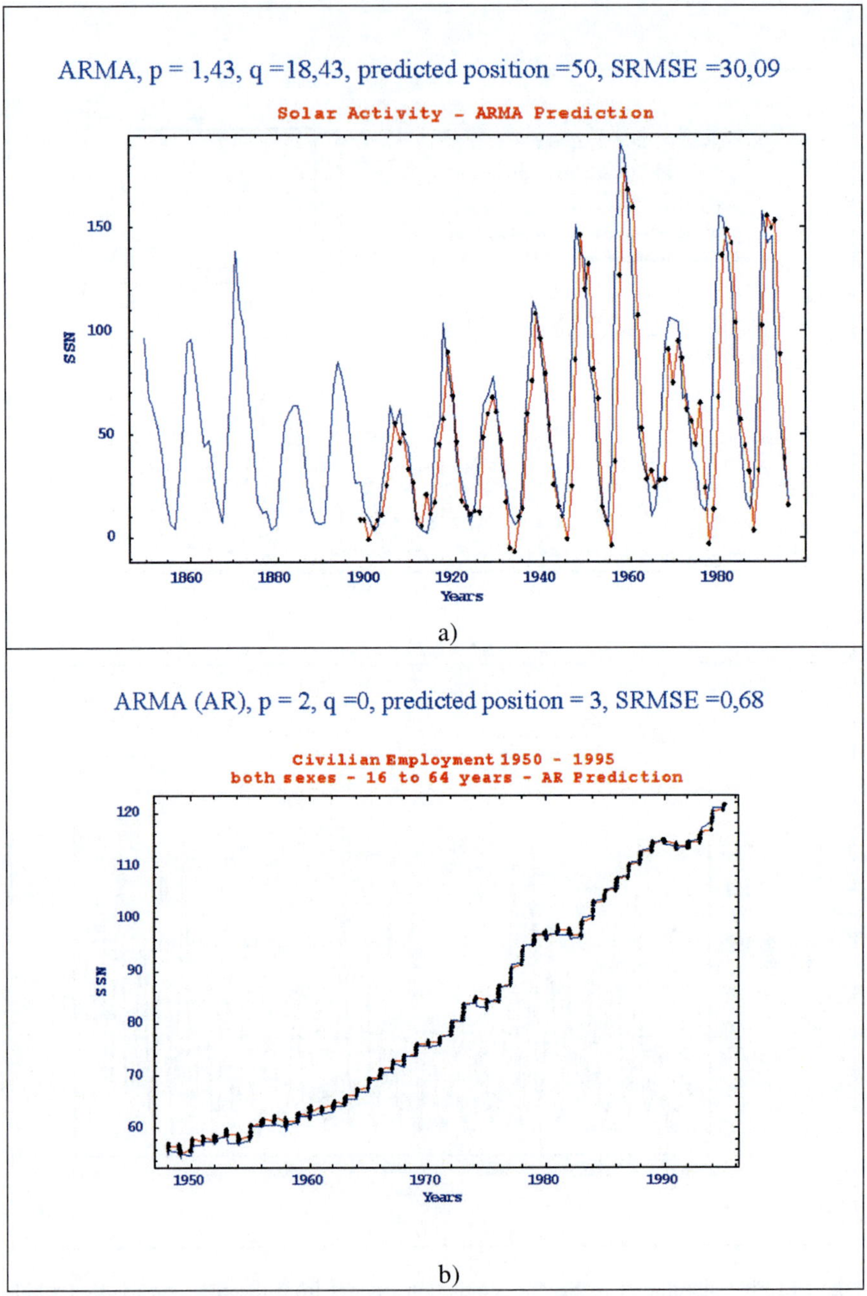

Fig. 3 a) Third simulation (EA was looking for suitable coefficients in last 50 data points) b) Fourth simulation (AR model on different time series)

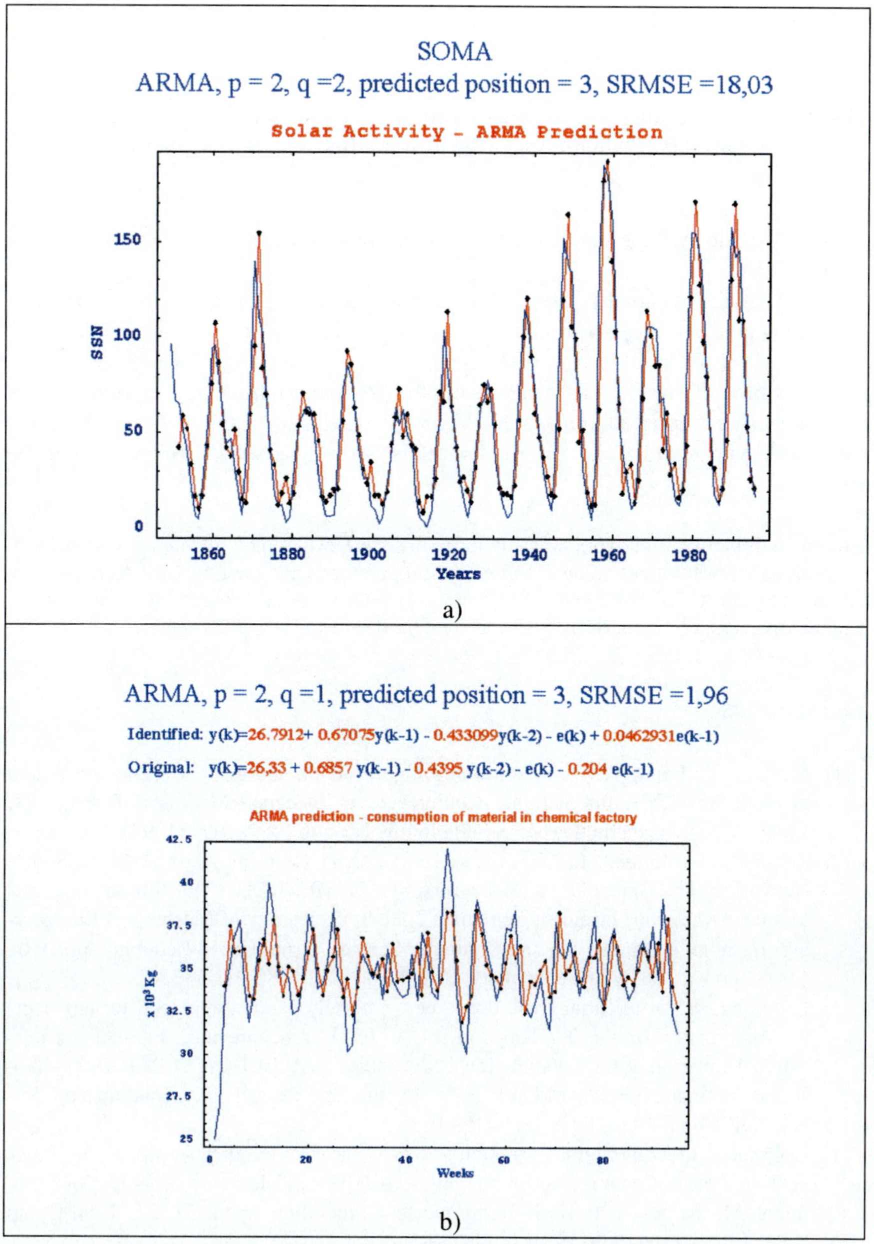

Fig. 4 a) Fifth simulation by SOMA (result is comparable with DE), b) Sixth simulation (EA was looking for coefficients of ARMA model from book [14] and then compared with coefficients from this book)

6 Conclusions

In this contribution was roughly described, how can be used various optimizing algorithms for finding or identification of suitable predictive models. There were made two sets of simulations where in the first set were made three kinds of simulations:

1. EI of model with given "non removable" position of its coefficients (Fig. 3, Fig. 4 b)
2. EI of model where EI was looking for suitable combinations of coefficients and its position (Fig. 4a))

As is visible, all simulations were successful. All simulations are temporal - they will be further improved and repeated in wider scale. It means that more time series, more various test and different models (neural networks, fuzzy,...) can be tried.

Acknowledgments. This work was supported by the Development of human resources in research and development of latest soft computing methods and their application in practice project, reg. no. CZ.1.07/2.3.00/20.0072 funded by Operational Programme Education for Competitiveness, co-financed by ESF and state budget of the Czech Republic.

References

[1] Masters, T., Land, W.: A new training algorithm for the general regression neural network. In: IEEE International Conference on Systems, Man, and Cybernetics, Computational Cybernetics and Simulation, vol. 3, pp. 1990–1994 (1997)

[2] Zelinka, I., Lampinen, J.: DELA – an Evolutionary Learning Algorithms for Neural Networks. In: Ošmera, P. (ed.) Proceedings of MENDEL 1999, 5th International Mendel Conference on Soft Computing, June 9-12. Faculty of Mechanical Engineering, Institute of Automation and Computer Science, Brno (Czech Republic), pp. 410–414. Brno University of Technology, Brno (1999) ISBN 80-214-1131-7

[3] Lampinen, J.: Differential Evolution – new naturally parallel approach for engineering design optimization. In: Topping, B.H.V. (ed.) Euroconference: Parallel and Distributed Computing for Computational Mechanics 1999 EURO-CM-PAR 1999 – Abstracts, Weimar, Germany, March 20-25. Lecture and Research Presentations, pp. 35–36. Civil-Comp Press, Edinburgh (1999)

[4] Lampinen, J.: Differential Evolution – New Naturally Parallel Approach for Engineering Design Optimization. In: Topping, B.H.V. (ed.) Developments in Computational Mechanics with High Performance Computing, pp. 217–228. Civil-Comp Press, Edinburgh (1999) ISBN 0-948749-59-8

[5] Lampinen, J., Zelinka, I.: Mechanical Engineering Design Optimization by Differential Evolution. In: Corne, D., Dorigo, M., Glover, F. (eds.) New Ideas in Optimisation. Mc-Graw-Hill, New York (1999) ISBN: 0077095065

[6] Lampinen, J., Zelinka, I.: Mixed Integer-Discrete-Continuous Optimization By Differential Evolution, Part 1: the optimization method. In: Ošmera, P. (ed.) Proceedings of MENDEL 1999, 5th International Mendel Conference on Soft Computing, June 9-12. Faculty of Mechanical Engineering, Institute of Automation and Computer Science, Brno (Czech Republic), pp. 71–76. Brno University of Technology, Brno (1999) ISBN 80-214-1131-7
[7] Zelinka, I.: SOMA – Self Organizing Migrating Algorithm. In: Babu, B.V., Onwubolu, G. (eds.) New Optimization Techniques in Engineering. Springer (2004) ISBN 3-540-20167X
[8] Price, K.: An Introduction to Differential Evolution. In: Corne, D., Dorigo, M., Glover, F. (eds.) New Ideas in Optimization, pp. 79–108. McGraw-Hill, London (1999)
[9] Price, K., Storn, R.: Differential evolution homepage (2001), `http://www.icsi.berkeley.edu/~storn/code.html` (accessed May 15, 2012)
[10] Varacha, P., Jasek, R.: ANN Synthesis for an Agglomeration Heating Power Consumption Approximation. In: Recent Researches in Automatic Control, pp. 239–244. WSEAS Press, Montreux, ISBN 978-1-61804-004-6
[11] Varacha, P., Zelinka, I.: Distributed Self-Organizing Migrating Algorithm Application and Evolutionary Scanning. In: Proceedings of the 22nd European Conference on Modelling and Simulation ECMS, pp. 201–206 (2008) ISBN 0-9553018-5-8
[12] Soeterboek, A.R.M.: Predictive Control, Proefschrift. Technische Universiteit Delft, Rotterdam (1990)
[13] Masters, T.: Neural, Novel & Hybrid Algorithms for Time Series Prediction. John Wiley & Sons, Inc.
[14] Cipra, T.: Time series analysis with applications in economics. SNTL/ALFA (1986) ISBN 04-012-86
[15] Grassberger, P., Procaccia, I.: Estimation of the Kolmogorov Entropy From a Chaotic Signal. Phys. Rev. 29 A, 2591 (1983b)
[16] Halsey, T.C., Jensen, M.H., Kadanoff, L.P., Procaccia, I., Schraiman, B.I.: Fractal Measures and Their Singularities: the Characterization of Strange Sets. Phys. Rev. 33 A, 1141 (1986)
[17] Eckmann, J.P., Procaccia, I.: Fluctuation of Dynamical Scaling Indices in Non-Linear Systems. Phys. Rev. 34 A, 659 (1986)
[18] Back, T., Fogel, D.B., Michalewicz, Z.: Handbook of Evolutionary Computation. Institute of Physics, London (1997)
[19] Koza, J.R.: Genetic Programming II. MIT Press (1998) ISBN 0-262-11189-6
[20] Koza, J.R., Bennet, F.H., Andre, D., Keane, M.: Genetic Programming III. Morgan Kaufnamm pub. (1999) ISBN 1-55860-543-6
[21] O'Neill, M., Ryan, C.: Grammatical Evolution. Evolutionary Automatic Programming in an Arbitrary Language. Kluwer Academic Publishers (2002) ISBN 1402074441
[22] Ryan, C., Collins, J.J., O'Neill, M.: Grammatical Evolution: Evolving Programs for an Arbitrary Language. In: Banzhaf, W., Poli, R., Schoenauer, M., Fogarty, T.C. (eds.) EuroGP 1998. LNCS, vol. 1391, pp. 83–96. Springer, Heidelberg (1998)
[23] Koza, J.R., Keane, M.A., Streeter, M.J.: Evolving Inventions. In: Scientific American, pp. 40–47 (February 2003) ISSN 0036-8733

[24] O'Sullivan, J., Conor, R.: An Investigation into the Use of Different Search Strategies with Grammatical Evolution. In: Foster, J.A., Lutton, E., Miller, J., Ryan, C., Tettamanzi, A.G.B. (eds.) EuroGP 2002. LNCS, vol. 2278, pp. 268–277. Springer, Heidelberg (2002)

[25] Johnson, C.G.: Artificial Immune Systems Programming for Symbolic Regression. In: Ryan, C., Soule, T., Keijzer, M., Tsang, E.P.K., Poli, R., Costa, E. (eds.) EuroGP 2003. LNCS, vol. 2610, pp. 345–353. Springer, Heidelberg (2003)

[26] Zelinka, I., Chen, G., Celikovsky, S.: Chaos Synthesis by Means of Evolutionary Algorithms. International Journal of Bifurcation and Chaos 18(4), 911–942 (2008)

[27] Zelinka, I., Celikovsky, S., Richter, H., Chen, G.: Evolutionary Algorithms and Chaotic Systems, 550 p. Springer, Germany (2010)

[28] Zelinka, I., Davendra, D., Senkerik, R., Jasek, R., Oplatkova, Z.: Analytical Programming - a Novel Approach for Evolutionary Synthesis of Symbolic Structures. In: Kita, E. (ed.) Evolutionary Algorithms. InTech, ISBN 978-953-307-171-8

[29] Takens, F.: Detecting strange attractors in turbulence. In: Rand, D.A., Young, L.-S. (eds.) Dynamical Systems and Turbulence. Lecture Notes in Mathematics, vol. 898, pp. 366–381. Springer

Data Mining by Symbolic Fuzzy Classifiers and Genetic Programming

Suhail Owais[1], Pavel Krömer[2], Jan Platoš[2], Václav Snášel[2], and Ivan Zelinka[2]

[1] Department of Computer Science, IT College, ASU– Applied Science University, Amman, Jordan
dr_suhail@asu.edu.jo
[2] Department of Computer Science, FEECS, VŠB – Technical University of Ostrava, 17. listopadu 12, Ostrava, Czech Republic
{pavel.kromer,jan.platos,vaclav.snasel,ivan.zelinka}@vsb.cz

Abstract. There are various techniques for data mining and data analysis. Among them, hybrid approaches combining two or more algorithms gain importance as the complexity and dimension of real world data sets grows. In this paper, we present an application of evolutionary-fuzzy classification technique for data mining. Genetic programming is deployed to evolve a fuzzy classifier and an example of real world application is presented.

1 Introduction

The recent time has seen a rise in the demand for advanced data mining algorithms. Many real world application domains generate huge amounts of data. Information hidden in such a data can be extracted and help in optimization of processes, designs, and algorithms. The growing dimension and complexity of said data sets represents a challenge for traditional search and optimization methods while the increase of power of widely available computers encourages the deployment of soft computing methods such as the populational meta-heuristic algorithms, artificial neural networks and fuzzy systems.

Fuzzy sets and fuzzy logic provide means for soft classification of data. In contrast to crisp classification, which states crisp decisions about data samples, fuzzy classification allows to analyze the data samples in a more sensitive way [2]. Fuzzy decision trees and if-then rules are examples of efficient, transparent, and easily interpretable fuzzy classifiers [2, 20].

Genetic programming is a powerful machine learning technique from the wide family of evolutionary algorithms. In contrast to the traditional evolutionary algorithms, it can be used to evolve complex hierarchical tree structures and symbolic expressions.

In this work, we use genetic programming for data mining by fuzzy classifier evolution. In particular, genetic programming is used to evolve symbolic fuzzy classifiers that are able to describe classes in a data set by means of its features.

I. Zelinka et al. (Eds.): Nostradamus: Mod. Meth. of Prediction, Modeling, AISC 192, pp. 273–282.
springerlink.com

Such a fuzzy classifier evolved over a training data set can be later used for efficient and fast classification of data samples e.g. for predicting quality of products, and generally to assign labels to data.

Artificial evolution of fuzzy classifiers is a promising approach to data mining because genetic programming has proven very good ability to find symbolic expressions in various application domains. The general process of classifier evolution can be used to evolve classifiers for different data classes and data sets with different properties. The resulting classifiers can be used as standalone data labeling tools or participate in collective decision in an ensemble of data classification methods.

2 Fuzzy Information Retrieval

The evolution of fuzzy classifiers for data mining was implemented on top of a framework for search query optimization designed for efficient information retrieval [6, 18]. For fuzzy data mining, we map data samples to documents and data features to index terms. In this section, we describe the principles of fuzzy information retrieval for better understanding of the proposed method.

The area of information retrieval (IR) is a branch of computer science dealing with storage, maintenance, and searching in large amounts of data [5]. It defines and studies IR systems and models. An IR model is a formal background defining the document representation, query language, and document-query matching mechanism of an IR system (i.e. a search engine).

2.1 Extended Boolean IR model

The proposed fuzzy classification algorithm builds on the extended Boolean IR model, which is based on the fuzzy set theory and fuzzy logic [5, 14]. In the extended Boolean IR model, documents are interpreted as fuzzy sets of indexed terms.

In each document, every indexed term has a weight from the range [0,1] expressing the degree of significance of the term for document representation. Many different weighting approaches can be used to assign weights to index terms, e.g. the tf×idft term weighting scheme [17].

A formal description of a document collection in the extended Boolean IR model is shown in (1) and (2), where d_i represents i-th document and t_{ij} j-th term in i-th document. The entire document collection can be represented by an index matrix D.

$$d_i = (t_{i1}, t_{i2} \dots, t_{im}), \qquad t_{ij} \in [0,1] \tag{1}$$

$$D = \begin{pmatrix} t_{11} & \cdots & t_{1m} \\ \vdots & \ddots & \vdots \\ t_{n1} & \cdots & t_{nm} \end{pmatrix} \tag{2}$$

The query language in the extended Boolean model of IR is improved with the possibility of weighting query terms in order to attribute different levels of importance to those in a search request and by weighting (parameterizing) aggregation operators (most often AND, OR, and NOT) to soften or blur their impact on query evaluation [5, 14].

Consider Q to be the set of user queries over a collection; then the weight of term t in query q is denoted as $a(q,t)$ satisfying a: $Q \times T \rightarrow [0,1]$. To evaluate the atomic query of one term representing single search criterion the function g: $[0,1] \times [0,1] \rightarrow [0,1]$ will be used. The value of $g(F(d,t),a)$ is called the retrieval status value (RSV). For RSV evaluation the interpretation of the query term weight a is crucial. The most commonly used interpretations see the query term weight as the importance weight, threshold or ideal document description [5, 14].

The theorems for the evaluation of RSV in the case of importance weight interpretation and threshold interpretation are shown in (3) and (4) respectively [5, 14], where $P(a)$ and $Q(a)$ are coefficients used for tuning the threshold curve. An example of $P(a)$ and $Q(a)$ could be as follows: $P(a) = \frac{1+a}{2}$ and $Q(a) = \frac{1-a^2}{4}$. The RSV formula (4) is illustrated in Fig. 1a. Adopting the threshold interpretation, an atomic query containing term t of the weight a is a request to retrieve documents having $F(d,t)$ equal or greater to a. Documents satisfying this condition will be rated with high RSV and contrariwise documents having $F(d,t)$ smaller than a will be rated with a small RSV.

$$g(F(d,t),a) = \begin{cases} \min(a, F(d,t)), & \text{for the } or \text{ op.} \\ \max(1-a, F(d,t)), & \text{or the } and \text{ op.} \end{cases} \tag{3}$$

$$g(F(d,t),a) = \begin{cases} P(a)\dfrac{F(d,t)}{a}, & \text{for} F(d,t) < a \\ \mathrm{P(a)} + \mathrm{Q(a)}\dfrac{F(d,t)-a}{1-a}, & \text{for} F(d,t) \geq a \end{cases} \tag{4}$$

Query term weight a could be understood as an ideal document term weight prescription. In that case, RSV will be evaluated according to (5), enumerating the distance between $F(d;\ t)$ and a in a symmetric manner as shown in Fig. 1b. This means that a document with a lower term weight will be rated with the same RSV as a document with a higher term weight, considering the same differences. An asymmetric version of (5) is shown in (6) and illustrated in 1c.

$$g(F(d,t),a) = e^{\kappa(F(d,t)-a)^2} \tag{5}$$

$$g(F(d,t),a) = \begin{cases} e^{\kappa(F(d,t)-a)^2}, & \text{for} F(d,t) < a \\ \mathrm{P(a)} + \mathrm{Q(a)}\dfrac{F(d,t)-a}{1-a}, & \text{for} F(d,t) \geq a \end{cases} \tag{6}$$

The operators *and*, *or*, and *not* can be evaluated with the help of fuzzy set operations. Fuzzy set operations are extensions of crisp set operations on fuzzy sets [21]. A characteristic function uniquely defines a fuzzy set and hence fuzzy set operations are defined using characteristic functions [8]. In [21] Lotfi Zadeh defined basic methods for the complement, union and intersection of fuzzy sets. Besides standard (Zadeh's) fuzzy set operations, entire classes of prescriptions for defining the complements, intersections and unions on fuzzy sets were later designed.

In this study, we use the threshold interpretation of RSV, standard t-norm $t(x,y) = \min(x,y)$ and t-conorm $s(x,y) = \max(x,y)$ for the implementation of *and* and *or* operators and fuzzy complement $c(x) = 1 - x$ for the evaluation of *not* operator .

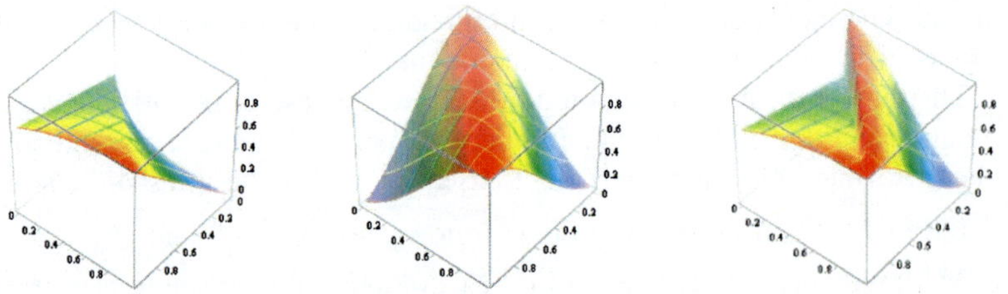

(a) *g(F(d; t); a)* according to (4) (b) *g(F(d; t); a)* according to (5) (c) *g(F(d; t); a)* according to (6)

Fig. 1 RSV functions

2.2 *IR Evaluation*

The effectiveness of an information retrieval system can be evaluated using the measures precision *P* and recall *R*. Precision corresponds to the probability of retrieved document to be relevant and recall can be seen as the probability of retrieving relevant document. Precision and recall in the extended Boolean IR model can be defined using the Σ-count $\|A\|$:

$$\rho(X|Y) = \begin{cases} \dfrac{\|X \cap Y\|}{\|Y\|}, & \|Y\| \neq 0 \\ 1, & \|Y\| = 0 \end{cases} \tag{7}$$

$$P = \rho(REL|RET),\ R = \rho(RET|REL) \tag{8}$$

where *REL* stands for the fuzzy set of all relevant documents, *RET* for the fuzzy set of all retrieved documents, and $\|A\|$ is the Σ-count, i.e. the sum of the values of characteristic function μ_A for all members of the fuzzy set *A* [22]:

$$\|A\| = \sum_{x \in A} \mu_A(x) \tag{9}$$

For an easier IR effectiveness evaluation, measures combining precision and recall into one scalar value were developed. The F-score *F* is among the most used scalar combinations of *P* and *R*:

$$F = \frac{(1 + \beta^2)PR}{\beta^2 P + R} \tag{10}$$

The index matrix D can be seen as a general data matrix with m rows (data samples) and n columns (data features). The evaluation of Extended Boolean query over the document collection generates an ordering of the documents (i.e. it assigns a real value from the range [0,1] to each document). The ordering can be also interpreted as a fuzzy set of documents. If we abandon the IR terminology, we can call the extended Boolean query a general fuzzy classifier and use it to describe fuzzy sets or fuzzy sub sets of data by its features

3 Genetic Programming for Classifier Evolution

Genetic programming (GP) is an extension to genetic algorithms (GA), which is a popular and wide spread member of the class of evolutionary algorithms (EA). EAs found a group of iterative stochastic search and optimization methods based on mimicking successful optimization strategies observed in nature [1]. The essence of EAs lies in their emulation of Darwinian evolution, utilizing the concepts of Mendelian inheritance for use in computer science [3]. Together with fuzzy sets, neural networks, and fractals, evolutionary algorithms are among the fundamental members of the class of soft computing methods.

EAs operate with a population (also known as a pool) of artificial individuals (also referred to as items or chromosomes) encoding possible problem solutions. Encoded individuals are evaluated using a carefully selected objective function which assigns a fitness value to each individual. The fitness value represents the quality (ranking) of each individual as a solution to a given problem. Competing individuals explore the problem domain towards an optimal solution [10].

The evolutionary algorithm for fuzzy classifier evolution was introduced in [19] and builds on the principles of fuzzy information retrieval [4, 5] and evolutionary optimization of search queries [7].

3.1 Genetic Programming

Genetic programming is an extension to genetic algorithms, allowing work with hierarchical, often tree-like, chromosomes with an unlimited length [11].

GP was introduced as a tool to evolve whole computer programs and represented a step towards adaptable computers that could solve problems without being programmed explicitly [11, 1].

In GP the chromosomes take the form of hierarchical variably-sized expressions, point-labeled structure trees. The trees are constructed from nodes of two types, terminals and functions. More formally, a GP chromosome is a symbolic expression created from terminals t from the set of all terminals T and functions f from the set of all functions F satisfying the recursive definition [1]:

- $\forall t \in T$: t is the correct expression
- $\forall f \in F$: $f(e1, e2, ..., en)$ is the correct expression if $\forall f \in F$ and $e1, ..., en$ are correct expressions. The function *arity*(f) represents the arity of f.
- there are no other correct expressions

GP chromosomes are evaluated by the recursive execution of instructions corresponding to tree nodes [1]. Terminal nodes are evaluated directly (e.g. by reading an input variable) and functions are evaluated after left-to-right depth-first evaluation of their parameters.

Genetic operators are applied to the nodes in tree-shaped chromosomes. A crossover operator is implemented as the mutual exchange of randomly selected sub-trees of the parent chromosomes. Mutation aims to modify the chromosomes by pseudo-random arbitrary changes in order to prevent premature convergence and broaden the coverage of the fitness landscape. Mutation could be implemented as: removal of a sub-tree at a randomly chosen node; replacement of a randomly chosen node by a newly generated sub-tree; replacement of node instruction by a compatible node instruction (i.e. a terminal can be replaced by another terminal, a function can be replaced by another function of the same arity); a combination of the previous.

3.2 *Fuzzy Classifier*

The fuzzy classifier takes form of a symbolic expression with data features (data set attributes) as terminals and operators as non-terminal nodes. Both terminals and non-terminals are weighted.

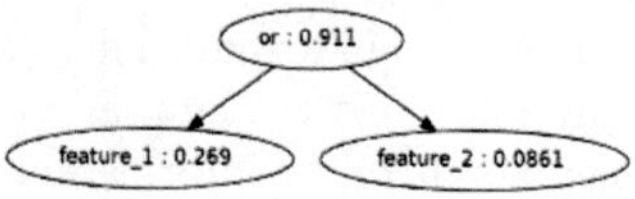

Fig. 2 An example of fuzzy classifier

Fuzzy classifier is evaluated for each data sample in the training collection. For each terminal, the value of corresponding feature is taken. The operators are implemented with the help of fuzzy set operators. The standard implementation of fuzzy set operators was used but any other pair of t-norm and t-conorm could be used. We also note that additional operators (e.g. various ordered weighted averaging aggregation operators) could be added.

4 Experiments

Genetic programming was used to evolve fuzzy rules describing faulty products in an industrial plant. During the production, a number of sensory inputs are read to record material properties, production flow and product features. The features include the chemical properties of the raw material, density, temperature at several processing stages, and many other values recorded several times during the production process. At the end, the product is classified as either valid or corrupt. The data and classification for a number of product samples is known and the goal of the genetic programming is to find a fuzzy classifier that could be used for product quality prediction.

We have obtained data sets from 5 different production lines of a production plant. The data sets contained readings from 508 sensors for each product. For each production line, the data was divided into training (40%) and test (60%) collection.

We label the data sets D1, D2, D3, D4 and D5 respectively. Selected properties of the data sets are shown in Table 1. All five data sets have the same number of features but since they come from different processing lines, their internal structure differs and the patterns describing faulty products are unique for each of them.

Table 1 Description of the data sets

Name	D1	D2	D3	D4	D5
Features	508	508	508	508	508
Training samples	562	154	755	4881	2022
Test samples	844	233	1134	73226	3034

Table 2 Results of classification of test data collections

	Data set				
	D1	D2	D3	D4	D5
OA	97.63	97.00	99.50	96.99	99.60
FP	1.30	3.00	0	0.43	0.07
FN	1.7	0	0.53	2.58	0.33

The proposed algorithm for classifier evolution was implemented and classifiers were sought for all five training sets. The results of the classification on collections are shown in Table 2. The table shows the overall accuracy (OA), percent of false positives (FP) and percent of false negatives (FN) obtained by the best classifiers.

5 Further Possibilities and Techniques

There are other alternative techniques of genetic programming nature. Generally, there are two well-known methods, which can be used for synthesis of various programs by means of computers. The first one is called genetic programming or GP [11, 1], and the other is grammatical evolution [12, 16].

The idea as to how to solve various problems using symbolic regression (SR) by means of EA was introduced by John Koza, who used genetic algorithms (GA) for GP. Genetic programming is basically a symbolic regression, which is done by the use of evolutionary algorithms, instead of a human brain. The ability to solve very difficult problems is now well established, and hence, GP today performs so well that it can be applied, e.g. to synthesize highly sophisticated electronic circuits [13].

In the last decade of the 20th century, C. Ryan developed a novel method for SR, called grammatical evolution (GE). Grammatical evolution can be regarded as an unfolding of GP due to some common principles, which are the same for both algorithms. One important characteristic of GE is that it can be implemented in any arbitrary computer language compared with GP, which is usually done (in its canonical form) in LISP. In contrast to other evolutionary algorithms, GE was used only with a few search strategies, for example with a binary representation of the populations in [15]. Another interesting investigation using symbolic regression was carried out by [9] working on Artificial Immune Systems or/and systems, which are not using tree structures like linear genetic programming.

Put simply, evolutionary algorithm simulates Darwinian evolution of individuals (solutions of given problem) on a computer and are used to estimate-optimize numerical values of defined cost function. Methods of GP are able to synthesize in an evolutionary way complex structures like electronic circuits, mathematical formulas etc. from basic set of symbolic (nonnumeric) elements.

Another alternative method of symbolic regression is in [23] called Analytic programming (AP). This method can bc used as well as GP or GE for various model syntheses and has been successfully compared to many standard problems with GP, with very good results. Our next step is to use AP and other alternative methods to create more complex comparative study, focused on problems, mentioned here.

6 Conclusions

This work presents an application of evo-fuzzy data mining technique to the classification of data samples in a real world industrial data set. Genetic programming has been used to evolve fuzzy classifiers in form of weighted symbolic expressions aggregating data features with the help of a set of operators. In contrast to previous efforts in this area, this approach is inspired by information retrieval. The information retrieval inspired fuzzy rules containing the operators *and*, *or*, and *not* provided a rich and flexible tool to express detailed soft classification criteria.

Data classes were interpreted as membership degrees of a fuzzy set and the algorithm sought for a classifier that would describe such a fuzzy set. In this sense, the described approach also differs from most of the traditional rule-based fuzzy classifiers that aim to mine the if-then relations from data.

The evolution of fuzzy classifier takes a number of parameters. The set of classifier operators, the interpretation of classifier weights and the fitness function can all be altered.

Acknowledgments. This paper has been elaborated in the framework of the IT4Innovations Centre of Excellence project, reg. no. CZ.1.05/1.1.00/02.0070 supported by Operational Programme 'Research and Development for Innovations' funded by Structural Funds of the European Union and state budget of the Czech Republic. This work was also supported by the Development of human resources in research and development of latest soft computing methods and their application in practice project, reg. no. CZ.1.07/2.3.00/20.0072 funded by Operational Programme Education for Competitiveness, co-financed by ESF and state budget of the Czech Republic.

References

[1] Affenzeller, M., Winkler, S., Wagner, S., Beham, A.: Genetic Algorithms and Genetic Programming: Modern Concepts and Practical Applications. Chapman & Hall/CRC (2009)

[2] Bezdek, J.C., Keller, J., Krisnapuram, R., Pal, N.R.: Fuzzy Models and Algorithms for Pattern Recognition and Image Processing (The Handbooks of Fuzzy Sets). Springer-Verlag New York, Inc., Secaucus (2005)

[3] Bodenhofer, U.: Genetic Algorithms: Theory and Applications. Lecture notes, Fuzzy Logic Laboratorium Linz-Hagenberg (Winter 2003/2004)

[4] Cordón, O., de Moya, F., Zarco, C.: Fuzzy logic and multiobjective evolutionary algorithms as soft computing tools for persistent query learning in text retrieval environments. In: IEEE Int. Conference on Fuzzy Systems 2004, Budapest, Hungary, pp. 571–576 (2004)

[5] Crestani, F., Pasi, G.: Soft information retrieval: Applications of fuzzy set theory and neural networks. In: Kasabov, N., Kozma, R. (eds.) Neuro-Fuzzy Techniques for Intelligent Information Systems, pp. 287–315. Springer, Heidelberg (1999)

[6] Húsek, D., Owais, S.S.J., Snášel, V., Krömer, P.: Boolean queries optimization by genetic programming. Neural Network World, 359–409 (2005)

[7] Húsek, D., Snášel, V., Neruda, R., Owais, S.S.J., Krömer, P.: Boolean queries optimization by genetic programming. WSEAS Trans. on Inf. Sci. and Applications 3(1), 15–20 (2006)

[8] Jantzen, J.: Tutorial On Fuzzy Logic. Technical Report 98-E-868 (logic), Technical University of Denmark, Dept. of Automation (1998)

[9] Johnson, C.G.: Artificial Immune System Programming for Symbolic Regression. In: Ryan, C., Soule, T., Keijzer, M., Tsang, E.P.K., Poli, R., Costa, E. (eds.) EuroGP 2003. LNCS, vol. 2610, pp. 345–353. Springer, Heidelberg (2003)

[10] Jones, G.: Genetic and evolutionary algorithms. In: von Rague, P. (ed.) Encyclopedia of Computational Chemistry. John Wiley and Sons (1998)

[11] Koza, J.: Genetic programming: A paradigm for genetically breeding populations of computer programs to solve problems. Technical Report STAN-CS-90-1314, Dept. of Computer Science, Stanford University (1990)

[12] Koza, J.R., Andre, D., Bennett, F.H., Keane, M.A.: Genetic Programming III: Darwinian Invention & Problem Solving, 1st edn. Morgan Kaufmann Publishers Inc., San Francisco (1999)

[13] Koza, J.R., Keane, M.A., Streeter, M.J.: Evolving inventions. Scientific American (2003)

[14] Kraft, D.H., Petry, F.E., Buckles, B.P., Sadasivan, T.: Genetic Algorithms for Query Optimization in Information Retrieval: Relevance Feedback. In: Sanchez, E., Shibata, T., Zadeh, L. (eds.) Genetic Algorithms and Fuzzy Logic Systems, World Scientific, Singapore (1997)

[15] O'Sullivan, J., Ryan, C.: An Investigation into the Use of Different Search Strategies with Grammatical Evolution. In: Foster, J.A., Lutton, E., Miller, J., Ryan, C., Tettamanzi, A.G.B. (eds.) EuroGP 2002. LNCS, vol. 2278, pp. 268–277. Springer, Heidelberg (2002)

[16] Ryan, C., Collins, J.J., O'Neill, M.: Grammatical Evolution: Evolving Programs for an Arbitrary Language. In: Banzhaf, W., Poli, R., Schoenauer, M., Fogarty, T.C. (eds.) EuroGP 1998. LNCS, vol. 1391, pp. 83–96. Springer, Heidelberg (1998)

[17] Salton, G., Buckley, C.: Term-weighting approaches in automatic text retrieval. Information Processing and Management 24(5), 513–523 (1988)

[18] Snasel, V., Abraham, A., Owais, S., Platos, J., Kromer, P.: User Profiles Modeling in Information Retrieval Systems. In: Emergent Web Intelligence: Advanced Information Retrieval. Advanced Inf. and Knowledge Proc., pp. 169–198. Springer, London (2010)

[19] Snášel, V., Krömer, P., Platoš, J., Abraham, A.: The Evolution of Fuzzy Classifier for Data Mining with Applications. In: Deb, K., Bhattacharya, A., Chakraborti, N., Chakroborty, P., Das, S., Dutta, J., Gupta, S.K., Jain, A., Aggarwal, V., Branke, J., Louis, S.J., Tan, K.C. (eds.) SEAL 2010. LNCS, vol. 6457, pp. 349–358. Springer, Heidelberg (2010)

[20] Verikas, A., Guzaitis, J., Gelzinis, A., Bacauskiene, M.: A general framework for designing a fuzzy rule-based classifier. Knowledge and Information Systems, 1–19

[21] Zadeh, L.A.: Fuzzy sets. Information and Control 8, 338–353 (1965)

[22] Zadeh, L.A.: Test-score semantics dor natural languages and meaning representation via Pruf. In: Empirical Semantics, Quantitative Semantics, vol. 12(1), pp. 281–349. Studienverlag Brockmeyer, Bochum (1981)

[23] Zelinka, I., Davendra, D., Senkerik, R., Jasek, R., Oplatkova, Z.: Analytical programming - a novel approach for evolutionary synthesis of symbolic structures. In: Kita, E. (ed.): Evolutionary Algorithms. InTech

Author Index